Voids in Materials

Voids in Materials

From Unavoidable Defects to Designed Cellular Materials

Second Edition

Gary M. Gladysz

Department of Materials Science and Engineering,
University of Alabama at Birmingham, Birmingham,
AL, United States

Krishan K. Chawla

Department of Materials Science and Engineering,
University of Alabama at Birmingham, Birmingham,
AL, United States

ELSEVIER

Elsevier
Radarweg 29, PO Box 211, 1000 AE Amsterdam, Netherlands
The Boulevard, Langford Lane, Kidlington, Oxford OX5 1GB, United Kingdom
50 Hampshire Street, 5th Floor, Cambridge, MA 02139, United States

British Library Cataloguing-in-Publication Data

A catalogue record for this book is available from the British Library

Library of Congress Cataloging-in-Publication Data

A catalog record for this book is available from the Library of Congress

ISBN: 978-0-12-819282-5

For Information on all Elsevier publications
visit our website at https://www.elsevier.com/books-and-journals

Publisher: Susan Dennis
Acquisitions Editor: Kostas KI Marinakis
Editorial Project Manager: Sara Valentino
Production Project Manager: Joy Christel Neumarin Honest Thangiah
Cover Designer: Alan Studholme

Typeset by MPS Limited, Chennai, India

Dedication

Gary M. Gladysz would like to dedicate this to his parents, Edward and Kathy; and his wife and daughters April, Amelia, and Claire for their constant encouragement!

Krishan K. Chawla would like to dedicate this to his wife, Nivi, for always being there!

Contents

About the authors xiii
Preface to the first edition xv
Preface to the second edition xvii

1. Introduction 1
   - 1.1 Overview 1
   - 1.2 Descriptions 2
     - 1.2.1 Intrinsic and intentional voids 2
     - 1.2.2 Closed and open cell porosity 3
     - 1.2.3 Unreinforced and reinforced voids 5
     - 1.2.4 Porosity in natural and synthetic materials 7
     - 1.2.5 Stochastic, nonstochastic, and Voronoi foams 9
     - 1.2.6 Material versus digital design of voids 12
   - 1.3 Voids through the length scale 13
   - References 15

2. Intrinsic voids in crystalline materials: Ideal materials and real materials 17
   - 2.1 Introduction 17
   - 2.2 Crystalline materials 18
     - 2.2.1 Ideal materials and properties 18
     - 2.2.2 Defects and real properties 21
     - 2.2.3 Density 25
   - 2.3 Mechanical properties 25
     - 2.3.1 Modulus 25

2.3.2 Effect of voids on strength 28
2.3.3 Griffith theory of brittle fracture 29
2.4 Processing and service-induced voids 30
2.4.1 Casting 30
2.4.2 Powder processing of materials 31
2.4.3 Voids in solders 31
2.5 Time-dependent properties 31
2.5.1 Diffusion of vacancies and voids 32
2.5.2 Clustering and failure 34
2.5.3 Kirkendall voids in crystalline materials 35
References 37

3. Intrinsic voids in polymeric networks 41
3.1 Polymer structure 41
3.2 Free volume and thermomechanical behavior 44
3.3 Kinetic theory of polymer strength 46
3.4 Thermal conductivity 48
3.5 Role of voids in physical aging in polymers 49
3.6 Measurement of free volume 49
References 50

4. Nanometer scale porous structures 53
4.1 Introduction 53
4.2 Nanotubes 54
4.3 Zeolites 59
4.4 Nanoporous polymers 61
4.5 Nanoporous organic networks 64
4.5.1 Covalent organic frameworks 65
4.5.2 Covalent triazine frameworks 67
4.5.3 Polymers of intrinsic microporosity (PIM) 67
4.5.4 Conjugated microporous polymers 68

4.6 Nanoporous noble metals 71
References 73

5. Hollow and porous structures utilizing the Kirkendall effect 77
5.1 Introduction 77
5.2 Generalized Kirkendall mechanism for formation of hollow particles 78
5.2.1 Symmetric hollow particles 80
5.2.2 Asymmetric hollow particles 81
5.3 Tubes 82
5.4 Porous and hollow structures 85
References 91

6. Techniques for introducing intentional voids into materials 95
6.1 Introduction 95
6.2 Commonalities of foam formation processes 96
6.3 Introduction of a gas 97
6.3.1 Mixing 97
6.3.2 Physical blowing agent 99
6.3.3 Chemical blowing agent 101
6.4 Templating or sacrificial pore former 105
6.4.1 Aerogels 106
6.5 Bonding together of spheres, fibers, powders, or particles 107
6.6 Additive manufacturing of cellular structures 109
6.7 Mechanical stretching 123
6.8 Exploiting chemically selective weakness in solids 124
6.9 Hierarchical design with voids 129
References 132
Further reading 137

7. Techniques of introducing intentional voids into particles and fibers 139

7.1 Introduction 139

7.2 Hollow and porous particles 139

7.2.1 Introduction 139

7.2.2 Processing of porous particles 140

7.2.3 Hollow particles 143

7.2.4 Hollow, porous particles 147

7.2.5 Porous and hollow macrometer scale particles 150

7.3 Hollow and porous fibers 154

7.3.1 Carbon nanotubes 158

7.4 Nonspherical hollow particles 160

References 162

8. Void characterization techniques 167

8.1 Introduction 167

8.2 Microscopy 167

8.2.1 Optical microscopy 167

8.2.2 Electron microscopy 168

8.3 Positron annihilation lifetime spectroscopy (PALS) 172

8.4 Three-dimensional imaging 174

8.5 Gas adsorption 178

8.6 Chromatographic porosimetry 180

8.6.1 Introduction 180

8.6.2 Inverse gas chromatography (IGC) 181

8.6.3 Inverse size exclusion chromatography 183

References 186

9. Characteristics and properties of porous materials 189

9.1 Introduction 189

9.2 General characterization 190

9.2.1 Cell size 190
9.2.2 Open versus closed cell 191
9.2.3 Reinforced versus unreinforced voids 193
9.2.4 Density 195
9.2.5 Relative density 196
9.2.6 Energy absorption 197
9.2.7 Cell size distribution and regularity 199
9.3 Conventional foams 202
9.3.1 Stress–strain behavior in compression 202
9.3.2 Elastic constants 203
9.3.3 Dielectric constant 206
9.4 Syntactic foams 208
9.4.1 Growth and performance 208
9.4.2 Compressive stress–strain relationship 210
9.5 Thermal properties 212
9.6 Finite element analysis (FEA) 217
9.7 Geopolymer foams 220
9.8 Metallic foams 222
References 226

10. Applications 231
10.1 Introduction 231
10.2 Syntactic foams 233
10.2.1 Deep-sea buoyancy 233
10.2.2 Hollow composite macrospheres and composite syntactic foams 238
10.2.3 Deep-sea thermal insulation 239
10.2.4 Syntactic foams and explosive formulations 240
10.2.5 Other application 242
10.3 Aerospace 243

10.3.1 Carbon nanotubes (CNT) 243
10.3.2 Honeycombs 245
10.3.3 Thermal protection systems and heat shields 246
10.3.4 Silica aerogel for a comet dust collector 249
10.3.5 Thermal barrier coatings (TBCs) 254
10.4 Energy 257
10.4.1 Lithium-ion battery 257
10.4.2 Electrochemical energy storage with porous metals 259
10.4.3 Guest–host complexes 261
10.4.4 Solar power 263
10.5 Titania and photocatalysis 263
10.6 Biomaterials and healthcare 265
10.6.1 Introduction 265
10.6.2 Biomaterials scaffold 267
10.6.3 Nerve regeneration 268
10.6.4 Drug delivery 272
10.7 Menger sponges 272
References 275

Glossary 281
Author index 285
Subject index 297

About the authors

Gary Gladysz is an adjunct associate professor of materials science and engineering at the University of Alabama at Birmingham, United States and founder at X-Link 3D at Youngstown, OH, United States. He received his PhD from the New Mexico Institute of Mining and Technology, where he participated in the NATO Collaborative Program with the German Aerospace Institute (DLR). Since receiving his PhD, he has led research efforts in university, government, and industrial settings. He has extensive research experience designing and characterizing thermoset composite materials for 3D printing, fibrous composites, ceramic composites, polymers, composite foams, and thin films. As a technical staff member at Los Alamos National Laboratory (LANL), he was technical lead for rigid composites and thermoset materials. In 2005 he was awarded the LANL Distinguished Performance Group Award for his work leading materials development on the Reliable Replacement Warhead Feasibility Project. Additionally, while the US Army, he developed composite materials and test protocols for ballistic head protection. He has served on funding review boards for LANL, National Science Foundation, ACS, and the Lindbergh Foundation. He has been guest editor on many issues of leading materials science journals, including *Journal of Materials Science* and *Materials Science & Engineering*. He has organized many international conferences/symposia on syntactic foams, composite materials, and innovative materials for additive manufacturing. He started and chairs the ECI international conference series on Syntactic and Composites Foams. He currently lives in Boston, Massachusetts, United States.

Professor **Krishan Chawla** obtained his BS from Banaras Hindu University and his MS and PhD degrees from the University of Illinois *at* Urbana-Champaign, United States. He has taught and/or done research at (in alphabetical order) Arizona State University, Tempe, AZ (United States); Ecole Polytechnique Federale de Lausanne (Switzerland); Federal Institute for Materials Research and Testing (BAM), Berlin (Germany); German Aerospace Research Institute (DLR), Cologne (Germany); Instituto Militar de Engenharia (Brazil); Laval University (Canada); Los Alamos National Lab (United States); New Mexico Tech (United States); Northwestern University (United States); University of Alabama at Birmingham (United States); and University of Illinois *at* Urbana-Champaign (United States).

He has published extensively in the areas of processing, microstructure, and mechanical behavior of materials, in general, and composite materials and fibers, in particular. Besides being a member of various professional societies, he is Editor of *International Materials Review* (published jointly by ASM International, United States and the Institute of Materials, London) and a member of the Editorial Board of various journals. During 1989-1990, he served as a Program Director for metals and ceramics in the Division of Materials Research, National Science Foundation, Washington, DC, United States. He serves as a consultant to the industry, US national laboratories, and various US federal government agencies. In 1992 he was the recipient of the *Eshbach Society Distinguished Visiting Scholar Award* from Northwestern University. During the period of June, 1994 through June, 1995 he held the *US Dept. of Energy Faculty Fellowship* at Oak Ridge National Lab. In 1996 he was given the *Distinguished Researcher Award* by the New Mexico Tech. In 1997 he was made a *Fellow of ASM international.* In 2000 he was awarded the *Distinguished Alumnus* award by Banaras Hindu University. He received the *President's Award for Excellence in Teaching,* University of Alabama at Birmingham in 2006. In 2018 he was awarded the *Albert Nelson Marquis Lifetime Achievement Award.*

Preface to the first edition

The title of this book, *Voids in Materials: From Unavoidable Defects to Designed Cellular Materials* says everything. All materials have voids at some scale. Sometimes the voids are ignored, sometimes they are taken into account, and other times they are the focal point of the research. In this book, however, we take due notice of all of these occurrences of voids, whether designed or unavoidable defects, we define these voids (or empty spaces in materials), categorize them, characterize them, and describe the effect they have on material properties.

After an introductory chapter, we devote a chapter each on intrinsic voids in crystalline materials (such as metals and ceramics) and in polymers. We explain the differences between ideal and real materials as rooted in the voids and defects. We discuss the origins, diffusion, and coalescence of voids/defects and the relation to phenomena such as creep, physical aging, diffusion, glass transition temperature, thermal expansion, how material properties change with size, distribution, and amount of voids, and the implications that voids have on product design. This is followed by a chapter on intentional voids in materials. Oftentimes, the methods and the vocabulary related to foams are material-specific. Similar methods can be called by different names when working with a metal or ceramic. We point out the commonalities in the way the voids are introduced in different materials, highlight the similarities, and point out the different terms used to describe them. In addition to a chapter on intentional voids in bulk materials, we devote a chapter on the introduction of voids into dispersed phases such as particles and fibers. Structures such as nanotubes, hollow and porous spheres, membranes, and nonspherical particles are technologically important in fields as diverse as catalysis, biomaterials, ablation, composite materials, and pharmaceuticals/medicine. A chapter is devoted to cellular materials or foams, wherein we highlight the commonalities in material properties of voids in polymers, metals, and ceramics. Finally broad applications of such cellular materials are described along with techniques used to characterize voids.

Throughout the book we have taken the approach of highlighting the physics and chemistry of the subject matter under consideration while minimizing the mathematical part. Extensive use is made of line drawings and micrographs to bring home to the reader the importance of voids as unavoidable structural defects as well as voids being an element of design to obtain the desired properties in a material. The intended audience for this book are students, researchers, practicing engineers in the fields of materials science and engineering, physics, chemistry, and mechanical engineering.

Finally, we would like to acknowledge our colleagues without whose help we would not have been able to do this project. Gary M. Gladysz would like to thank A. Boccaccini, K. Carlisle, W. Congdon, L. Dai, S. Emets, N. Godfrey, M. Koopman, M. Lewis, J. Lula, U. Mann, G. McEachen, D. Mendoza, B. Perry, W. Ricci, S. Rutherford, C. Sandoval, and V. Shabde.

Krishan K. Chawla would like to thank A. Boccaccini, N. Chawla, K. Carlisle, M. Koopman, M. Lewis, and A. Mortensen. Thanks are due to Kanika Chawla and A. Woodman for help with the figures.

Gary M. Gladysz[1,2], Krishan K. Chawla[1]

[1]DEPARTMENT OF MATERIALS SCIENCE AND ENGINEERING, UNIVERSITY OF ALABAMA AT BIRMINGHAM, BIRMINGHAM, AL, UNITED STATES [2]X-LINK 3D, YOUNGSTOWN, OH, UNITED STATES

Preface to the second edition

Right from the stage of conception of this book, *Voids in Materials: From Unavoidable Defects to Designed Cellular Materials*, our goal was to create a book focusing on voids, independent of the material containing the voids. We believe this is what distinguishes our book from other books on foams or defects, etc. Since the first edition in 2014 there has been significant progress in this field. The expansion is driven by the maximizing functionality while minimizing mass of the part and the time to needed to make the part. We have expanded on these themes in the second edition, adding new content that includes the latest topics and applications of cutting edge research and development. We hope that the students, scientists, researchers, and engineers will find the new content and organization of the book compelling and in accord with the overall concept. Another positive aspect of this book is the wealth of references and further reading suggestions that we provide at the end of each chapter. This will direct readers who wish to explore topics in more depth to the relevant literature.

The cover design of this edition visually encapsulates many of the key concepts discussed. Though there is no scale marker, one can immediately recognize the bottom graphic as a carbon nanotube. That fact that the hollow portion is an intentional void on the single nanometer length scale is a testament to the precision we now have on the design of materials. Even on a smaller length scale is the arrangement of the atoms creating the graphene shell of the carbon nanotube creating a honeycomb like pattern with nonstochastic intrinsic voids. The top image shows a blown stochastic foam with a gradient cell structure made of micrometer scale intentional voids. In contrast, the interface regions contain intrinsic voids; intrinsic since the level of control over the process make these defects unintentional yet inevitable. The span of length scale of voids just discussed, subnanometer, nanometer, and micrometer suggest that ability for hierarchical design of porosity.

A topic that has that been expanded with new content in multiple chapters is additive manufacturing (AM) or 3D printing. This reflects the rapid, worldwide expansion of AM in research and development, manufacturing, and applications. We discuss the seven distinct technologies within AM and highlight interesting porous structures for each. We discuss the new concept of the digital design of voids and the role it has in mass customization; tailoring adidas shoes and biomaterial scaffolds to suit an individual's need. Digital design of porosity has made it possible to tailor regularity (R) from completely stochastic and random ($R = 0$) to nonstochastic and ordered ($R = 1$) and everything in between. We discuss regularity as it relates to Voronoi structures, designing and tailoring voids within a structure to achieve certain functionalities. Needless to say, this is only possible with AM.

In the first edition we had a section or two discussing the Kirkendall effect. The Kirkendall effect when initially identified in the 1940s was thought of as detrimental in metals. However, in the early 2000s, methods and procedures were developed to constructively use this "detrimental" effect to design and create hollow structures. Since becoming a process for creating intentional porosity, the research has dramatically increased. We now devote an entire Chapter 5, Hollow and porous structures utilizing the Kirkendall effect, to cover the topic in detail. We cover hollow particle and tube formation on the nanometer and micrometer scale and well as porous particles and fibers. We discuss *symmetric* and *asymmetric* Kirkendall mechanisms for creating hollow particle and fibers and explain the underlying phenomena that lead to each.

Another entirely new chapter is Chapter 4, Nanometer-scale porous structures. Nanotechnology has already made its way into our daily lives in the optics, chemical filtration, desalinization, and sensors to name a few. We discuss many types of cutting edge research on nanocellular foams, nanoporous noble metals, zeolites as well as the various nanoporous organic networks (covalent organic frameworks, covalent triazine frameworks, polymers of intrinsic microporosity, and conjugated microporous polymers).

We have expanded the *Applications* chapter to include a wide variety of new applications from using aerogel to collect comet dust, electrochemical energy storage, to thermal barrier coating of jet engines.

Finally, we would like to acknowledge our colleagues whose contributions to this second edition have been invaluable. Gary M. Gladysz would like to thank A. Boccaccini, K. Carlisle, A. Campanella, J. J. Castellón, W. Congdon, M. Gromacki, N. Gupta, C. Hershey, M. Koopman, M. Lewis, J. Lindahl, O. Manoukian, D. Mendoza, V. Mishra, B. Pillay, D. Schmidt, C. Sandoval, and K. Shah. Krishan K. Chawla would like to thank A. Boccaccini, K. Carlisle, N. Chawla, M. Koopman, M. Lewis, and A. Mortensen. Thanks are due to Kanika Chawla and M. Armstrong for help with figures.

Gary M. Gladysz[1,2], Krishan K. Chawla[1]

[1]*DEPARTMENT OF MATERIALS SCIENCE AND ENGINEERING, UNIVERSITY OF ALABAMA AT BIRMINGHAM, BIRMINGHAM, AL, UNITED STATES* [2]*X-LINK 3D, YOUNGSTOWN, OH, UNITED STATES*

September 1, 2020

1 Introduction

1.1 Overview

So why write a book just on voids in materials as the topic? The answer is simple and twofold—first, the juxtaposition of "empty space" adjacent to solid material seemed, to us, an interesting dichotomy. Second, depending on one's perspective or desired outcome, voids can limit or enhance the performance of materials. Is the target a consolidated material or a foam? If the target is a foam, how do material properties change with the type and amount of voids; open cell versus closed cell, the mean size, size distribution, volume fraction of voids, etc.? If the target is a consolidated material, some important questions might be how do properties change with the volume percent, location, geometry, and size of unwanted voids? Another interesting question is how do those voids, on the subnanometer and nanometer scales, which are typically not characterized by foam researchers, play into the final properties.

Voids are also very important to understand from a practical engineering standpoint. Even in the most highly engineered densified materials, defects, such as voids, will limit the design of real structures. So along with the theoretical exploration of voids in materials, in this book, we provide examples of real-world applications wherein voids are prevalent in structures and affect their properties.

In this book we explore such dichotomies; solid versus empty and desired versus undesired aspects of voids in materials. Furthermore, we would like to shed light on a "middle ground" of the smart use of voids to help in the optimization of performance of a part. By middle ground we mean a neutral look at the impact voids have on material/parts and use of voids as a design parameter for optimizing performance in multifunctional materials. There is much published work available on foams (Gibson & Ashby, 1997; Shutov, 2004). Even more numerous are those that provide a passing mention of voids when they are incidental/unwanted during the fabrication of nominally dense materials. This book, however, is not just about foams or residual porosity in materials, important though these contributions are; instead it focuses on the *void* itself. The fact is that all materials have voids, that is, they are pervasive in all materials at some length scale. So, in addition to voids in foams, this book brings in information from a number of different fields of study such as material science and engineering, physics and chemistry of materials, and mechanics of materials. This book treats all of these "different" types of voids equally and

Voids in Materials. DOI: https://doi.org/10.1016/B978-0-12-819282-5.00001-6

highlights their commonalities in all aspects—from processing, formation, and characterization to the resulting material properties.

For the purpose of this book, a void has two essential properties, it must be (1) a volume measured in cube of some unit of length and (2) occupied by a vacuum or gas (i.e., solid/liquid materials are absent). In general, there is no size or shape requirement on a void; so it ranges from subnanometers to millimeters, sometimes even larger, in equivalent diameter. We should make it clear that voids in a liquid and gaseous medium will not be covered in this book. We devote the rest of this chapter to a general discussion on voids.

1.2 Descriptions

1.2.1 Intrinsic and intentional voids

Intrinsic voids appear in materials because of inherent structure, natural processes, processing limitations, and/or aging in service environments. At some length scale, all real materials have intrinsic voids. At the atomic level, if we examine the Bohr atomic model, we see that most of the volume occupied by an atom is empty space. We will not be going into details of the Bohr model in this book, but it is important to mention. A general chemistry text is sufficient to review the general structure of atoms.

When voids are thought of as defects, they are viewed as having a detrimental effect on material properties. However, sometimes defects can be beneficial and essential to specific material behavior, such as color centers and semiconducting properties. The intrinsic voids generally range from 10^{-15} to 10^{-3} m; examples include atomic vacancies, free volume, lattice holes, and process induced porosity.

Intentional voids are incorporated by design into a solid material. Such materials are usually, but not always, referred to as foams. This is especially true when the voids are on a micrometer scale. In the early 21st century, technology has evolved to an extent that we can control voids in the single nanometer size range. Materials with intentional voids can be classified in many ways. Some important examples are *single-phase foams, composite* and *syntactic foams* (Gladysz & Chawla, 2002). There are many ways to introduce an intentional void into a material, the method of introducing these voids is highly dependent on the type of material they are introduced into as well as the desired properties needed in the finished material. Details of these processes will be covered in Chapters 5–7.

When we discuss intentional voids in a material, it is important to remember that the intrinsic voids on the atomic and/or nanometer length scale may still be present. Fig. 1–1 is an example of an intrinsic void in the wall of a hollow (intentional void) glass microspheres formed during the spray drying formation process. The hollow core of the sphere is in the micrometer range and the unintentional voids in the

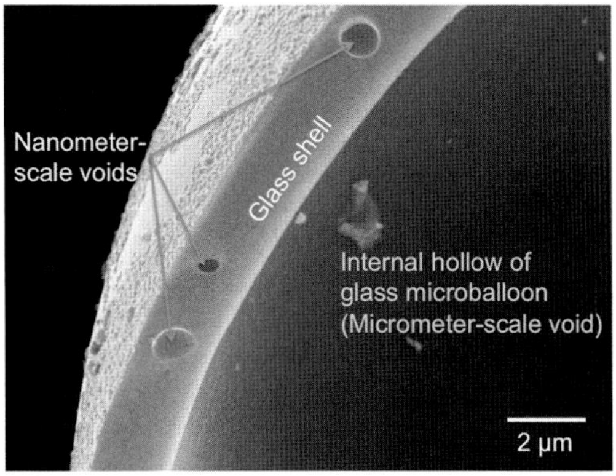

FIGURE 1–1 A hollow glass sphere illustrating nanometer-scale intrinsic voids caused by processing and the intentional void, the hollow core of the sphere.

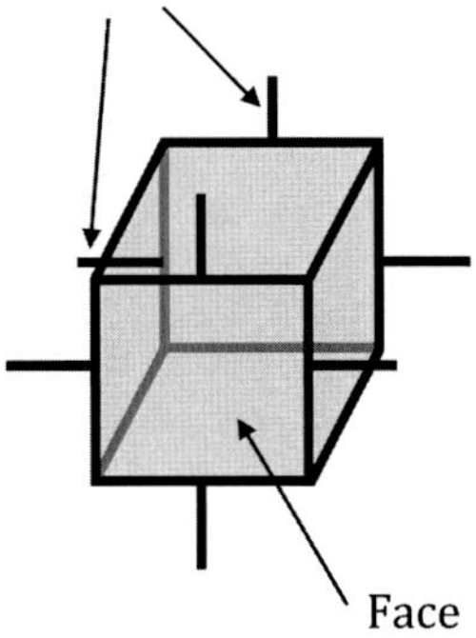

FIGURE 1–2 An idealized structure of a cell consisting of struts and faces.

shell are in nanometer/submicrometer range. This intrinsic void weakens the shell of the sphere and can be the cause of failure during service.

Independent of the material, voids can be categorized as *reinforced* or *unreinforced* and *open* or *closed cell.* We will discuss the concepts of open cell versus closed cell and reinforced versus unreinforced in Sections 1.2.2 and 1.2.3, respectively.

1.2.2 Closed and open cell porosity

Whether discussing cells in foam or just general porosity, they are simply voids dispersed in a solid phase. Cells are made up of struts and faces, as shown schematically in Fig. 1–2, that surround the void space. In a closed cell, the face of the cell wall consists of a continuous solid phase. In an open cell material, a part of that wall is missing.

In a flexible open cell foam, see Fig. 1–3A and B, gas can freely flow in or out of the cells when the structure is extended or compressed. Because the cell faces are

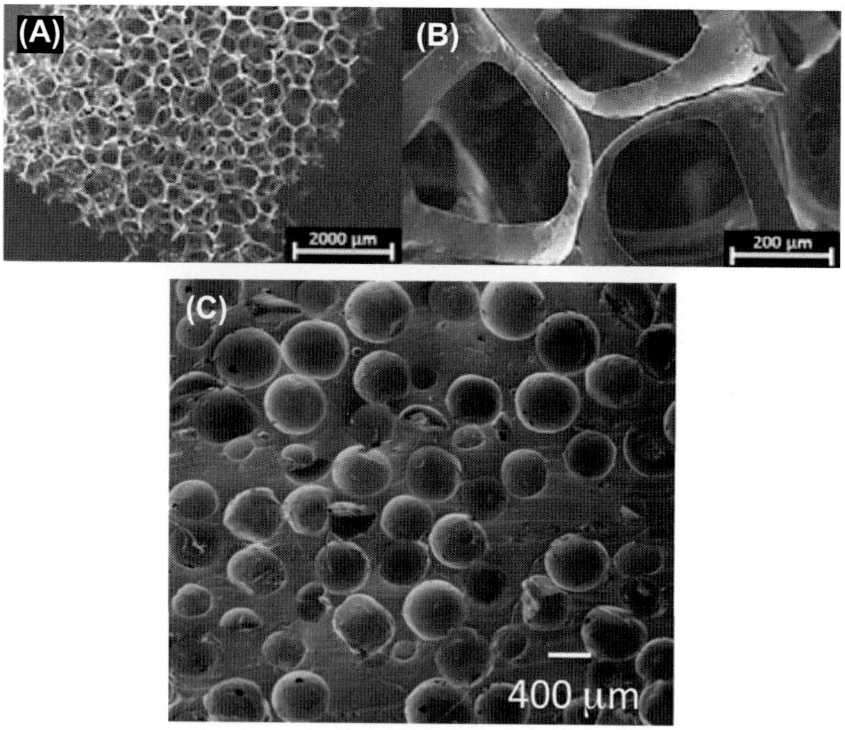

FIGURE 1–3 Examples of the structure: (A) low magnification and (B) higher magnification of open cell reticulated foam (C) a closed cell silicone foam. *Source: Lepage, G., Albernaz, F. O., Perrier, G., & Merlin, G. (2012). Characterization of a microbial fuel cell with reticulated carbon foam electrodes. Bioresource Technology, 124, 199–207.*

discontinuous, materials containing open cells typically have a lower modulus and strength than those containing closed cells. The reticulated (meaning weblike) foam in Fig. 1–3 is an extreme example of an open cell foam as it is composed entirely of struts without faces. This material is a candidate for an electrode in microbial fuel cell (Lepage, Albernaz, Perrier, & Merlin, 2012). There are some general conditions needed for a material to be open cell. According to Shutov (2004), the following two criteria must be met for a predominantly open cell structure:

- Each polygonal cell must have at least two discontinuous or broken faces.
- An overwhelming majority of the cell struts must be shared by at least three cells.

From the above criteria, it is clear that the physical structure of open cell foams and resulting properties can vary widely. We discuss these structure–property relationships in more detail in Chapter 9, Characteristics and properties of Porous Materials. In general, open cell foams exhibit good absorption capacity for water and good acoustic damping properties (Zhang, Li, Hu, Zhu, & Huang, 2012) compared to closed cell foams.

In closed cell foams (Fig. 1–3C) the faces are continuous, which leaves gases inside individual cells isolated from the surrounding cells. Because the faces are intact, closed cell foams typically have higher strength and modulus than open cell foams. In addition to superior mechanical properties, they are used extensively for their insulating properties (Jelle, 2011), because the air trapped in the cells reduce thermal conductivity.

1.2.3 Unreinforced and reinforced voids

Unreinforced voids are present in most materials that we deal with on a day-to-day basis. Conventional, single-phase foams are the most recognizable materials that have unreinforced voids. A common example of a single-phase foam containing an unreinforced void phase is the polyurethane foam used, for example, for cushioning in furniture and expanded polystyrene (PS) used for insulation. Examples of unreinforced voids are shown in Fig. 1–3A–C.

Foams containing unreinforced voids, as we mentioned above, make a very large class of materials and find wide applications. There are many books and journals dedicated to the behavior of such foams, so what we present in this book on this topic will be of a general nature and the reader will be directed to the suggested reading listed for more details.

Reinforced voids are mostly encountered in a class of materials called *syntactic foams* or *composite foams* (Gladysz & Chawla, 2002). They occur when one of the reinforcing phases is hollow or porous. Examples of hollow reinforcing phases are glass microballoons and hollow fibers, such as carbon nanotubes.

The need to distinguish between reinforced and unreinforced voids became evident with the development of syntactic foams in the 1960s and 1970s. The first widespread use of syntactic foams was for use in deep-sea buoyancy and insulation applications. Voids are introduced in a syntactic foam by bonding together of a hollow material, typically in the form of microballoons, with a binder phase. The hollow particle or microballoon is the reinforced void phase; the shell is commonly made of glass; however, the shell can be made of phenolic, carbon, ceramic, or metal also. The binder phase can be a polymer, metal, or ceramic.

Syntactic foams can be further categorized as two- or three-phase syntactic foams. A three-phase syntactic foam is made from microballoons, a binder phase, and interstitial voids. This interstitial void is an unreinforced void; it can be either open or closed cell and can be engineered into the syntactic material in order to minimize density. Although not referred to as such, hollow fibers, such as nanotubes when embedded in a matrix, can be viewed as reinforced void and a syntactic foam. Syntactic foams are used where high specific strength and modulus materials are needed.

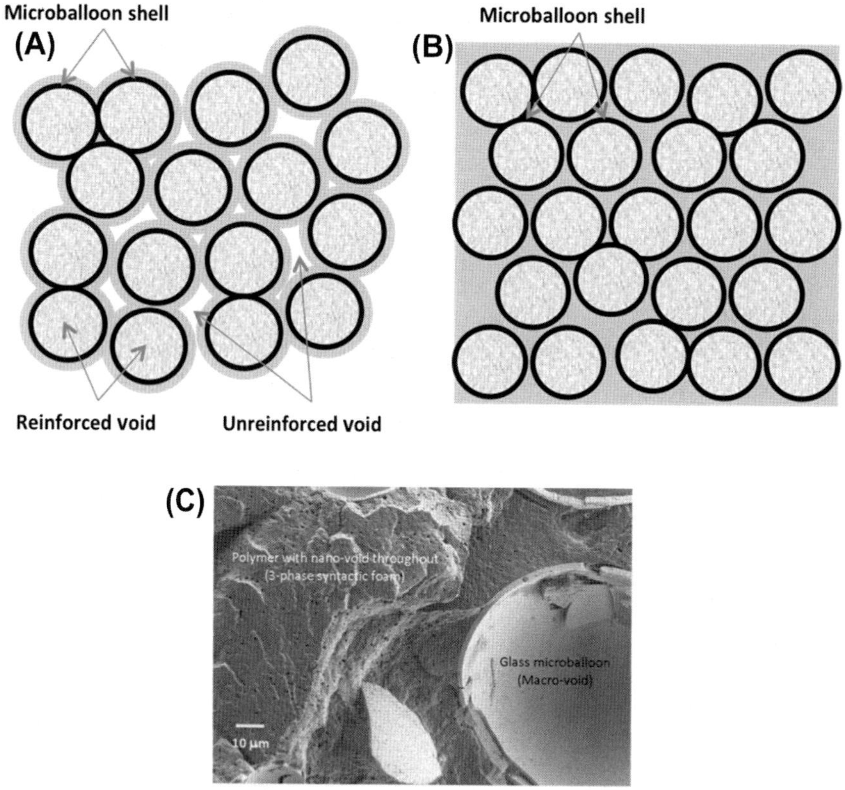

FIGURE 1–4 (A) A schematic of a three-phase syntactic foam where interstitial void is engineered into the material and (B) a two-phase syntactic foam where there is sufficient binder to fill the interstices. (C) Scanning electron micrograph of a cross section of a syntactic foam, illustrating unreinforced (matrix nanoporosity) and reinforced (glass microballoon) void on two different length scales, micrometer and nanometer.

Fig. 1–4 compares two- and three-phase syntactic foams. Three-phase syntactic foams (Fig. 1–4A) are designed such that the microballoon and binder phase volume fraction is less than one; the remainder being the unreinforced, interstitial void. Two-phase syntactic foams (Fig. 1–4B) are designed so that the volume fractions of the binder phase and microballoons add up to unity, that is, there is enough binder to fill the interstices between the microballoons. Typically, the dimensions of the unreinforced voids in a syntactic foam are on a micrometer scale. Fig. 1–4C shows the microstructure of a three-phase syntactic foam, made of glass microballoons and porous polymer binder.

There are several examples of hollow microspheres (micrometer scale in diameter) available commercially. Glass, phenolic, and ceramic particles are the most common ones. In large scale manufacturing, hollow particle formation processes rely on a *blowing agent.* The internal gas expands the skin of the hollow particle which then cools and hardens into a particle with a central void.

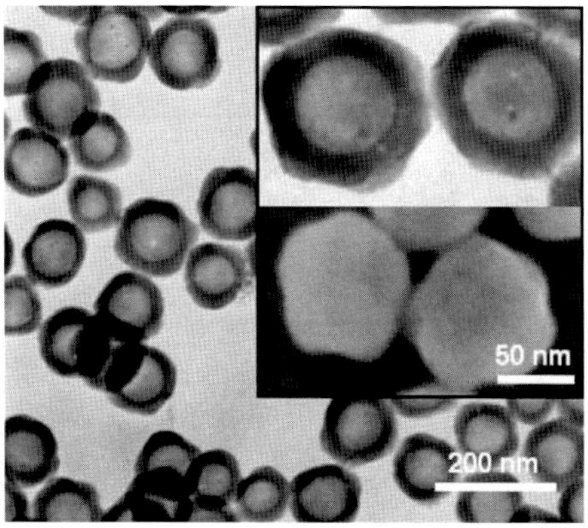

FIGURE 1–5 Silica nanospheres demonstrate the control of voids down to 50 nm range. The insets show higher magnification of the nanospheres (SEM and TEM). *Source: Huang, S., Yu, X., Dong, Y., Li, L., & Guo, X. (2012). Spherical polyelectrolyte brushes: Ideal templates for preparing pH-sensitive coreshell and hollow silica nanoparticles. Colloids and Surfaces A: Physicochemical and Engineering Aspects, 415, 22–30.*

It is possible to make reinforced voids on nanometer, micrometer, and macrometer scales. On a nanometer scale, hollow spherical shells have been made by templating nanoparticles on a sacrificial core material (Minami, Kobayashi, & Okubo, 2005) or bubble (Hadiko, Han, Fuji, & Takahashi, 2005). Another technique that can be used is plasma polymerization (Cao & Matsoukas, 2004). Fig. 1–4C shows a three-phase syntactic foam; the binder phase having nanometer-scale unreinforced porosity, with a glass microballoon reinforced void. Fig. 1–5 shows an example of hollow particle fabricated at the submicrometer range and nanometer range (Huang, Yu, Dong, Li, & Guo, 2012). These are silica hollow particles produced via the templating method. The sacrificial core was PS with poly acrylic acid (PAA) attached to the surface. The functionality of the PAA was to direct and control the coating of silica. The PS/PAA core was dissolved using chloroform leaving behind a hollow particle.

1.2.4 Porosity in natural and synthetic materials

There are many naturally occurring materials that are used today, for example, granite and marble countertops and lumber and stone used in construction. In fact, from a supply chain and raw materials standpoint all of our advanced materials come from nature. As the stone and lumber industry illustrates, "[I]f it is not grown, then it is mined." For example, polymers can be made from fossil fuel or plant-based feed stocks. Natural materials are the building blocks of all engineered materials. There are numerous examples of porous materials that are found in nature. Sometimes these materials are mined and directly used as functional materials, whereas at other times a natural material is studied, and its microstructure is replicated in the laboratory. The replication in the lab is generally an attempt to narrow the inevitable variability seen in a natural material. We do this by implementing process and raw material controls to reduce variability; thus creating an engineered product.

We would like to use Castella de San Marcus, located in Florida, United States as an example. Its construction dates back to 1672, centuries before the emergence of modern materials science and engineering. Since 1672 the structure has survived wars and hurricanes, at least partly, because of the durability of the walls (Fig. 1–6). The porous walls of this structure are known for the ability to absorb the energy of projectiles, like bullets and cannon balls, without large crack emanating from the point of impact that would lead to catastrophic failure (Subhash, Jannotti, & Subhash, 2016). Interestingly, because of the materials selection and construction of the walls, this fort was never taken by force. The natural material of wall construction is a porous rock formation call coquina. The ability to absorb the energy of projectiles has to do with the unique microstructure). Coquina is classified as a sedimentary rock, that is, it is made up of fragments of other preexisting rocks. It is a composite of fragmented

FIGURE 1–6 A photograph of the Castella de San Marcus located in St. Augustine, FL, United States. The original construction started in 1672. The choice of material for the exterior facing walls is coquina, a porous sedimentary rock with a noted ability to absorb the energy from bullets and cannon balls of that time without experiencing brittle fracture.

marine shells, fossils and coral, limestone, sand, minerals and clay, with significant void space between these solid components. The voids are located at the interstitial positions between the shell fragments. Following is a summary of the natural process that results in coquina and sedimentary rocks, in general:

- Weathering of existing rock, marine shells,
- transportation of the weathered shells,
- deposition of shells in layers,
- compaction, and
- cementation.

The last two items in the above list are called *lithification,* meaning "convert into solid rock." Since there are interstitial voids present, this type of rock is considered partially lithified. This natural process of introducing voids in sedimentary rock has analogous processing steps in engineered synthetic materials. However, to make comparable engineered material, the time scale needs to be significantly compressed. We discuss these processes in detail in Chapter 6, Techniques for Introducing Voids into Bulk Materials, Techniques of Introducing Intentional Voids into Particles and Fibers.

A synthetic material that has microstructural and processing similarities to coquina is frit material; frit has a network of particles connected at contact points with interstitial porosity. For example, Cho and Kim (2016) used zinc aluminum borosilicate glass ceramic to investigate the dependence of particle size on the densification and mechanical behavior of the final frit material. The process has similarities with the partially lithified sedimentary rock formation, the process done by Cho and Kim is as follows, with the analogous sedimentary rock formation steps given in parentheses:

- Jet mill and ball mill to crush glass into desired particle size (Weathering).
- Sedimentation to size-separate particles (Transportation/deposition).
- Centrifuge to pack the particles (Compaction).
- High temperature sintering (Cementation).

Fig. 1−7 is an SEM micrograph of the materials made up of different particle sizes after the centrifuge step. At this point the loose powder is consolidated but there is no bonding between the discrete particles. These materials were processed further at 900°C to fuse the particles together (Fig. 1−8). Just as the coquina has interstitial porosity so does this glass frit material, ranging from 4% to 14% by volume.

1.2.5 Stochastic, nonstochastic, and Voronoi foams

Before the advent of additive manufacturing (AM) it was virtually impossible to fabricate, in a reproducible manner, the exact microstructure of a chemically blown foam. Studies would normalize data related to foams based on density with the implicit

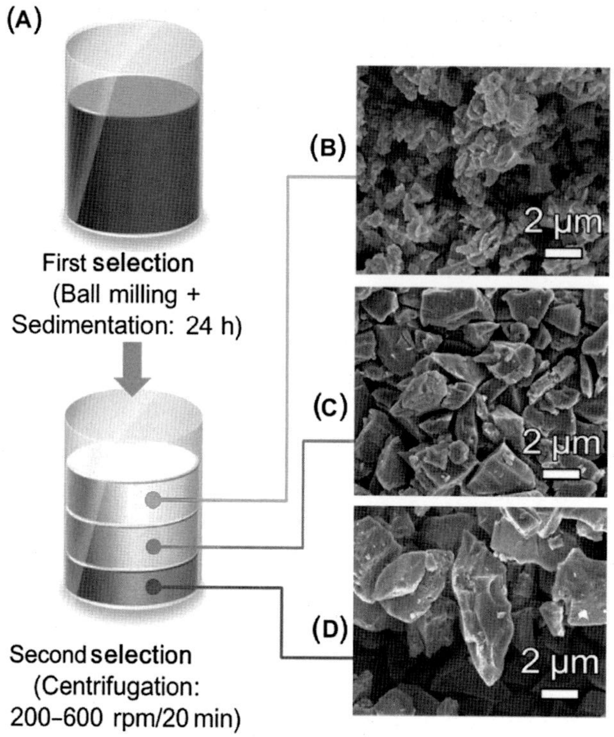

FIGURE 1–7 Process schematic showing the (A) particle size reduction and sedimentation (via sieve screens) of zinc aluminum borosilicate glass ceramic. The micrographs in (B–D) are the unsintered microstructure of the compacted particles with average particle size of 1.2, 2.9, and 4.8 μm, respectively. *Source: Cho, I.S., & Kim, D. (2016). Glass-frit size dependence of densification behavior and mechanical properties of zinc aluminum calcium borosilicate glass-ceramics. https://doi.org/10.1016/j.jallcom.2016.06.008.*

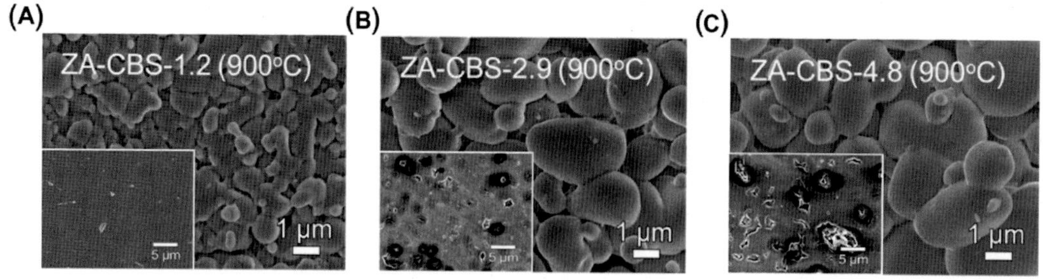

FIGURE 1–8 The sintered zinc aluminum borosilicate glass ceramic particles with average particle size (A) 1.2 μm (B) 2.9 μm, and (C) 4.8 μm. Void percentages range from 4% to 14% by volume. *Source: Cho, I.S., & Kim, D. (2016). Glass-frit size dependence of densification behavior and mechanical properties of zinc aluminum calcium borosilicate glass-ceramics. https://doi.org/10.1016/j.jallcom.2016.06.008.*

understanding that the cell structures from different samples were not identical. We classify such foams as *stochastic foams*. Foams with identical repeating units are classified as *nonstochastic foams*. Nonstochastic foams, other than honeycomb, were also not widespread. Additive manufacturing has led to the ability to design and reproducibly fabricate completely nonstochastic foams as well as the stochastic or disordered foams. In addition, the degree of disorder or randomness can be digitally designed into the structure.

In this regard, it would be appropriate to introduce here the concept of Voronoi foams. Voronoi tessellation, Voronoi decomposition, Voronoi partitioning, or Dirichlet tessellation are terms used to describe partitioning of a plane into regions. The partitioning of a plane into regions is done based on distances from specific points. The set of points is called seeds or sites. Voronoi tessellation finds applications in a variety of fields, science and technology, visual arts, etc.

The word tessellation means covering of a plane using a limited number of different shapes, usually polygons, although rectangles, hexagons, or triangles can also be used, with no overlaps and gaps. It would appear that the Voronoi based evaluation would work fine with analysis of foams. The foam structures can be designed from completely ordered and repeatable cell structure to one that is completely random.

Voronoi (also Kelvin) tessellation was once a topic of theoretical and numerical investigations, but with additive manufacturing this field has opened to empirical research. Lord Kelvin, who in 1887 asked the question, "[W]hat is the best way that space could be partitioned into cells of equal volume with the least area of surface between them?", proposed the Kelvin structure in 1887, a 14-faced polyhedron having 6 square faces and 8 hexagonal faces. It has become known as the Kelvin Problem. The Kelvin structure is called a Voronoi foam with a regularity, $R = 1$. Regularity can range from completely random, $R = 0$, to that of the Kelvin structure with complete regularity, $R = 1$. In a volume of space, V_o, if the number of nuclei, N, contained in it are equally spaced, d_o, then we get the Kelvin structure, that is, $R = 1$ is attained. The distance, d_o, between any two center points, that is, nuclei of adjacent polyhedrons is given by (Duan et al., 2019):

$$d_0 = \frac{\sqrt{6}}{2}\left(\frac{V_0}{N\sqrt{2}}\right)^{\frac{1}{3}}$$

One can use numerical methods to create structures of $R > 1$ by placing a nucleus at a distance, $\delta > d_o$ from any existing nucleus. The regularity, R, can then be defined as the ratio:

$$R = \frac{\delta}{d_0}$$

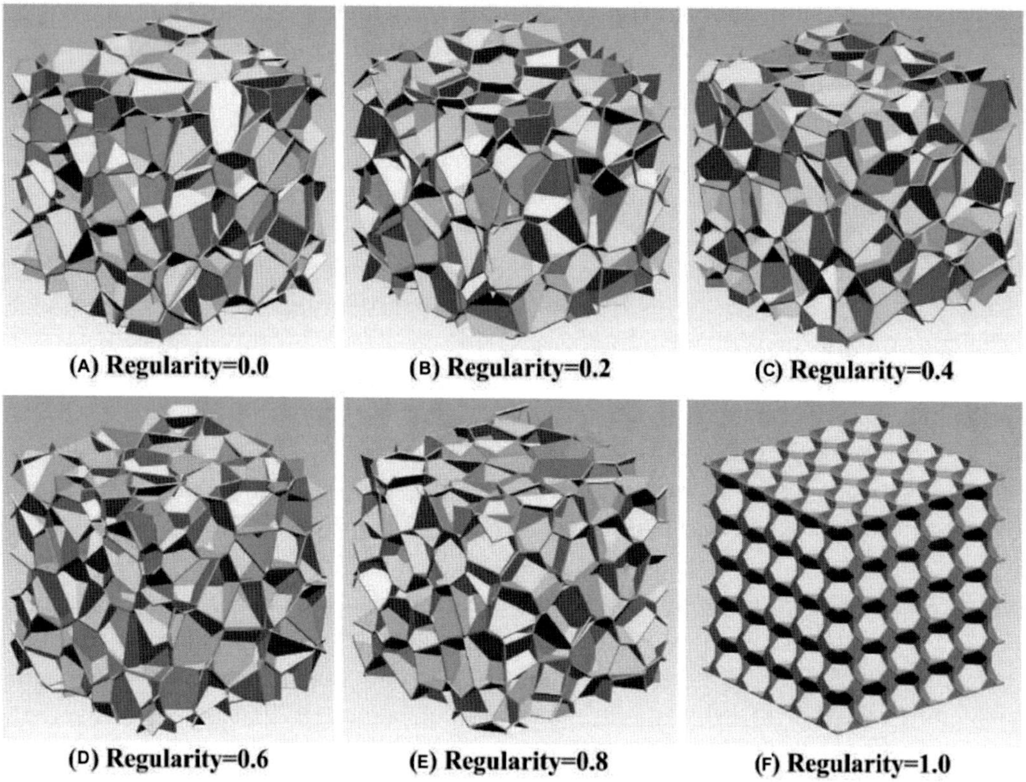

FIGURE 1–9 Voronoi foams designed with regularities, R, from 0 (totally random) to 1 (regular and repeating). *Source: Duan, Y., Du, B., Zhao, X., Hou, N., Shi, X., Hou, B., & Li, Y. (2019). The cell regularity effects on the compressive responses of additively manufactured Voronoi foams. International Journal of Mechanical Sciences, 164, 105151.*

Computer generated structures (Duan et al., 2019) with $R = 0.2$, 0.4, 0.6, 0.8, and 1.0 are shown in Fig. 1–9.

Duan and colleagues (2019) used fused deposition modeling, a material extrusion type of AM to create and test Voronoi foams from computer models. Their samples had a range of R from 0 to 1. We will go into more detail on processing and mechanical properties as a function on R in Chapter 9, Characteristics and Properties of Porous Materials.

1.2.6 Material versus digital design of voids

A combination of techniques can be used to create different size, shape, volume fraction, etc. of voids in a material. One can choose an AM technique via computer control, or one can choose a technique from a variety of conventional manufacturing processes described in Chapter 6, Techniques for Introducing Intentional Voids into

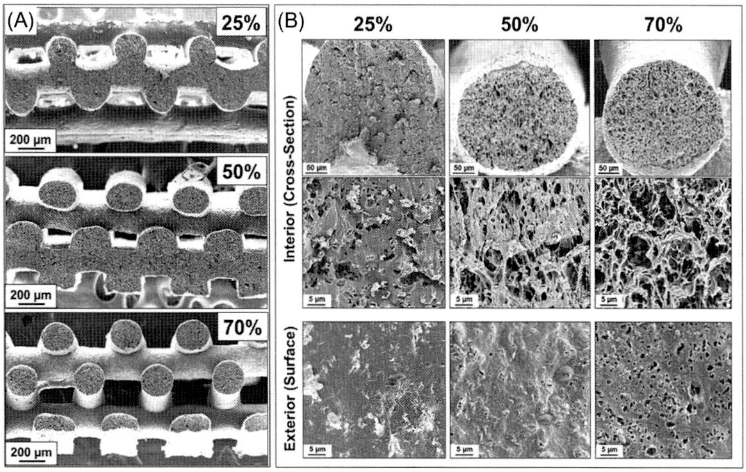

FIGURE 1–10 Cross section of the 3D-printed porous PLGA illustrating both digital and material designed porosity as well as hierarchical porosity. (A) Printing the PLGA in a grid-like pattern leaves pores on the millimeter scale. (B) Micrometer sized porosity was designed by incorporating $CuSO_4$ particles into the PLGA. After printing, the $CuSO_4$ was dissolved with appropriate solvents. *Source: Jakus, A.E., Geisendorfer, N.R., Lewis, P.L., & Shah, R.N. (2018). 3D-printing porosity: A new approach to creating elevated porosity materials and structures. https://doi.org/10.1016/j.actbio.2018.03.039.*

Bulk Materials. When using AM, it is important to make use of ideas of material design to create a porous structure. As an example, we cite the work of Jakus, Geisendorfer, Lewis, and Shah (2018) who combined material and digital design to create a material with elevated (hierarchical) porosity. They used an extrusion AM technique, printing a gridline pattern, see Fig. 1–10, to create voids on the millimeter length scale. The material design was using a polylactide-co-glycolide (PLGA)-based printing ink with copper sulfate ($CuSO_4$) salt as a pore-forming agent. Once the grid structure was printed, the salt was dissolved in ethanol followed by immersion in a water bath leaving voids with the same shape and volume fraction as the $CuSO_4$ that was loaded into the printing ink. They were able to have a total porosity between 66.6% and 94.4% by volume.

1.3 Voids through the length scale

From the structure of a single chemical element to complex multielement structures, the arrangement of atoms is the key, although not the only contribution, in defining the resultant macroscale properties. If one were to examine the breadth of materials-related industries, one would immediately see that a large variety of materials are processed for many different applications. Whether these are manufactured ceramics, metals, polymers, composites, or naturally occurring materials (e.g., wood or stone),

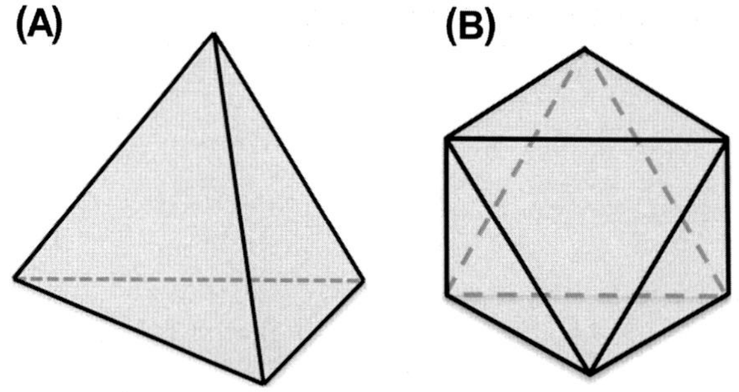

FIGURE 1–11 Geometry of a (A) tetrahedral void made from four spheres touching and (B) octahedral void made from six spheres touching. *Source: Krishna, P., & Pandey, D. (1981). Close-packed structures. (Pamphlet). Cardiff: University Collage Cardiff Press for the International Union on Chrystallographic Teaching. Retrieved from http://www.iucr.org/__data/assets/pdf_file/0015/13254/5.pdf.*

they all have very distinct and different range of structures and properties. The structure–processing–properties relationships are critically important in this whole endeavor spanning from synthesis to processing to applications. Understanding the processing and the effects it has on thermal, mechanical, electrical, etc., properties is very important so one can take advantage of the effects in material design. In this book, we choose to describe these effects on material properties on different length scales through the medium of voids.

When building atoms/ions into crystal specific structures of metals and ceramics, an important concept is that of theoretical packing efficiency, which leads us to certain mechanical and physical properties of these crystals. These packing efficiencies for crystals composed of a single type of atoms range from 74% for hexagonal close packed and face centered cubic crystals to 52% for simple cubic crystals. The voids or empty spaces in the various close-packed crystalline structures are categorized as octahedral and tetrahedral holes (Krishna & Pandey, 1981) (Fig. 1–11). A tetrahedral hole occurs when four spheres are in contact while an octahedral hole is in the interstitial space, where six spheres are touching (Fig. 1–11).

In noncrystalline materials the space between inorganic networks or polymer chains is referred to as *free volume*. The most important factor in determining the characteristic of free volume in a polymer is the structure of the polymer or polymers in the system. At the subnanometer length scale, there is hindrance between polymer chains. Hindrance is a term used to describe the arrangement of polymer chains based on a polymer system's inherent spatial and topological constraints (Micheletti, Marenduzzo, & Orlandini, 2011). All polymers have a measureable free volume and it can be determined experimentally by a technique called Positron Annihilation Lifetime Spectroscopy (Pethrick, 1997). Positron is a positively charged subatomic particle, a positive electron. Positrons can interact with materials and form positronium. Positronium is attracted to

Table 1–1 A general outline of the voids.

Length scale (m)	Intrinsic	Intentional
10^{-15}	Atomic defects	
10^{-11}	Electronic defects	
10^{-10}	Cubic, octahedral, and tetrahedral holes	
10^{-10}–10^{-8}	Free volume	Tuning of free volume
10^{-9}	Processing induced voids	Tubes and hollow particles
10^{-8}–10^{-6}	Processing induced voids	Conventional foams/hollow particles
10^{-5}–10^{-3}		Hollow spheres/foams
Macro		Hollow spheres/foams honeycomb, digital design through AM

voids in the material. The lifetime of a positronium within a void, when compared to the characteristic lifetime, can be translated into the size of the free volume as well as free volume percent. Free volume has been correlated with many physical and chemical properties in polymers. This technique has been used successfully in polymers, both thermoplastic and thermosetting, glasses, metals, and semiconductors.

For further assembly of crystalline or noncrystalline materials into bulk samples, it is important to remember that the intrinsic, subnanomter voids of either free volume or lattice holes and associated properties are still present. This assembly requires processing. Processing induced scale of intrinsic and intentional voids can then reach cube of nanometers to micrometers in volume. The amount and extent of processing voids depend on the type of process, extent of process control, and processing conditions chosen. The introduction of these voids, whether intrinsic or intentional, can have a measurable effect on material properties.

We point out here that these unintentional voids at this level are artifacts of the process that is used to make the bulk materials and are considered intrinsic. Although they are unavoidable and appear because of limitations of the process, the porosity can be altered or minimized by either optimizing processing parameters or choosing another processing method, which is more capable.

A general outline of the length scales and the types of voids that are encountered are summarized in Table 1–1. These voids will be discussed in detail throughout this book.

References

Cao, J., & Matsoukas, T. (2004). Synthesis of hollow nanoparticles by plasma polymerization. *Journal of Nanoparticle Research*, *6*(5), 447–455.

Cho, I.S., & Kim, D. (2016). *Glass-frit size dependence of densification behavior and mechanical properties of zinc aluminum calcium borosilicate glass-ceramics*. https://doi.org/10.1016/j.jallcom.2016.06.008.

Duan, Y., Du, B., Zhao, X., Hou, N., Shi, X., Hou, B., & Li, Y. (2019). The cell regularity effects on the compressive responses of additively manufactured Voronoi foams. *International Journal of Mechanical Sciences, 164*, 105151.

Gibson, L. J., & Ashby, M. F. (1997). *Cellular solids: Structure and properties*. Cambridge: Cambridge University Press.

Gladysz, G. M., & Chawla, K. K. (2002). *Composite foams. In Encyclopedia of polymer science and technology* (pp. 267–297). New York: John Wiley & Sons.

Hadiko, G., Han, Y. S., Fuji, M., & Takahashi, M. (2005). Synthesis of hollow calcium carbonate particles by the bubble templating method. *Materials Letters, 59*(19–20), 2519–2522.

Huang, S., Yu, X., Dong, Y., Li, L., & Guo, X. (2012). Spherical polyelectrolyte brushes: Ideal templates for preparing pH-sensitive core–shell and hollow silica nanoparticles. *Colloids and Surfaces A: Physicochemical and Engineering Aspects, 415*, 22–30.

Jakus, A.E., Geisendorfer, N.R., Lewis, P.L., & Shah, R.N. (2018). *3D-printing porosity: A new approach to creating elevated porosity materials and structures*. https://doi.org/10.1016/j.actbio.2018.03.039.

Jelle, B. P. (2011). Traditional, state-of-the-art and future thermal building insulation materials and solutions—Properties, requirements and possibilities. *Energy and Buildings, 43*(10), 2549–2563.

Krishna, P., & Pandey, D. (1981). Close-packed structures. *(Pamphlet)*. Cardiff: University Collage Cardiff Press for the International Union on Chrystallographic Teaching. Retrieved from http://www.iucr.org/__data/assets/pdf_file/0015/13254/5.pdf.

Lepage, G., Albernaz, F. O., Perrier, G., & Merlin, G. (2012). Characterization of a microbial fuel cell with reticulated carbon foam electrodes. *Bioresource Technology, 124*, 199–207.

Micheletti, C., Marenduzzo, D., & Orlandini, E. (2011). Polymers with spatial or topological constraints: Theoretical and computational results. *Physics Reports, 504*(1), 1–73.

Minami, H., Kobayashi, H., & Okubo, M. (2005). Preparation of hollow polymer particles with a single hole in the shell by SaPSeP. *Langmuir, 21*(13), 5655–5658.

Pethrick, R. A. (1997). Positron annihilation—A probe for nanoscale voids and free volume? *Progress in Polymer Science, 22*(1), 1–47.

Shutov, F. A. (2004). Cellular structure and properties of foamed polymers. *In Handbook of polymeric foams and foam technology* (pp. 17–53). Munich:: Hanser.

Subhash, S., Jannotti, P., & Subhash, G. (2016). Impact response of coquina. *In Dynamic behavior of materials* (1, pp. 1–4). Springer.

Zhang, C., Li, J., Hu, Z., Zhu, F., & Huang, Y. (2012). Correlation between the acoustic and porous cell morphology of polyurethane foam: Effect of interconnected porosity. *Materials and Design, 41*(0), 319–325.

2 Intrinsic voids in crystalline materials: Ideal materials and real materials

2.1 Introduction

In this chapter, we explore the concepts of *ideal materials* and *real materials*, that is, conceptually building materials from atomic scale to bulk and probing their properties. Although most of us only encounter, characterize, and manufacture real materials, it is important to understand the ideal behavior in materials. The link between the ideal and real materials is rooted in defects and voids within and on their surface. An ideal material has a *theoretical strength*, which is a function of the atomic or molecular bonding between atoms and temperature (Kelly & Macmillan, 1986).

There are two practical reasons to study the concept of ideal materials. First, the control of materials is reaching ever-smaller dimensions and miniaturization is widespread in many fields. As components get smaller, their material properties become size dependent, trending toward more ideal properties. Knowledge of this effect of size is essential in order to design these devices properly. Examples of areas where low-dimensional devices are finding applications include thin film coatings, micromachines, and biomedical fields. Second, in order to overcome the current limitations of real materials in large scale parts, a thorough understanding of the link between ideal and real materials is essential.

The properties of the ideal material are mainly a function of the type of bonding as well as the arrangement of atoms. We will now analyze the changes that *intrinsic voids* and void like defects have on material properties and discuss why bulk materials have significantly lower properties than ideal materials. To understand intrinsic voids, one must investigate atomic structure, bonding, and processing that lead to voids in materials. As such, concepts such as free volume, holes, defects, and vacancies are important. Also described are the mechanisms of void nucleation, growth, and coalescence leading to three-dimensional void formation, ductile failure, creep, and *Kirkendall voids*.

One can experimentally probe the theoretical strength of a material by placing only a small volume of material under strain. This *size effect* is demonstrated by applying

Voids in Materials. DOI: https://doi.org/10.1016/B978-0-12-819282-5.00002-8

intrinsic or *extrinsic constraints* (Zhu, Bushby, & Dunstan, 2008); that is, by creating very small sample size volumes (extrinsic) or by testing only small volumes of material (intrinsic). It is known that as sample size or test volume decreases the material properties are enhanced.

On a macroscale, the stress–strain follows the elastic–plastic models and is independent of specimen size. As sample dimensions become smaller, one starts seeing the size effect on material properties. Common low-dimensional sample geometries are thin films, fibers, or whiskers. An instrument that can probe small sample volumes is the nanoindenter. We give examples of size effects on material performance involving both solid and void domain sizes. For example, when investigating solid domain size effects, Fu et al. (2018) found that the creep resistance of alternating lamina of the titanium (Ti) and aluminum (Al) increased as the laminae thickness decreased from 250 to 10 nm.

2.2 Crystalline materials

The atoms in metals and ceramics can be modeled as hard spheres. A crystal is characterized by packing of these spheres in various repeatable arrangements in three dimensions.

2.2.1 Ideal materials and properties

When packing atoms, represented by hard spheres, into crystal structures the packing efficiency is never 100%. There are always voids or holes in the crystal lattice. They allow the crystal to incorporate other atoms, lead to formation of solid solutions, diffusion of atoms, and allow formation of more complex structures.

From a length-scale standpoint, *lattice holes* are the smallest voids that we will highlight in this book. The size of a hole in a crystal depends on the atomic radius, r_{atom}, as well as the packing mode. The size of the hole can be described by the radius of a hypothetical sphere, r_{hole}, that can be contained within the void created by the surrounding atoms. We can define a quantity called the *radius ratio* as follows:

$$\text{radius ratio} = \frac{r_{\text{hole}}}{r_{\text{atom}}}. \tag{2.1}$$

This radius ratio is a constant for each crystal structure. Thus, the larger the atomic radius of the atom making up the crystal, the larger the radius of the void will be in order to keep the radius ratio constant. Common lattice voids and values for the corresponding radius ratios are given in Table 2–1.

Table 2–1 Common types of lattice voids or holes.

Void type	Description	Size of void (radius ratio)	Schematic representation
Linear void	Two large atoms are arranged linearly. A small atom can occupy this void.		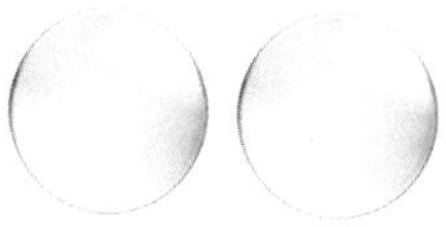
Trigonal planar	Three atoms of the same size joined in a plane.	0.155	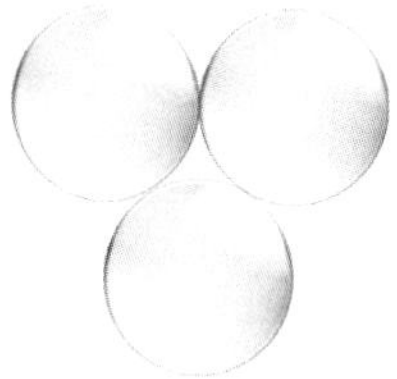
Tetrahedral	One atom placed over three other atoms which are in contact.	0.225	
Octahedral	Four atoms in a plane which are in contact, then putting one atom above and one below this plane.	0.414	
Cubic	Four atoms in one plane which are in contact and four atoms in the plane above.	0.732	

In close packed crystals, there are two types of voids, tetrahedral and octahedral, as shown in Fig. 2–1. In a face centered cubic crystal there are eight tetrahedral and four octahedral voids, whereas in a hexagonal close packed (HCP) crystals there are four tetrahedral and two octahedral voids.

Ionic crystals are made of positively charged anions and negatively charged cations. The ionic radii of the anion and cation must be considered when calculating the radius ratio. In

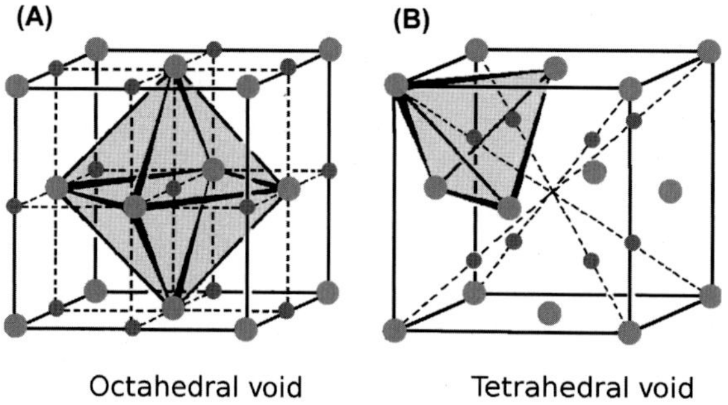

FIGURE 2–1 Schematic of the geometry of (A) octahedral and (B) tetrahedral voids.

Table 2–2 The relationship between the radius ratio and the preferred hole type.

Radius ratio	Coordination number	Void type for cation packing
0.225–0.414	4	Tetrahedral
0.414–0.732	6	Octahedral
0.732–1	8	Cubic
1	12	Close packed structure

these crystals, larger anions form a close-packed structure. The tetrahedral, octahedral, and cubic holes, having radius ratios of 0.225, 0.414, and 0.732, respectively, will incorporate cations. The octahedral holes, because of the larger radius ratio, incorporate larger cations that cannot be accommodated in tetrahedral sites. The situation is similar in alloys, say of metals A and B, that is, their *atomic radii* must be considered. Table 2–2 summarizes radius ratio, coordination number, and the type of hole. This table highlights the transition from one hole type to the next with increasing radius ratio. As radius ratio approaches 1, the ions are effectively the same size and incorporate into a close packed structure.

2.2.1.1 Density

The density of a material is often correlated to other material properties such as strength, modulus, and thermal conductivity. As such, it is an excellent predictor of a whole host of material properties and is relatively easy to measure. Knowing the theoretical and experimental density values, one can calculate the porosity or void content in any material.

The theoretical density, ρ, of crystalline material is described by the following expression:

$$\rho = \frac{\sum_{i=1}^{y} (nA)_i}{VN}, \tag{2.2}$$

where

n = number of "i" atoms/ions per unit cell
i = identity of element in the unit cell
y = number of distinct elements in the unit cell
A = relative atomic mass, g/mol
V = volume of the unit cell, cm^3
N = Avogadro's number (6.023×10^{23}/mol)

The numerator in this density expression is the summation of the number of each atom in a unit cell multiplied by its corresponding relative atomic mass. The denominator is the volume of the unit cell multiplied by Avogadro's number.

2.2.2 Defects and real properties

It is almost axiomatic to say that "the perfect solid does not exist in nature and its properties are determined, to a great extent, by the defects present in its structure." As mentioned in Chapter 1, Introduction, the focus of this book is the characterization and functionality of voids in materials, regardless of their size or shape. As the quote above implies, physical and chemical properties are not solely determined simply by the chemical structure of solid. Many of their properties are derived from the voids and empty spaces present in the material.

As such, intrinsic point defects in crystalline materials are of interest because they impart interesting and diverse functionality to materials. Some of these properties are color centers or F-centers, diffusion, and semiconducting properties. Point defects can also diffuse, agglomerate, and their amount or concentrations can be measured and to a certain extent, controlled.

2.2.2.1 Types of point defects

Point defects in a crystal lattice are nothing but a kind of void, commonly referred to as a vacancy. They are caused by missing atoms at a crystal lattice site. A *Schottky defect* is a special type of vacancy in an ionic crystal where cations and anions are missing from a crystal lattice. We should emphasize that the cations and anions are missing in a stoichiometric ratio in order to maintain the charge neutrality. In metals, which consist of metallic ions distributed in a sea of electrons, a vacancy is simply a missing atom; there is no restriction of charge neutrality.

Schottky defects are present in all ionic crystalline materials and thus are a type of intrinsic void. Vacant sites are present in metals as well. There are several parameters that can influence the amount or concentration of these vacancies, temperature being the major factor. The following expression gives us the number of point defects in thermal equilibrium:

$$\frac{N_D}{N} = B\exp\left(\frac{-AE}{kT}\right), \tag{2.3}$$

where N_D is the number of defects, N is the number of atomic sites, A and B are dimensionless factors ~ 1, E is the energy required to create the defect, k is Boltzmann's constant $(=1.381\times 10^{-23}$ J/K), and T is the temperature in kelvin. Typical energy of formation for a Schottky defect is between 0.7 and 1.5 eV, translating to vacancy concentrations, N_D ranging from 10^{-4} to 10^{-6} (Eyre & Matthews, 2001). The left-hand side of this expression is a dimensionless concentration of point defects. This concentration of point defects, it should be noted, is in thermodynamic equilibrium at a given temperature. The concentration of point defects increases exponentially with increasing temperature. Heating a material to high temperatures will increase the concentration of point defects. If this is followed by rapid cooling (quenching) to room temperature, we will "freeze-in" the defect concentration of the higher temperature down to room temperature.

Here it is appropriate to discuss a few techniques to experimentally determine the defect or vacancy concentration. One can determine the equilibrium vacancy concentration in a crystal structure by experimentally monitoring the change in sample length per unit length, $\Delta L/L$, and the change in lattice parameter, $\Delta a/a$, as a function of temperature. Simmons and Balluffi (1960) measured the vacancy concentration experimentally in aluminum. In general, an expression for vacancy or defect concentration can be written as:

$$N_D = 3\left(\frac{\Delta L}{L} - \frac{\Delta a}{a}\right).$$

An indirect experimental method is to quench a material heated to a desired temperature and measure a physical property such as electrical conductivity. This measurement could then be correlated with a known defect concentration obtained via a direct method. Another direct method for characterizing vacancies and vacancy clusters involves the use of *positron annihilation lifetime spectroscopy* (Su et al., 2012). A positron is the antiparticle of the electron and has a charge of +1. When positron enters a crystalline material, it can form positronium and is attracted to areas of low charge density such as vacancies, voids, and surfaces. Within this area, the positronium undergoes annihilation when it interacts with an electron. This interaction results in two photons being emitted. The magnitude of the positronium lifetime gives indirect data on free defect concentrations in metals and semiconductors as well as void diameter and distribution in porous materials.

2.2.2.2 Mechanical properties

In crystalline materials, atoms or ions are arranged in a periodic repeating pattern. Different repeating patterns lead to different arrangements or crystal structures, and the different crystal structures thus have atoms or ions occupying certain lattice sites with certain coordination number with other sites. The crystal structures span the range from simple cubic to more complex structures such as HCP, orthorhombic, etc., with lower elements of symmetry. As the crystal symmetry decreases, properties such as strength and modulus become more complex. Ignoring details of the specific crystal structure, it can be shown by using a model by Orowan (1948) that the theoretical stress of a perfect crystal is approximately E/π, where E is the Young's modulus of the material. The theoretical cleavage or tensile strength is the stress required to separate two planes of atoms perpendicular to the applied stress as shown in Fig. 2–2.

A maximum stress occurs when the layers are cleaved at the peak of the curve.

Fig. 2–2 shows a sinusoidal variation of stress, σ, as a function of interatomic distance, x. The curve can be represented by the following expression:

$$\sigma = K \sin\frac{\pi}{x}(x - d_0). \tag{2.5}$$

In this expression, K is a constant, x is the interatomic distance, and d_0 is the equilibrium separation between the planes. The stress is zero at the equilibrium separation between the planes, d_0. It can be shown (Meyers & Chawla, 2009) that by using such a sinusoidal curve model, the theoretical strength of a perfect crystal is given by

$$\sigma_{\max} \approx \frac{E}{\pi}.$$

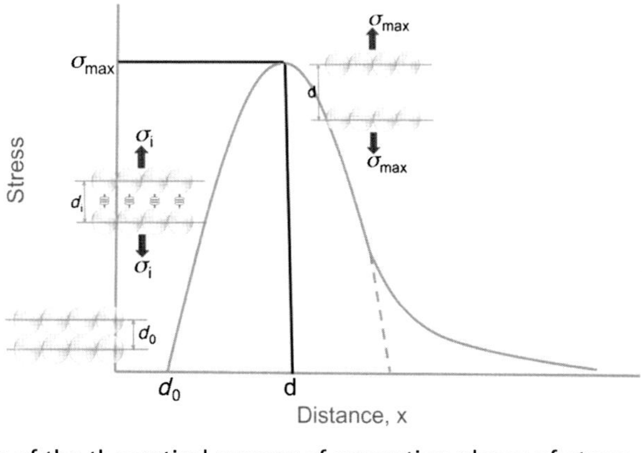

FIGURE 2–2 Schematic of the theoretical process of separating planes of atoms.

This theoretical strength is very high; it is not realized in practice in real materials. The reason for that is the real materials contain cracklike defects which reduce their strength.

2.2.2.3 Thermal properties

The primary means of thermal transport include electrons and/or phonons. Although phonons play a role in the thermal conductivity in metals, the process is dominated by electron transport. There are usually one or two *conduction electrons* moving freely around the metal nuclei, and it is these free electrons that are primarily responsible for the transport of thermal energy. In ionic and covalent materials, the electrons are shared. In these types of materials, phonon transport dominates thermal transport. Phonon transport is reduced by phonon–phonon collisions, phonon collisions with the surface of the crystals, and phonon interactions with impurities and defects. An approximation from the kinetic theory of gases gives the following expression for thermal conductivity (Kittel, 2004):

$$k_p = \frac{1}{3} C_p v l, \tag{2.7}$$

where C_p is the heat capacity per unit volume, v is average particle velocity, and l is the mean free path between phonon collisions. Lattice voids play an important role in limiting thermal conductivity in crystals. At higher temperatures, where lattice voids increase in concentration and the phonon–impurity interactions limit the mean free path, k is proportional to l, the mean free path. Furthermore, $l \sim 1/T^x$ where T is the temperature and the exponent x is between 1 and 2. At low temperatures where lattice voids have low concentrations, the mean free path, l, is limited by the sample dimensions and k is proportional to C_p. Since C_p is proportional to T^3, it follows that at low temperatures the thermal conductivity is also proportional to T^3.

In addition to the presence of lattice voids, filling lattice sites with smaller atoms gives one the ability to tune thermal conductivity. Nolas, Cohn, and Slack (1998) showed that by randomly incorporating smaller La^+ ions in skutterudites (a mineral that is mainly cobalt and nickel arsenide, $(Co,Ni)As_3$) and only partially filling all potential sites, one can minimize the thermal conductivity of skutterudites. The importance of the relative size of the lattice void compared to the ion residing therein is the key factor in tailoring thermal conductivity. The smaller ion leads to rattling, disrupting phonon transfer, which, in turn, leads to a smaller mean free path for phonon collisions. Not filling 100% of the lattice voids can also be important. Having lattice voids in the materials leads to a more random structure that will increase scattering of phonons. This illustrates that controlling the amount or concentration of intrinsic void on an atomic scale can be used to tailor the thermal conductivity.

Eq. (2.7) can also be used to describe thermal conductivity of highly disordered crystals and amorphous materials. The mean free path term is adjusted to account for the high defect concentration in crystals. In the case of an amorphous material, the mean free path is very short because of the absence of long range order.

2.2.3 Density

One practical consequence of a material having atomic vacancies is that it has a lower theoretical density. Presence of atomic vacancies means there is an absence of ions or atoms from the lattice; thus, the material will have a lower density than that calculated based on the unit cell dimensions. For example, for titanium (II) oxide, even when its composition is a stoichiometric TiO, about one-sixth of the Ti^{2+} and O^{2-} sites are vacant (Swaddle, 1997), which reduces the density by 16% when compared to the theoretical density.

2.3 Mechanical properties

2.3.1 Modulus

In general, the presence of voids in any material will affect the mechanical properties, be that a high modulus, engineered ceramic material such as alumina, a soft, low modulus, tin-based electronic solder, or a composite material. A very good example of voids affecting mechanical properties is that of the elastic modulus; irrespective of how the voids are introduced. The deviation of properties from the ideal value, therefore, depends on the presence and characteristics of voids within the material.

In general, the effect of voids on modulus is a more important problem in powder processed materials. Ceramic components are commonly fabricated by powder processing methods and generally have micrometer-scale porosity. This porosity affects the modulus and other mechanical properties of the ceramic.

There are various empirical expressions that allow one to predict the strength or the elastic modulus of a material as a function of the amount of voids or porosity (Mackenzie, 1950; Ramakrishnan & Arunachalam, 1993; Wachtman, 1963). We provide some examples.

Mackenzie (1950) proposed the following expression for the Young's modulus of a material containing spherical voids:

$$E = E_0(1 - f_1 p + f_2 p^2), \tag{2.8}$$

where E_0 is the modulus of the material with no voids and E is the modulus of the material having a porosity p, and f_1 and f_2 are constants. For spherical voids, f_1 and f_2 are equal to 1.9 and 0.9, respectively, for a material with a Poisson's ratio of 0.3.

Fig. 2–3 shows a plot of Eq. (2.8) in the form of a fractional modulus versus volume fraction of porosity. The plot shows two lines; one for $(1 - 1.9p + 0.9p^2)$ and the other for $(1-1.9p)$, where p is the porosity volume fraction. For low volume fraction of porosity, the two curves coincide. It turns out that the experimental data of Coble and Kingery (1956) on the effect of porosity on the modulus of alumina correlate well with Eq. (2.8).

It is worth checking if we can treat the voids as a simple second component, that is, could we treat the material as a composite of two components A and B, where the component B is nothing but void with a modulus of zero. For a composite material such as a carbon fiber reinforced polymer, we commonly use a rule of mixtures to obtain the modulus of the composite (Chawla, 2019). Assuming the rule of mixtures for the effect of porosity on the elastic modulus, we can write

$$E = E_A(1 - f_B) + E_B f_B, \tag{2.9}$$

where f is the volume fraction of a phase and the subscripts A and B denote the two phases.

In the present case, phase B is the pore with modulus equal to zero, therefore we can write

$$E = E_0(1 - p), \tag{2.10}$$

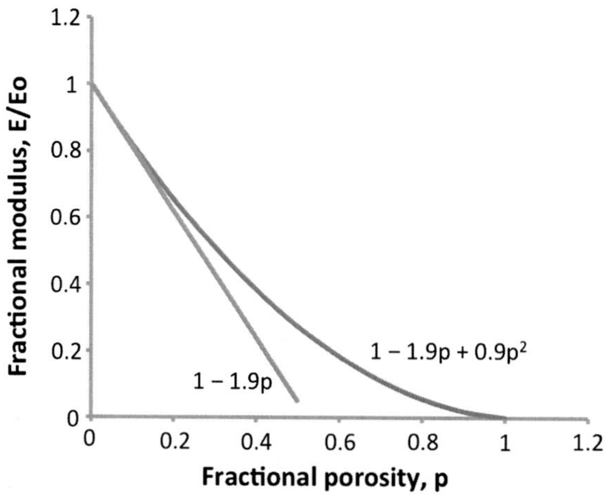

FIGURE 2–3 Effect of void volume fraction on modulus.

where E_0 is the modulus of the material with no voids and E is the modulus of the material having a porosity, p. For relatively low porosity, the quadratic term in Eq. (2.8) can be neglected, and the Mackenzie equation becomes

$$E = E_0(1 - 1.9p). \tag{2.11}$$

If E varied linearly with p, we would have E given by Eq. (2.10)

$$E = E_0(1 - p). \tag{2.12}$$

Thus, the physical significance of Mackenzie's equation is that porosity has an effect on E equals to approximately double the volume fraction of pores.

We take another example of the effect of porosity on modulus from solders used in electronic components. The Young's modulus of an Sn–3.5Ag solder was measured by two techniques: (1) loading–unloading measurements in tension and (2) nondestructive Resonant Ultrasound Spectroscopy (RUS) (Chawla et al., 2004). This work showed that the Young's modulus was significantly affected by the porosity of the solder. As expected, the Young's modulus of the solder decreased with increasing fraction of porosity. The decrease in modulus with porosity in solders was modeled using the approach of Ramakrishnan and Arunachalam (1993). Fig. 2–4 shows relationship between modulus and fraction of porosity in tin. The Young's modulus of a material, E, with a given fraction of porosity, p, is given by the following expression due to Ramakrishnan and Arunachalam (1993) (R–A model):

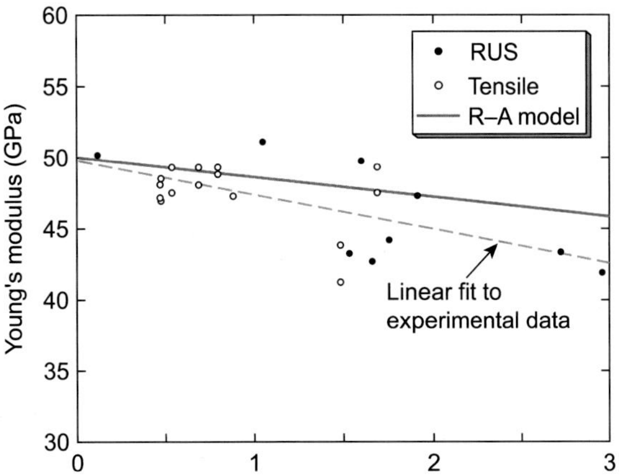

FIGURE 2–4 Effect of porosity on Young's modulus. Young's modulus measured during tensile testing and by Resonant Ultrasound Spectroscopy (RUS). *Source: Adapted from Chawla N.; Ochoa F.; Ganesh V.; Deng X.; Koopman M.; Chawla K.; Scarritt S. (2004) Measurement and prediction of Young's modulus of a Pb-free solder. Journal of Materials Science: Materials in Electronics 15 (6) 385–388.*

$$E = E_0 \frac{(1-p)^2}{1+kp}, \tag{2.13}$$

where E_0 is the Young's modulus of the fully dense solder (taken by extrapolating the experimental data to zero porosity, which yielded a value of approximately 50 GPa), and κ is a constant in terms of the Poisson's ratio of the fully dense material, ν_0:

$$\kappa = 2 - 3\upsilon_0. \tag{2.14}$$

For a tin-rich solder the Poisson's ratio, ν_0 is approximately 0.36. Fig. 2–4 shows the experimental data and the R–A model prediction, showing reasonable agreement at porosity levels less than 1%, but a somewhat larger deviation from the model at porosity greater than 1%. This may be attributed to interaction between pores at larger porosity levels.

When voids take the form of a slit, that is, when they become flattened, we call them *microcracks*. Microcracks can form during the cooling of a ceramic, due to thermal expansion (or contraction) anisotropy, which is shown by noncubic structures. Different grains contract by different amounts along different orientations, resulting in a buildup of thermoelastic stresses. Yet another source of microcrack formation is the anisotropy of elastic constants, which can generate elastic stress concentrations at the grain boundaries where the neighboring grains undergo different strains (due to differences in crystallographic orientation).

The change in the Young's modulus with microcracking has been analyzed by a number of investigators. The formulations give predictions that vary with the orientation of the microcracks with respect to the tensile axis, among other parameters. As an example we give the following expression developed by Salganik (1973):

$$\frac{E}{E_0} = \frac{1}{1+ANa^3}, \tag{2.15}$$

where E is the Young's modulus of the cracked material, and E_0 is Young's modulus of the uncracked material, respectively, a is the crack length, and N is the number of cracks per unit volume. The factor A varies between 1.77 and 1.5 when the Poisson's ratio, υ_0 varies between 0 and 0.5. To a first approximation, one can arrive at

$$\frac{E}{E_0} = \frac{1}{1+1.63\ \mathrm{Na}^3}. \tag{2.16}$$

2.3.2 Effect of voids on strength

As mentioned in earlier in this chapter, the theoretical cohesive strength of a brittle solid is of the order of E/π, where E is the Young's modulus of the material. This high strength is generally not realized in practice because of the presence of defects such as voids, cracks, and/or dislocations. Inglis (1913) recognized this and explained the loss of strength

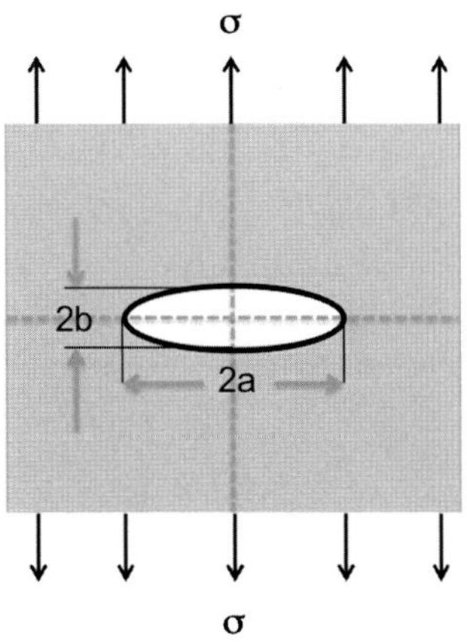

FIGURE 2–5 Schematic representation of an elliptical crack in a plate under an applied stress.

due to stress concentration caused by the presence of defects. Fig. 2–5 shows a material containing a void or crack under an applied stress. Inglis's expression for this situation is

$$\sigma_{\max} = \sigma\left(1 + 2\frac{a}{b}\right), \tag{2.17}$$

where $\sigma_{\max}$ is the maximum stress at the tip of the crack, σ is the applied stress or far-field stress, and a and b refer to the semimajor and semiminor axes of the ellipse or void. For a tip radius, ρ, which is much smaller than the length of the defect, Eq. (2.17) is reduced to

$$\sigma_{\max} = \sigma\left(1 + 2\sqrt{\frac{a}{\rho}}\right) \cong 2\sigma\sqrt{\frac{a}{\rho}} \quad \text{for } \rho \ll a. \tag{2.18}$$

2.3.3 Griffith theory of brittle fracture

Around 1920, Griffith did some remarkable experimental work with glass fibers and came to a very important conclusion, namely, the glass fibers are weakened by microscopic flaws (notches or voids) located on the surface or interior. He went a step further than Inglis. He recognized that in a material under load, stress increases at the crack tip. However, whether or not the void or crack will propagate under the applied stress will depend on a fundamental thermodynamic criterion: the balance between the strain energy released when the crack extends and surface energy created by the new crack surfaces. This energy balance results in the following expression for the critical stress for crack propagation:

$$\sigma_c = \left(\frac{2E\,\gamma_S}{\pi\,a}\right)^{\frac{1}{2}}, \tag{2.19}$$

where E is the modulus of elasticity, γ_s is the specific surface energy, and a is one-half of the internal crack length.

When plastic deformation is predominant, we must add a term of plastic work done, γ_p to γ_s, the surface energy term. Generally, we use a term, G, called the *strain energy release rate* or the crack extension force. Thus,

$$G = 2\left(\gamma_s + \gamma_p\right). \tag{2.20}$$

Crack propagation occurs when $G > G_c$, a critical crack extension force. We will not go into details of fracture mechanics here; the reader can consult a standard text on mechanical behavior of materials (Meyers & Chawla, 2009).

2.4 Processing and service-induced voids

Voids can form in materials during processing and service. As before, we use the terms voids, pores, cavities, and cracks as synonyms. We describe some situations where voids can occur, how they form, and the effect of such void formation on the behavior of material.

Void formation in a material during some forming processes is quite common. Usually such voids are on micrometer scale, and there is plentiful literature in the fields of mechanics and materials science and engineering highlighting a particular material property, such as Young's modulus, and the effects that void content has on it. We have given some examples earlier. The voids that form in a material as a result of some processing condition do not allow the achievement of the theoretical density. Below we give some common situations that result in void formation in materials.

2.4.1 Casting

Casting of materials, that is, pouring of a liquid material (metal, polymer, or ceramic) in a mold frequently results in some residual porosity. Such a processing technique is more common in metals and alloys than in other material types. The source of porosity can be normal shrinkage porosity (a typical liquid metal shrinks between 4% and 6% when it solidifies) or any gas evolution during solidification. At low gas concentrations, porosity due to gas evolution can be described by Sievert's law

$$S = k\,p^{0.5}, \tag{2.21}$$

where S is the solubility, k is a constant, and p is the partial pressure of gas over the melt. This equation implies that one can reduce the porosity by reducing the partial pressure of gas. Indeed, one can reduce the amount of dissolved gas by making some additions to the liquid melt, which will react with the gases. A good example is the practice of making what are called *killed* steels where we use Al or Si to deoxidize the melt and thus reduce the partial pressure of the gas and the high shrinkage associated with the gas removal.

Despite the common practice of degassing of metals, cast metallic parts do contain some residual porosity. If a cast part is deformed later, that is, made into a wrought product, the voids can be closed, essentially densifying the metal, during secondary processing such as rolling or forging. Sometimes, hydrostatic pressure is applied to close the pores or cavities formed during service, for example, at high service temperatures encountered in a turbine. Under these conditions creep strain can become significant because of the presence of pores.

2.4.2 Powder processing of materials

Ceramic materials are generally processed from starting materials in the form of a powder. Some metals and alloys are also processed with powders as the starting materials. Materials in powder form are consolidated by sintering or hot pressing. Invariably, these processes result in some residual porosity, more so in sintering and less in hot pressing.

2.4.3 Voids in solders

Soldering or brazing is the process of joining two metals or ceramics by using a solder or brazing alloy (generally a tin-based, low melting point alloy) as the filler. Even though the absolute temperatures involved in soldering are low, they are high relative to the melting point of the solder alloy, that is, in terms of the homologous temperature (temperature in kelvin/melting point in kelvin), the temperature is high.

We have discussed the effect of porosity on modulus in Section 2.3.

2.5 Time-dependent properties

There are some mechanical properties that are time dependent. Creep is an important example of such a mechanical property. In jet aircraft engines or land based turbines, blades can undergo creep at low stresses. Typically, such creep deformation processes involve diffusion of voids and defects. We will describe briefly the role of voids and atomic defects in creep and other time dependent properties. However, before we do that, we describe the important role of diffusion processes in materials.

2.5.1 Diffusion of vacancies and voids

Diffusion is very important in our daily life. If we have two species of materials, A and B, put together in a container, the process of diffusion will make them spread out over time. Common examples include the diffusion of air fresheners in the air and dissolution of sugar in a cup of tea or coffee. We are interested in the diffusion that takes place in solid materials and the very important role that vacancies and defects play in the process. There are two important laws in diffusion called Fick's first and second laws, which govern the process of diffusion.

2.5.1.1 Fick's first law of diffusion

This law says that the diffusional flux of a species in a given direction is proportional to the concentration gradient of the species in that direction. The assumption underlying Fick's first law is that the concentration gradient does not change with time. Mathematically, Fick's first law can be written as:

$$J = -D\frac{dC}{dx}, \tag{2.22}$$

where J is the diffusional flux in the x direction and is proportional to the concentration gradient, dC/dx. The proportionality constant, D, also called as the diffusivity or coefficient of diffusion, has the units of m^2/s. The negative sign indicates that the diffusion occurs down the concentration gradient.

2.5.1.2 Fick's second law of diffusion

Assuming that diffusivity, D, is independent of composition, Fick's second law of diffusion takes the following form:

$$\frac{dC}{dx} = -D\frac{\partial^2 C}{\partial x^2}. \tag{2.23}$$

A very important result that follows from these laws is the following approximate expression (the detailed expression involves error function, which we skip) that relates the diffusion distance, x, and time, t:

$$x \approx \sqrt{Dt}, \tag{2.24}$$

where D is the diffusion coefficient given by the following expression:

$$D = D_o \exp\left[\frac{-Q}{RT}\right]. \tag{2.25}$$

In the above expressions, x is the effective diffusion distance, D_o is a constant that depends on the diffusion couple, Q is the activation energy for the diffusion process, R is the universal gas constant, and T is the temperature in kelvin.

2.5.1.3 Vacancy diffusion

The process of diffusion in crystalline solids involves movement of vacancies. When a vacancy diffuses, say in a piece of aluminum, it exchanges places with an adjacent aluminum atom. In order to accomplish this movement a certain potential energy barrier must be overcome. This process is called *self-diffusion*, and it involves the flow of vacancies and atoms in opposite directions.

The movement of vacancies is an important process in creep deformation. Creep is defined as the time-dependent deformation of a solid, that is, it is the strain that occurs as a function of time under a constant stress at a given temperature. We define the homologous temperature for creep strain (T_H) as the temperature under considerations, T, in kelvin divided by the melting point, T_m, of the material in kelvin. Thus,

$$T_H = \frac{T}{T_m}. \tag{2.26}$$

Cracks and voids play an important role in creep deformation of materials. Voids may form continually during creep, but total rupture is associated with growth and coalescence of cracks and voids. Grain boundary cavitation is observed under creep conditions in metals and ceramics. Creep conditions become appreciable at low stresses and a homologous temperature, T_H, between 1/3 and 2/3.

The movement or diffusion of vacancies during creep leads to a clustering and coalescence of these point defects into three-dimensional voids (Wilkinson, 1987). Ultimately, the growth of voids can lead to failure of the material (Orsini & Zikry, 2001; Smallman & Westmacott, 1972; Wilkinson, 1987, 1988).

There are two mechanisms where vacancies diffuse through a crystalline material leading to clustering, creep, and eventual failure: *Nabarro–Herring creep* and *Coble creep*. Both Nabarro–Herring and Coble creep are stress-induced flow of vacancies leading to a clustering of voids in grain boundaries perpendicular to the tensile axis, called *r-type voids*; see Fig. 2–6.

Coble creep is the diffusion of the vacancies through the grain boundary, whereas Nabarro–Herring creep involves vacancy currents through the grain.

For Nabarro–Herring creep, the creep rate, is given by the following equation (Herring, 1950; Nabarro, 1948):

$$\dot{\epsilon}_{NH} = A_{NH}\frac{D_l \Omega \sigma}{d^2 kT}, \tag{2.27}$$

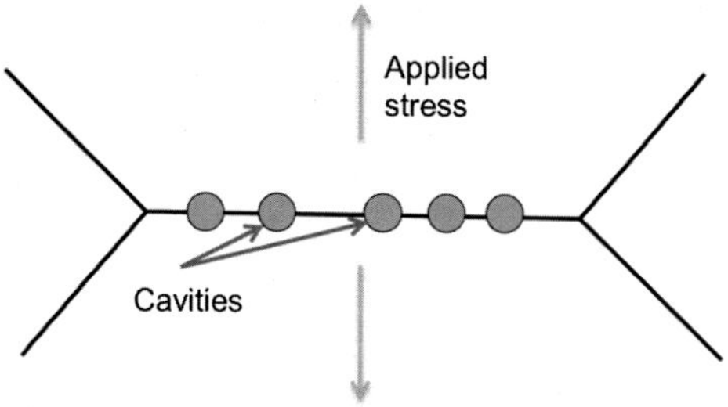

FIGURE 2–6 Cavities formed at a grain boundary perpendicular to the applied stress. Such cavities are called r-type cavities.

where D_l is the coefficient for lattice self-diffusion, Ω is the atomic volume, d is the grain size, k is Boltzmann's constant, T is the absolute temperature, and A_{NH} is a constant having a theoretical value between 12 and 40 depending on the shape of the grains and the grain size and load distribution. Typically, Nabarro–Herring creep dominates at low stresses and high homologous temperatures, $T_H > 0.8$.

For Coble creep, we can write the governing equation as (Coble, 1963):

$$\dot{\epsilon}_{Co} = A_{Co}\frac{\delta D_{gb}\Omega\sigma}{d^3kT}, \tag{2.28}$$

where δ is the effective width of the grain boundary, D_{gb} is the diffusion coefficient for grain boundary diffusion, and A_{Co} is a constant $\sim 150/\pi$. Since the Coble creep rate is dependent on vacancy diffusion through the grain boundary region rather than lattice self-diffusion, Coble creep becomes important at low homologous temperatures, typically $T_H < 0.7$.

2.5.2 Clustering and failure

In the later stage of creep phenomenon, called tertiary creep, voids are the main agents of deformation. They generally nucleate at grain boundaries that are normal to the tensile stress. Continued creep strain results from the growth of voids and eventual coalescence of voids leading to fracture. Fig. 2–7 shows such voids formed in copper.

Typically, the cavitation sites are grain boundary triple points, particles, or ledges at grain boundaries. Grain boundary sliding (GBS) can lead to void formation at the grain boundary triple points. Such voids are frequently called w-type cavities, see Fig. 2–8. GBS refers to relative displacement of adjacent grains at the grain boundary. It is an important mode of creep deformation. The total longitudinal creep strain can be represented as

$$\varepsilon_T = \varepsilon_B + \varepsilon_G, \tag{2.29}$$

where ε_B is the GBS contribution and ε_G is the intragranular contribution to strain.

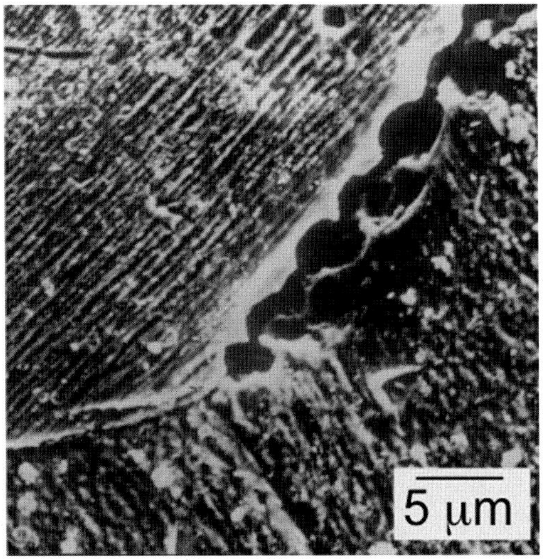

FIGURE 2–7 Cavities and grain boundary sliding at a triple point. *Source: Adapted from Chawla K.K. (1973) Grain boundary cavitation and sliding in copper/tungsten composites due to thermal stress. Philosophical Magazine 28 401.*

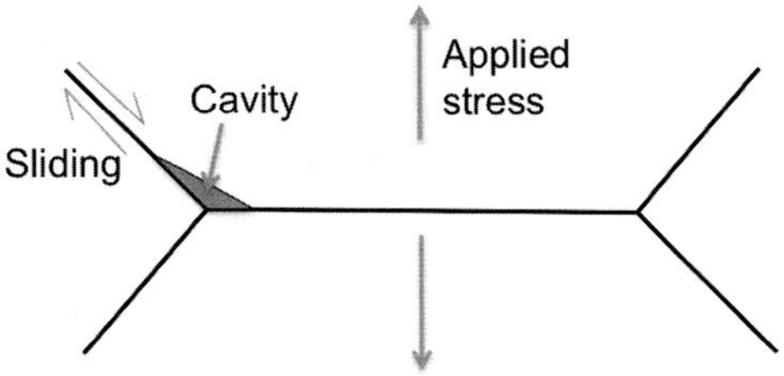

FIGURE 2–8 A schematic of a w-type cavity or crack formed at a grain boundary triple point. Such a cavity will form by sliding of two grains along a grain boundary.

2.5.3 Kirkendall voids in crystalline materials

Pore formation in a material during service can involve diffusion. Such porosity is called Kirkendall porosity (Smigelskas & Kirkendall, 1947). Kirkendall porosity appears when two metals, A and B, forming a diffusion couple have significantly different diffusivities or diffusion coefficients. An easy experiment to illustrate the phenomenon of Kirkendall porosity is as follows. We make a diffusion couple of two metals, A and B, that have different diffusivities, that is, the diffusion fluxes will be different. When exposed to a high temperature, significant diffusion will occur. If the flux of A is larger than that of B, there will occur a net flow of matter past the interface and an equal and opposite flow of vacancies in the opposite direction. A good illustration of this is the work done by Wang, Matlock, and Olson (1993) on Ni–Cu laminate composites.

Fig. 2–9 shows micrographs from their work. The samples were heat treated at 1000°C in argon for different times. Note the appearance of pores in Cu. The explanation for this void formation is as follows. Copper diffuses faster than nickel (the melting point of Cu is lower than that of Ni), which causes a net flow of vacancies from Ni to Cu. At a critical vacancy concentration, pores form in Cu. Note also in Fig. 2–10

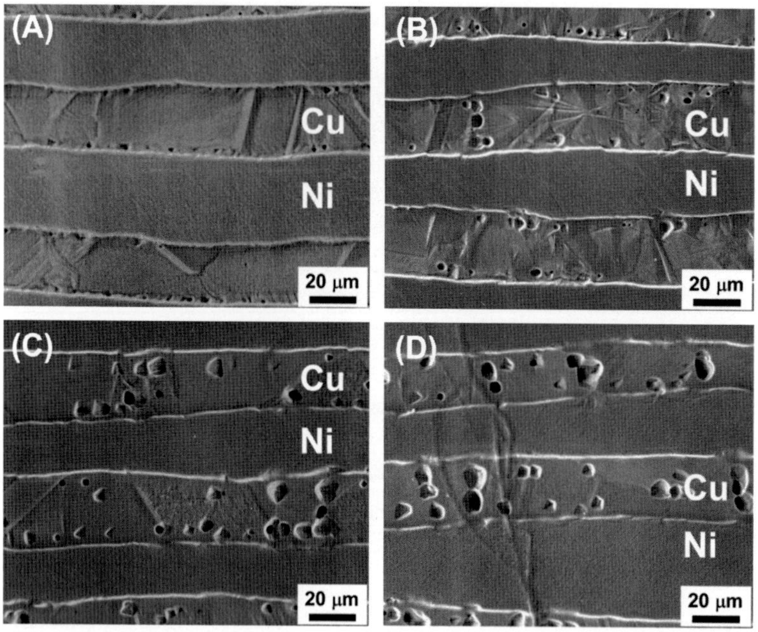

FIGURE 2–9 Kirkendall porosity in a Ni–Cu laminate. The Ni–Cu laminates were heated to 1000°C in vacuum for (A) 1 min, (B) 5 min, (C) 10 min, and (D) 15 min. Copper diffuses faster than nickel (the melting point of Cu is lower than that of Ni), which causes a net flow of vacancies from Ni to Cu. SEM. *Courtesy: D. Matlock.*

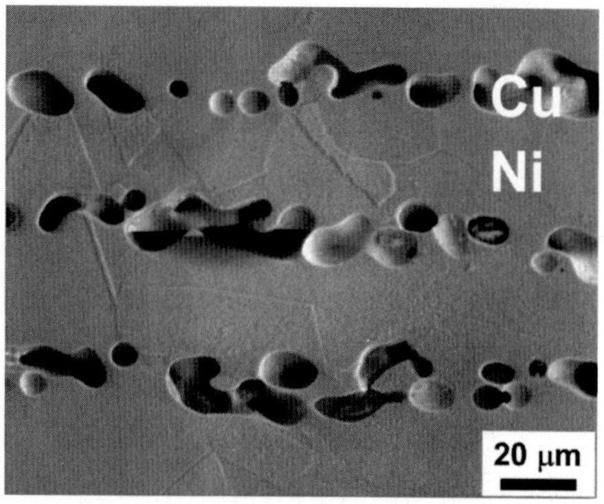

FIGURE 2–10 Coalescence of Kirkendall porosity in the copper layer in a Ni–Cu laminate that was heated to 1000°C in vacuum for 24 h. SEM. *Courtesy: D. Matlock.*

that with increasing time of heat treatment the average pore diameter increases and coalescence of pores occurs.

In practice one can find evidence of Kirkendall porosity in filamentary superconductors (Klein, Warshaw, Dudziak, Cogan, & Rose, 1981), nickel-based superalloys (Darolia, 2013), materials subjected to neutrons in a nuclear reactor (Roberts, 1975). Nickel-based superalloys are used in jet engines that operate at temperatures above 1100°C in very aggressive environment of hot gases, etc. Commonly nickel-based superalloys are coated with thermal barrier ceramic coatings, which provide resistance against oxidation but interdiffusion at the coating/substrate interface can lead to Kirkendall porosity.

The irradiation of a metal in a nuclear reactor can cause formation of vacancies and interstitial defects. Such mobile defects can vanish by migration to a sink such as a dislocation. Alternatively, if one of the defects migrates faster than the other, then we will have an excess of that defect. If this excess defect are vacancies, it can result in the formation of voids, which in turn can lead to dimensional instability.

References

Chawla, K. K. (1973). Grain boundary cavitation and sliding in copper/tungsten composites due to thermal stress. *Philosophical Magazine, 28*, 401.

Chawla, K. K. (2019). Composite materials: Science and engineering (4th ed.). Cham, Switzerland: Springer Nature.

Chawla, N., Ochoa, F., Ganesh, V., Deng, X., Koopman, M., Chawla, K., & Scarritt, S. (2004). Measurement and prediction of Young's modulus of a Pb-free solder. *Journal of Materials Science: Materials in Electronics, 15*(6), 385–388.

Coble, R. (1963). A model for boundary diffusion controlled creep in polycrystalline materials. *Journal of Applied Physics, 34*(6), 1679–1682.

Coble, R. L., & Kingery, W. D. (1956). Effect of porosity on physical properties of sintered alumina. *Journal of the American Ceramic Society, 39*(11), 377–385.

Darolia, R. (2013). Thermal barrier coatings technology: Critical review, progress update, remaining challenges and prospects. *International Materials Reviews, 58*(6), 315.

Eyre, B. L., & Matthews, J. R. (2001). Crystals: point defects. *Encyclopedia of materials: Science and technology* (pp. 1919–1924). Oxford: Elsevier.

Fu, K., Sheppard, L. R., Chang, L., An, X., Yang, C., & Ye, L. (2018). Length-scale-dependent nanoindentation creep behaviour of Ti/Al multilayers by magnetron sputtering, *Materials Characterization,* 139, 165–175.

Herring, C. (1950). Diffusional viscosity of a polycrystalline solid. *Journal of Applied Physics, 21*, 437.

Inglis, C. (1913). Stresses in a plate due to the presence of cracks and sharp corners. *Transactions of the Royal Institute of Naval Architects, 3*, 219.

Kelly, A., & Macmillan, N. H. (1986). *Strong solids* (3rd. ed.). New York: Oxford Science Publications.

Kittel, C. (2004). Thermal conductivity. *Introduction to solid state physics* (8th ed., p. 121) John Wiley & Sons.

Klein, J. D., Warshaw, G., Dudziak, N., Cogan, S. F., & Rose, R. M. (1981). On the suppression of Kirkendall porosity in multifilamentary superconducting composites. *IEEE Transactions on Magnetics, 17*(1), 380–382.

Mackenzie, J. K. (1950). The elastic constants of a solid containing spherical holes. *Proceedings of the Physical Society. Section B, 63*(1), 2.

Meyers, M. A., & Chawla, K. K. (2009). *Mechanical behavior of materials* (2nd ed., p. 407) Cambridge, UK: Cambridge University Press.

Miller, D., & Kumar, V. (2011). Microcellular and nanocellular solid-state polyetherimide (PEI) foams using sub-critical carbon dioxide II. Tensile and impact properties, *Polymer, 52*(13), 2910–2919.

Nabarro, F. R. N. (1948). Report of a conference on strength of solids. *Physical Society of London,* 75.

Nolas, G., Cohn, J., & Slack, G. (1998). Effect of partial void filling on the lattice thermal conductivity of skutterudites. *Physical Review B, 58*(1), 164.

Orowan, E. (1948). Fracture and strength of solids. *Reports on Progress in Physics, 12,* 185.

Orsini, V. C., & Zikry, M. A. (2001). Void growth and interaction in crystalline materials. *International Journal of Plasticity, 17*(10), 1393–1417.

Ramakrishnan, N., & Arunachalam, V. S. (1993). Effective elastic moduli of porous ceramic materials. *Journal of the American Ceramic Society, 76*(11), 2745–2752.

Roberts, J. T. A. (1975). Radiation effects problems in nuclear fuel rods. *IEEE Transactions on Nuclear Science,* 2219.

Salganik, R. L. (1973). Mechanics of solids with a large number of cracks. *Izv. Akad. Nauk SSSR, Mekh. Tverd. Tela, 4,* 149–158.

Simmons, R. O., & Balluffi, R. W. (1960). Measurements of equilibrium vacancy concentrations in aluminum. *Physical Review, 117*(1), 52–61.

Smallman, R. E., & Westmacott, K. H. (1972). The nature and behaviour of vacancy clusters in close-packed metals. *Materials Science and Engineering, 9*(0), 249–272.

Smigelskas, A. D., & Kirkendall, K. O. (1947). Zinc diffusion in alpha brass. *Transactions of AIME, 171,* 130.

Su, L. H., Lu, C., He, L. Z., Zhang, L. C., Guagliardo, P., Tieu, A. K., ... Li, H. J. (2012). Study of vacancy-type defects by positron annihilation in ultrafine-grained aluminum severely deformed at room and cryogenic temperatures. *Acta Materialia, 60*(10), 4218–4228.

Swaddle, T. W. (1997). *The defect solid state. Inorganic chemistry* (pp. 95–114). San Diego: Academic Press.

Wachtman, J. B. (1963). Elastic deformation of ceramics and other refractory materials. In J. B. Wachtman (Ed.), *Mechanical and thermal properties of ceramics* (p. 139). Washington, DC: Special Publication 303 National Bureau of Standards.

Wang, S. H., Matlock, D. K., & Olson, D. L. (1993). An analysis of the critical conditions for diffusion-induced void formation in Ni–Cu laminate composites. *Materials Science and Engineering, 167*(1–2), 139–145.

Wilkinson, D. S. (1987). The effect of time dependent void density on grain boundary creep fracture—continuous void coalescence. *Acta Metallurgica, 35*(6), 1251–1259.

Wilkinson, D. S. (1988). The effect of a non uniform void distribution on grain boundary void growth during creep. *Acta Metallurgica, 36*(8), 2055–2063.

Zhu, T., Bushby, A., & Dunstan, D. (2008). Materials mechanical size effects: A review. *Materials Technology: Advanced Performance Materials, 23*(4), 193–209.

3

Intrinsic voids in polymeric networks

3.1 Polymer structure

Polymers are held together by primary (covalent bonds) and secondary bonds (van der Waals and hydrogen bonds). In covalent bonding there is a sharing of valence electrons (the *s* and *p* shells) to complete an octet (a group of eight electrons) in an atom. The most notable exception to this octet is hydrogen, which needs only two electrons. A polymer chain consists of a covalently bonded repeating unit, *mer*, see Fig. 3–1. The mer is repeated until the polymer reaches a desired molecular weight, and it is then terminated with an end group on each side of the repeating unit.

Polyethylene, one of the simplest polymer structures, has a repeating *mer* of two covalently bonded carbon atoms and having two hydrogen atoms each, C_2H_4. The bonding between chains, however, is weak van der Waals type. Most common polymers are based on carbon, however, silicon-based polymers, known as polysiloxanes or silicones, are also common.

There are many different types of polymers and are classified as either *thermoplastic* or *thermoset*, see Fig. 3–2. These polymers can be linear or branched, see Fig. 3–2A and B. Polymers can have a rigid three-dimensional network structure, see Fig. 3–2C. Another type of polymers can be based on structure, amorphous and crystalline (see Fig. 3–2D and E). Polymers are rarely 100% crystalline. One can achieve close to 100% crystallinity only under special circumstances.

Semicrystalline thermoplastics consist of a mixture of amorphous and crystalline domains, see Fig. 3–2E. These polymers will experience both melting of crystalline domain and glass transition of amorphous domain upon heating and resolidification and recrystallization during cooling. The amorphous domain will have a glass transition temperature above which the mechanical properties fall rapidly, and the material goes from solidlike phase to a liquidlike phase. The crystalline part can be repeatedly solidified by cooling and reformed into a new shape. Because of this ability to melt and resolidify, these type of polymers are frequently recycled. Examples of thermoplastics are nylon (polyamide), polyethylene, polyvinyl chloride, polypropylene, and polystyrene.

In contrast, thermosetting polymers undergo a curing or crosslinking of polymer chains, and set into a rigid three-dimensional network structure, see Fig. 3–2C. Once cured, a thermoset cannot enter a melt phase and cannot be reformed as a

Voids in Materials. DOI: https://doi.org/10.1016/B978-0-12-819282-5.00003-X

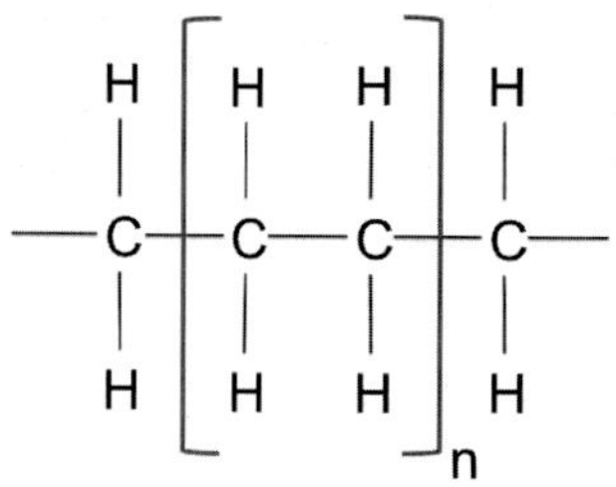

FIGURE 3–1 Illustration of a *mer* making ethylene.

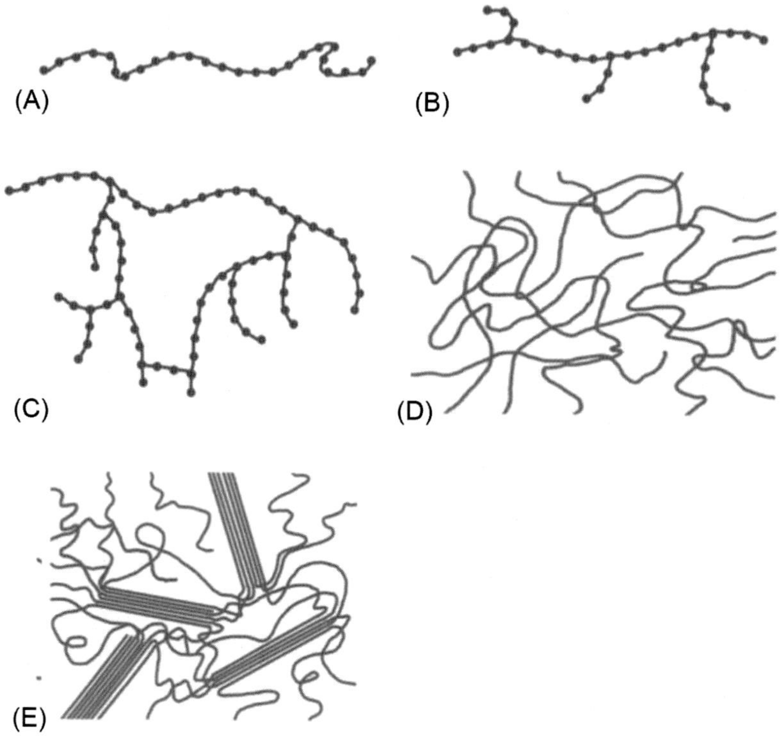

FIGURE 3–2 Illustration of the different types of polymers; (A) linear, (B) branched, (C) cross-linked, (D) amorphous, and (E) semicrystalline.

thermoplastic materials. Common examples of thermosetting polymers are epoxies, phenolics, unsaturated polyesters, and rubbers that have been vulcanized.

There are several types of intrinsic voids that are present in both thermoplastic and thermosetting polymers. These voids constitute *free volume*. The free volume is a very important concept when studying polymers. According to Sears and Darby (1982), increasing free volume permits increased motion of polymer molecules, and therefore a study of plasticization is nothing but a study of ways to increase free volume.

Actually, the concept of free volume dates back to Eyring (1936) who used it in an attempt to describe molecular motion. In the simplest sense, free volume is the empty space or void between polymer chains. Polymer chains are generally not packed like, say, the atoms in a metal, such as a piece of aluminum, as we described in Chapter 2, Intrinsic voids in crystalline materials: ideal materials and real materials, which results in lattice holes. This free volume in a polymer allows the chains to vibrate, rotate, and undergo molecular motion. For these modes to occur there must be a void of a critical size or volume. The understanding of the physical implications of the concept of free volume in polymer systems came about after several key studies were undertaken in the late 1930s to the early 1950s. Eyring studied free volume as it related to viscosity, plasticity, and diffusivity, while others studied the free volume as it related to temperature and viscosity (Van Wijk & Seeder, 1937, 1939) and the relationship of free volume with the compressibility of polymer liquids (Brinkman, 1940).

The term *specific volume* means volume per unit mass. In the simplest form, specific free volume (V_{f}) in a polymer is the difference between total specific volume (V_{T}) and specific occupied volume (V_{occ}) (Sears & Darby, 1982) (Fig. 3–3).

$$V_{\mathrm{f}} = V_{\mathrm{T}} - V_{\mathrm{occ.}} \tag{3.1}$$

Total specific volume, V_{T}, is simply $(1/\rho)$, where ρ is the density. V_{f} is the sum of two distinct free volumes: *hole specific free volume* (V_{fh}) and *interstitial specific free volume* (V_{fi}). Eq. (3.1) above becomes:

$$V_{\mathrm{T}} = (V_{\mathrm{fh}} + V_{\mathrm{fi}}) + V_{\mathrm{occ}}. \tag{3.2}$$

Eq. (3.2) shows the contributions of individual specific volumes that lead to the total specific volume in a material. Much of the literature surrounding this topic is about polymers and their behavior above and below the glass transition temperature, T_{g}, and the mechanisms of the addition of plasticizers to thermoplastic polymers.

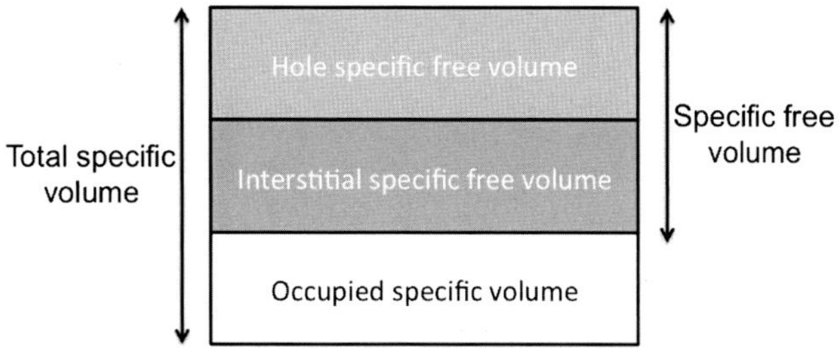

FIGURE 3–3 Graphical definition of the constituents of total specific volume and specific free volume in a polymer or glassy network system.

3.2 Free volume and thermomechanical behavior

There are two important but related thermal attributes to understand when discussing polymer free volume: melting point, T_m, and glass transition temperature, T_g. Pure, crystalline materials show a well-defined, thermodynamic parameter called melting point, T_m. It is at this fixed temperature that crystal structure is destroyed when we heat a crystalline material. Amorphous materials (most polymers) do not have a melting point. When an amorphous polymer is cooled, there occurs a contraction in its dimensions because of a decrease in the thermal vibrations of molecules and a corresponding reduction in the free volume. Fig. 3–4 shows a plot of specific volume (volume per unit mass) as a function of temperature. Many times in engineered products, we have a polymer system that consists of a mixture of crystalline and amorphous polymers. These are referred to semicrystalline polymers. The crystalline component will show a fixed melting point. The amorphous component does not show a fixed transition. Instead the amorphous polymer continues to contract slowly but with a marked change in slope. One can then define a glass transition temperature, T_g, where the slope of cooling curve (specific volume vs. temperature) shows a distinct change. It is important to recognize that although in practice the glass transition temperature and melting point have many similarities, they represent very different characteristics in terms of the internal structure. As the temperature decreases and reaches T_g, the liquid changes to a very viscous, glassy material. The transition is frequently quantified in terms of a sudden increase in viscosity, about 10^{-12} Pas. The internal structure, however, remains that of a liquid. The T_g of a material is not fixed like the

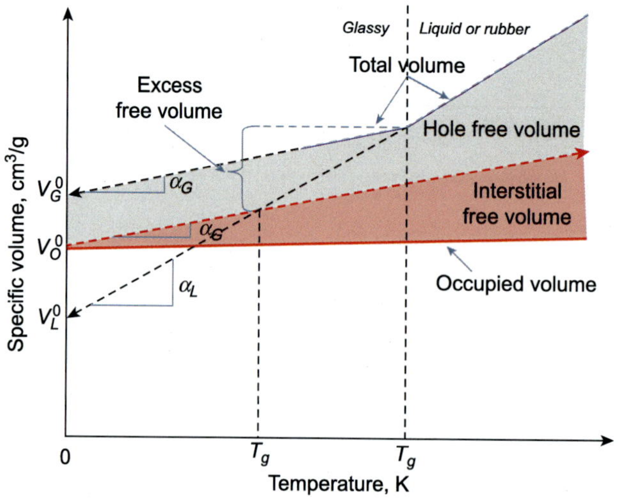

FIGURE 3–4 Graphical explanation of the total volume and free volume in a model glassy/rubbery system. **Source:** *Adapted from Sears J.K.; Darby J.R. (1982) Technology of plasticizers. John Wiley & Sons: New York.*

T_m. As we will describe below, the T_g of a polymer can change depending on, for example, processing conditions, aging, and thermal history.

Fox, Thomas, and Flory (1950) discovered that the glass transition temperature (T_g) of any polymer, independent of structure, occurs at a viscosity of $\sim 10^{-12}$ Pas. Williams, Landel, and Ferry (1950) extended the free volume concept to the T_g of rigid polymers. They concluded that physically all polymers at T_g, independent of structure, have similar fractional free volumes, between 0.11 and 0.17. Subsequent studies suggest a wider range in values of fractional free volume and the value depends on the type of polymer.

The interstitial free volume accounts for the majority of the polymer volume expansion with temperature, or *coefficient of thermal expansion* (CTE) at temperatures below the T_g. Linear CTE is typically determined by measuring the change in length of a material as a function of temperature. Most materials expand with an increase in temperature. CTE (α) is defined by the following relationship:

$$\alpha = \frac{\Delta L}{L_o \Delta T}, \tag{3.3}$$

where L_O is the initial length of the specimen, ΔL is the change in length when the temperature is changed by ΔT.

Since glassy materials are nonequilibrium structures, they contain free volume that can collapse during aging. Above T_g, properties of the polymer change dramatically as the free volume increases with temperature. This gives the polymer chains the freedom to move into adjacent areas, as they are now rubbery or even liquidlike. Above T_g, the CTE typically can increase 1000 times compared to that of the glassy phase below the T_g. As a polymer cools to T_g, and transits from a rubbery material or liquid to a glassy state, the specific volume of the material decreases.

As the polymer cools further below the T_g, the CTE is significantly decreased from its corresponding value above the T_g. Fig. 3–4 shows in a schematic manner, the effect of temperature and CTE on the total volume in a dilatometry experiment with a model glassy/rubbery system.

Here we need to define one other term, namely, *equilibrium glass transition temperature*, T_g^{∞}. This is a hypothetical value that identifies a glass transition temperature $T_g^{\infty} < T_g$ that corresponds to material experiencing an infinite relaxation time where the free volume a minimum.

According to Williams et al. (1950), the free volume of any polymer at the T_g is between the volume fractions of 0.11 and 0.17. One can extrapolate the slope of the line above (α_L) and below (α_G) the observed T_g to absolute zero (0K) to find the corresponding volumes, V_L^o and V_G^o, respectively, see Fig. 3–4. The actual occupied volume, V_O^o, is between V_L^o and V_G^o. The occupied volume is the volume taken up by the polymer molecules and is considered impenetrable. Extending a line of slope (α_G), from V_O^0 through the T_g, represents the trend of volume with increasing temperature, including the occupied volume and the interstitial free volume, shaded red. Above this line is the free volume hole, shaded gray. In a glassy material, this difference between the volume at T_g^{∞} and T_g is the volume available during the physical aging process. The mechanism of physical aging is the collapse of the free volume holes present in the nonequilibrium glassy structures. The volume difference between the equilibrium and nonequilibrium volume curves is referred to as *excess free volume.*

Occupied volumes of polymers were first described by Bondi (1964) using van der Waals volumes, V_{vdw}. According to Bondi the following relationship exists between van der Waals volume and specific volume extrapolated to 0K:

$$V_O^0 = 1.3\ V_{vdw,} \tag{3.4}$$

where V_O^0 is occupied volume via extrapolation to 0K and V_{vdw} is van der Waals volume. A modified additive method was proposed by (Park and Paul, 1997). This modification used permeability of various gases to predict density and estimate the occupied volume specific for each gas.

3.3 Kinetic theory of polymer strength

The theoretical strength of a polymer chain is a notional strength of the C–C bond connecting adjacent *mers* in the polymer. This value is significantly higher than that of a bulk polymer, for example, measured in a tensile test. This difference is due to the presence of thermal fluctuations leading to free volume and stress concentrators. The result of the kinetic theory relating to the fracture strength, σ_f, of both semicrystalline and amorphous polymer materials can be expressed as (Zhurkov, 1984):

$$\sigma_f = \frac{1}{V}\left[U - kT\ln\left(\frac{t}{t_0}\right)\right], \tag{3.5}$$

where V is the activation volume, U is the activation energy of breaking the C–C bond, k is the Boltzmann's constant, T is absolute temperature in kelvin, t is the time to failure under a constant stress, and t_0 is the reciprocal of the molecular oscillation frequency. As Eq. (3.5) indicates, the fracture strength is both temperature and time dependent.

One can find the maximum strength (σ_{max}) of a polymer by ignoring the temperature and time dependence, and substituting the following for activation volume, V (He, 1986):

$$V = V_d q, \tag{3.6}$$

where V_d is the activation volume of the polymer without considering the stress concentration factor, q. The stress concentration factor accounts for flaws, defects or voids, and free volumes in the bulk polymer.

The maximum strength of a polymer, $\sigma_{\max}$, is written with the stress concentration factor included while the theoretical strength, σ_{th}, excludes this term.

The theoretical strength of a polymer from He (1986) is given as:

$$\sigma_{\text{th}} = \frac{1}{s}\frac{\sqrt{4r\,E_c U}}{3n}, \tag{3.7}$$

where s is the cross-sectional area of the polymer chain, r is the interatomic distance, which is equal to the length of the C—C bond, E_c is the Young's modulus of the polymer chain, U is the activation energy needed for thermal decomposition, and n is the number of atoms.

When testing for the tensile properties of a polymer, generally the carbon chains are not aligned with the principal loading axis. However, there are forms of materials, such as fibers, where significant effort is made to align these polymer chains with the longitudinal axis of the fiber. When these fibers are tested in tension, the C—C backbone of the polymer may support a significant portion of the force.

Table 3—1 illustrates this point. Polyethylene can be processed into a highly aligned, linear, ultrahigh molecular weight polyethylene fiber. Even though it is highly aligned, the strength is an order of magnitude lower and the Young's modulus is less by a factor of 2.6 when compared to theoretical values. Most bulk polymers have randomly oriented chains. When testing the strength of the randomly oriented polymer, one rarely tests the strength of the primary C—C bond in the polymer. Instead, one engages with the secondary bonds, such as hydrogen and van der Waals bonds, which are present between adjacent polymer chains and any intrinsic stress concentrations included in the q term of Eq. (3.6).

Table 3–1 A comparison of tensile properties of polyethylene.

Polyethylene form	σ (GPa)	E (GPa)	Reference
Theoretical	32.5	340	He (1986)
UHMWPE fiber	3.2	130	Elices and Llorca (2002)
Bulk HDPE	0.021—0.037	0.59—1.1	Meyers and Chawla (2009)

Notes: A significant drop is noted between theoretical values of polyethylene and ultrahigh molecular weight polyethylene (UHMWPE), and also the bulk high-density polyethylene (HDPE).

3.4 Thermal conductivity

The thermal properties of polymers are also dependent on the structure. There are significant differences between thermal properties of amorphous and semicrystalline polymers. In addition, anisotropy caused by preferred orientation of the polymer chains, for example in fibers, will lead to a directional dependence. Preferential orientation can also occur in polymer thin films during the deposition process and during deformation on stretching (Kurabayashi, 2001).

In general, the principal thermal energy carriers are electrons and phonons. Phonons are thermal vibrations transmitted by the polymer chain. The dominant mechanism of thermal transport in all polymers is phonon transport. However, the thermal conductivity of amorphous polymers can be an order of magnitude lower than that of highly crystalline polymers. This difference in behavior is mainly attributed to scattering mechanisms stemming from mean free path. In highly crystalline polymers, the packing of the chains is more efficient leading to a large mean free path and higher thermal conductivity.

Energy transport by phonons is reduced by the following mechanisms:

1. Phonon–phonon collisions.
2. Collisions with the surface.
3. Interaction with impurities, that is, atomic scale defects, holes, atomic impurities.

An approximation from the kinetic theory of gases gives the following expression for thermal conductivity, k:

$$k = \frac{1}{3} C_{\mathrm{p}} v l, \tag{3.8}$$

where C_{p} is the heat capacity per unit volume, v is the average particle velocity, and l is the mean free path between phonon collisions.

Eq. (3.8) can also be used to describe thermal conductivity in highly ordered crystalline materials, highly disordered crystals, and amorphous solids. Recall from Chapter 2, Intrinsic voids in crystalline materials: ideal materials and real materials, the mean free path term can be adjusted to account for the defect concentration in the crystalline phase. Somewhat analogous are amorphous polymers and glasses where there is no long-range order and very small mean free path value.

Material design and processing have progressed to the point where phonon interactions and transport can be controlled, leading to an ability to tailor thermal properties. As such, a new field called *phononics* or *phonon engineering* (Balandin & Nika, 2012) has emerged.

3.5 Role of voids in physical aging in polymers

Physical aging in a polymer refers to the time dependent reduction of the free volume. This decrease in free volume causes changes in physical and mechanical properties. The difference in free volume and equilibrium free volume is the free volume available for reduction in the physical aging process. This volume reduction during physical aging is accomplished by the diffusion of voids out of the material. As voids diffuse from the interior to the surface in a polymer below the T_g there occurs a reduction in free volume. Aging is a time dependent process involving a slow decrease in volume, with very long relaxation times. This process is also characterized by a decrease in polymer chain mobility and increase in material density, both of which are directly related to the decrease in free volume. Accompanying the aging process there are also changes in mechanical, electrical, structural, and physical properties.

Alfrey, Goldfinger, and Mark (1943) described and modeled the densification or diffusion of voids from the bulk to the surface in a polymer. Thornton, Nairn, Hill, Hill, and Huang (2009) extended Alfrey's work in their model called empirically derived vacancy diffusion (EVD).

This EVD model leads to an empirically derived relationship between the diffusion coefficient for free volume transport, D, and fractional free volume, f:

$$D = \alpha \exp(\beta f), \tag{3.9}$$

where α and β are constants. Fig. 3–5 illustrates the role of free volume as it relates to polymer aging near its T_g. The top curve shows the decrease in free volume with time of a polymer equilibrated at 40°C and lowered to 35°C, whereas the bottom curve shows the increase in free volume of the polymer equilibrated at 30°C and raised to 35°C. In the top curve, because the temperature is lowered, there is a decrease in free volume until an equilibrium free volume is reached at 35°C. In the bottom curve, since the temperature is raised, the free volume increases until it reaches the equilibrium free volume. These two curves converge on the equilibrium free volume value at 35°C.

3.6 Measurement of free volume

A technique called Positron Annihilation Lifetime Spectroscopy (PALS) has been used since the 1960s to investigate defects and free volumes in materials (Oka, Jinno, & Fujinami, 2009). A positron is the antiparticle of the electron having a charge of +1. Positron lifetime, specifically the positronium lifetime, can be used to quantify free volume holes in polymers, and void diameter and distribution in

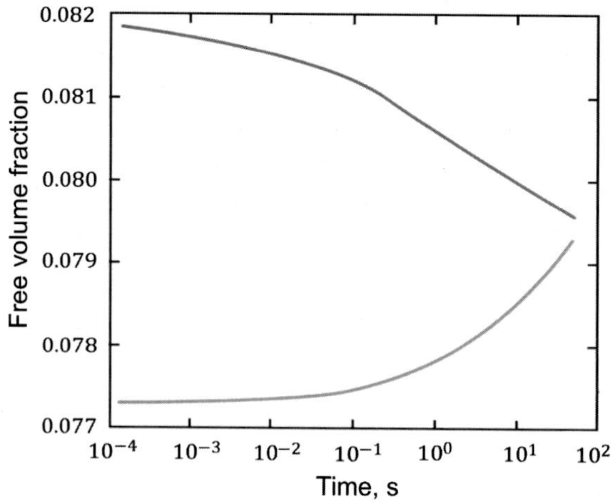

FIGURE 3–5 Graph illustrating the role of free volume as it relates to polymer aging near its T_g. The polymer in the bottom curve was equilibrated at 30°C and raised to 35°C, whereas the polymer in the top curve was equilibrated at 40°C and lowered to 35°C. Note that the two curves converge on the equilibrium free volume value at 35°C. **Source:** *Adapted from Thornton A.W.; Nairn K.M.; Hill A.J.; Hill J.M. (2009) New relation between diffusion and free volume: I. Predicting gas diffusion. Journal of Membrane Science 338 (1–2) 29–37; Thornton A.W.; Nairn K.M.; Hill A.J.; Hill J.M.; Huang Y. (2009) New relation between diffusion and free volume: II. Predicting vacancy diffusion. Journal of Membrane Science 338 (1–2) 38–42.*

porous materials. This technique has been used successfully on thermosetting and thermoplastic polymers. Free volume has been correlated to diffusivity (Choudalakis & Gotsis, 2012; Eceolaza, Iriarte, Uriarte, Del Rio, & Etxeberria, 2012; Laksmana, Hartman Kok, Vromans, & Van Der Voort Maarschalk, 2009; Shi, Nieh, & Chou, 2000; Thornton, Nairn, Hill, & Hill, 2009; Thornton et al., 2009), wear (Brostow, Hagg-Lobland, & Narkis, 2006), and thermal properties (Jia, Zheng, Zhu, Li, & Xu, 2007) in polymers. This technique has been used successfully for characterizing thermoplastics, thermosets, glasses, metals, and semiconductors. See Chapter 8, Void characterization techniques, for a detailed discussion of PALS.

References

Alfrey, T., Goldfinger, G., & Mark, H. (1943). The apparent second-order transition point of polystyrene. *Journal of Applied Physics* (14), 700.

Balandin, A. A., & Nika, D. L. (2012). Phononics in low-dimensional materials. *Materials Today, 15*(6), 266–275.

Bondi, A. (1964). van der Waals volumes and radii. *Journal of Physical Chemistry, 68*(3), 441–451.

Brinkman, H. C. (1940). On the theory of liquids. *Physica, 7*(8), 747–752.

Brostow, W., Hagg-Lobland, H. E., & Narkis, M. (2006). Sliding wear, viscoelasticity, and brittleness of polymers. *Journal of Materials Research, 21*, 2422–2428.

Choudalakis, G., & Gotsis, A. D. (2012). Free volume and mass transport in polymer nanocomposites. *Current Opinion in Colloid & Interface Science, 17*(3), 132–140.

Eceolaza, S., Iriarte, M., Uriarte, C., Del Rio, J., & Etxeberria, A. (2012). Influence of the organic compounds addition in the polymer free volume, gas sorption and diffusion. *European Polymer Journal, 48*(7), 1218–1229.

Elices, M., & Llorca, J. (2002). *Fiber fracture*. Oxford: Elsevier.

Eyring, H. (1936). Viscosity, plasticity, and diffusion as examples of absolute reaction rates. *Journal of Chemical Physics, 4*(4), 283–291.

Fox, J., Thomas, G., & Flory, P. J. (1950). Second-order transition temperatures and related properties of polystyrene. I. Influence of molecular weight. *Journal of Applied Physics, 21*(6), 581–591.

He, T. (1986). An estimate of the strength of polymers. *Polymer, 27*(2), 253–255.

Jia, Q. M., Zheng, M., Zhu, Y. C., Li, J. B., & Xu, C. Z. (2007). Effects of organophilic montmorillonite on hydrogen bonding, free volume and glass transition temperature of epoxy resin/polyurethane interpenetrating polymer networks. *European Polymer Journal, 43*(1), 35–42.

Kurabayashi, K. (2001). Anisotropic thermal properties of solid polymers. *International Journal of Thermophysics, 22*(1), 277–288.

Laksmana, F. L., Hartman Kok, P. J. A., Vromans, H., & Van Der Voort Maarschalk, K. (2009). Predicting the diffusion coefficient of water vapor through glassy HPMC films at different environmental conditions using the free volume additivity approach. *European Journal of Pharmaceutical Sciences: Official Journal of the European Federation for Pharmaceutical Sciences, 37*(5), 545–554.

Meyers, M. A., & Chawla, K. K. (2009). *Mechanical behavior of materials* (2nd ed.). Cambridge, UK: Cambridge University Press.

Oka, T., Jinno, S., & Fujinami, M. (2009). Analytical methods using a positron microprobe. *Analytical Sciences: International Journal of the Japan Society for Analytical Chemistry, 25* (837).

Park, J. Y., & Paul, D. R. (1997). Correlation and prediction of gas permeability in glassy polymer membrane materials via a modified free volume based group contribution method. *Journal of Membrane Science, 125*(1), 23–39.

Sears, J. K., & Darby, J. R. (1982). *Technology of plasticizers*. New York: John Wiley & Sons.

Shi, F. G., Nieh, T. G., & Chou, Y. T. (2000). A free volume approach for self-diffusion in metals. *Scripta Materialia, 43*(3), 265–267.

Thornton, A. W., Nairn, K. M., Hill, A. J., & Hill, J. M. (2009). New relation between diffusion and free volume: I. Predicting gas diffusion. *Journal of Membrane Science, 338*(1–2), 29–37.

Thornton, A. W., Nairn, K. M., Hill, A. J., Hill, J. M., & Huang, Y. (2009). New relation between diffusion and free volume: II. Predicting vacancy diffusion. *Journal of Membrane Science, 338* (1–2), 38–42.

Van Wijk, W. R., & Seeder, W. A. (1939). The influence of the temperature and the specific volume on the viscosity of liquids II. *Physica, 6*(2), 129–136.

Van Wijk, W. R., & Seeder, W. A. (1937). The influence of the temperature and the specific volume on the viscosity of liquids. *Physica, 4*(10), 1073–1088.

Williams, M. L., Landel, R. F., & Ferry, J. D. (1950). The temperature dependence of relaxation mechanisms in amorphous polymers and other glass-forming liquids. *Journal of the American Chemical Society, 14*, 3701.

Zhurkov, S. (1984). Kinetic concept of the strength of solids. *International Journal of Fracture, 26* (4), 295–307.

4

Nanometer scale porous structures

4.1 Introduction

In this chapter we discuss nanometer scale materials and voids. As such, it is appropriate to give a brief history of this technical field. The first mention of nanotechnology has been attributed to Richard Feynman in 1959 (Nunes et al., 2019) in a lecture titled "There's plenty of room at the bottom." However, the term nanotechnology properly was coined by Norio Taniguchi et al. in their 1974 paper "On the basic concept of nanotechnology" (Taniguchi, Arakawa, & Kobayashi, 1974).

A basic description of *nanomaterials* has the size domains that make up the material on a nanometer length scale. Macrostructures can technically be nanomaterials, if for example, the crystal grains of a metal or ceramic are on the nanometer-scale. Other nanomaterials, which are probably more familiar, are discrete structures at nanometer (nm) length scale such as nanotubes and nanoparticles. The prefix *nano-* is the Greek word that is used in the SI system of units meaning one billionth or 10^{-9}. The *meter* is the base unit of length, therefore 1 nm is one billionth of a meter or 1×10^{-9} meters. We use the nm dimension to describe an object or feature with a measurement ranging from 1 nm to less than 1000 nm. For an object that is of 1000 nm or greater, we would use micrometer as the dimension. We follow this convention in this chapter and throughout the book. However, when reviewing literature about *nanoporosity*, one will encounter a naming system due to International Union of Pure and Applied Chemists (IUPAC) for porosity. This system is based on three pore size regimes (Pierotti & Rouquerol, 1985):

1. Macropores—greater than 50 nm.
2. Mesopores—between 2 and 50 nm.
3. Micropores—less than 2 nm.

The pore size categories are somewhat arbitrary in that they were developed around the interaction of voids with nitrogen gas as it is adsorbed on the surface of a porous material. These are based on transport mechanisms of the adsorbed nitrogen gas at its normal boiling point through the porosity. The structures measured as per this IUPAC scale are referred to as being nanoporous and this porosity has particular functionality. We go into more detail about these mechanisms in Chapter 8, Void characterization techniques of this book.

Voids in Materials. DOI: https://doi.org/10.1016/B978-0-12-819282-5.00004-1

The field of nanometer scale porous materials is vast. We give an overview of this field and encourage interested readers to consult references for details. The main challenges in this field are the tailoring and controlling of the pore size and shape to suit a particular application. Nanoporous materials include numerous applications: fuel cell, H_2 and other gas storage, biomaterials and bioengineering, membranes, separations, nanoscaffolds for containing phase changed materials for heat storage, catalysts, environmental remediation, microelectromechanical systems, optical and photonic devices, chemical filtration, desalination, and chemical and biological sensors.

The ability to control the domains of a *solid* material in the nanometer range has led to important breakthroughs in performance of materials. Consider the important mechanical and electrical properties and outstanding performance demonstrated by graphene, carbon nanotubes (CNTs) and fullerenes, etc. Analogous to the control of solid domains are the dimensions and precision we now have of voids and porosity. This control is evidenced by Faucher et al. (2019) in a review on the mass transport through materials with single nanometer porosity. By describing a material with single nanometer porosity, Faucher indicates that it is intentionally designed porosity that is <10 nm in diameter.

In what follows we present important materials that have functional nanometer-scale porosity. As discussed, this is an extensive field with many important areas of research. We highlight research that focuses on controlling and designing nanometer scale voids within a material and the specific functionality that results.

4.2 Nanotubes

In Chapter 5, Hollow and porous structures utilizing the Kirkendall effect, we discuss the Kirkendall effect and how it can be used, in an advantageous way, as a route to create nanotubes. The Kirkendall effect has been exploited to make a wide variety of nanotubes including sulfides, oxides, selenides, tellurides, fluorides, phosphides, metallic alloys, and intermetallic compounds (El Mel, Nakamura, & Bittencourt, 2015). In Chapter 7, Techniques of introducing intentional voids into particles and fibers, we discuss the process of electrospinning as a versatile and efficient technique to fabricate inorganic hollow structured nanofibers of metal oxides, carbon, metals, and composites (Li et al., 2017) as well as natural and synthetic polymers (Barhoum et al., 2019). The combination of these two important techniques allows for fabrication of dozens of materials into nanotubular structures. However, by far the most studied nanotube is the carbon nanotube. We briefly discuss the CNTs in Chapter 7, Techniques for introducing intentional voids into particles and fibers but, because of its importance, a more detailed discussion is warranted.

In Chapter 7, Techniques for introducing intentional voids into particles and fibers, we discuss some of the processing techniques to make single walled carbon nanotubes (SWCNT), double walled carbon nanotubes (DWCNT), and multiwalled carbon nanotubes (MWCNT). These methods are shown in Fig. 4–1. Now we turn our attention to changing and controlling the diameter of the central void of CNTs. The building block of a SWCNT is graphene; a two-dimensional hexagonal array of carbon atoms. When this sheet of graphene is rolled up, it forms a hollow cylinder that is one-atom-thick carbon layer. The thickness has been measured for a single layer and up to 10 layers of graphene by Ni et al. (2007). A single layer of graphene was found to be 0.335 nm and the thickness of additional walls/layers is simply multiples of 0.335 nm. For example, the thickness of three layers of graphene is $3 \times (0.335)$ or 1.005 nm. We wish to highlight the central void, its functionality, and applications.

It should be noted that not all works in this area characterize the inner diameter (i.e., the void portion) of the CNT. The reader should be careful as to which CNT diameter an author is referring to. Many, in fact the majority, refer to the overall CNT diameter dimension, which theoretically is the sum of the diameter of the central void and two graphene layers (0.670 nm). We add two graphene layer thicknesses, one at each end of the central void. The diameter of CNT can range from just 1 nm for a single-walled carbon nanotube

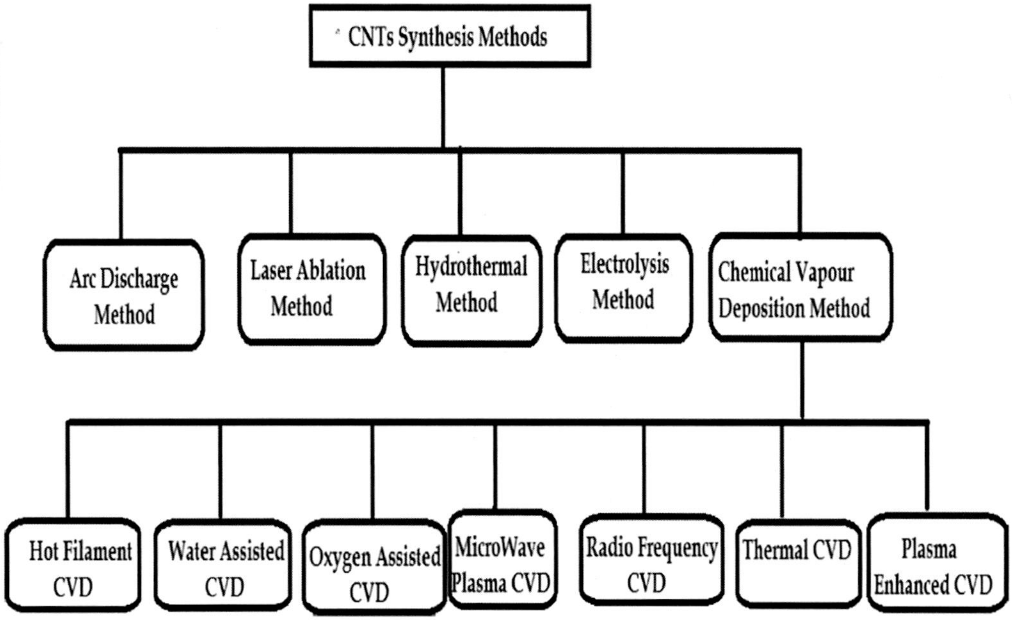

FIGURE 4–1 Various methods for making carbon nanotubes. Source: *Shanmugam, K., Manivannan, J., & Manjuladevi, M. (2019). Stupendous nanomaterials: Carbon nanotubes synthesis, characterization, and applications.* Nanomaterials-toxicity, human health and environment *IntechOpen.*

to 60 nm for multiwalled carbon nanotubes and even higher for ropes of carbon nanotubes (Ragab & Basaran, 2011). To determine the diameter and diameter distribution of CNTs, Raman spectroscopy can be used with a laser excitation wavelength of 633 nm. In the Raman spectra, there are three peaks or regions that are used in determining the diameter: the radial breathing modes (RBM ~ 100–300 cm^{-1}), D peak (~1350 cm^{-1}), and G peak (~1570 cm^{-1}).

The RBM peaks are the distinctive peaks of carbon nanotubes, analogous with the diameters of carbon nanotubes (Shanmugam et al., 2019). The details of determining the CNT diameter using Raman spectra are beyond the scope of this book, and readers should consult references (Ahmad & Silva, 2020; Levshov et al., 2017). Processing parameters that affect the diameter, number of walls are the carbon source, carrier gas, flow rate, chamber pressure, temperature, substrate material, catalyst material and its thickness, size and distribution of the catalyst nanoparticles, catalyst support material and temperature (Ahmad & Silva, 2020).

Data listed in CNT publications invariably include the CNT diameter. Knowing that the graphene making the tube wall is 0.355 nm, the central void diameter can be calculated even if the central void dimension is not specifically mentioned. There are times when the dimensions of the CNT void are just incidental to fabricating a specific CNT diameter. At other times a specific interior CNT void dimension is essential to the application.

An example of the void being incidental to the functioning of the CNT is that of the reinforcing capability of nanotubes in a polymer matrix. Many of the theories that explain the properties of composite materials on a micrometer level are not scalable to the nanometer level. Relevant to this example is the rule of mixtures and Halpin–Tsai theories. On a micrometer level, the behavior of a composite is based on the *volume fraction* of each phase that is present in the composite. Instead, Cadek et al. (2004) found that mechanical properties of CNT reinforced composites depend on the average diameter of the carbon nanotubes and therefore on the *total surface area* per unit volume. The surface area, *SA*, is related to the CNT diameter, *d*, CNT length, *l*, by the follow equation:

$$SA = \pi dl.$$

In other words, Cadek found that the reinforcement scales linearly with the total surface area of the CNT. The surface area is proportional to the diameter of the central void. The increase in mechanical properties was attributed to the interfacial interaction between the CNT and polymer causing the matrix to crystallize at the interphase region. Although the increased mechanical properties correlate to the average void dimension in the CNTs, the actual mechanism is unrelated.

There is a significant effort to study and devise processes that can create CNTs with controlled diameters. Tian et al. (2011) studied the relationship of the CNT diameter to optical properties. Tian was able to tailor the mean diameter of SWCNTs by the addition of appropriate amounts of CO_2 mixed with the gas-phase carbon source into an aerosol that enters a chemical vapor deposition reactor. The CO_2 selectively "etches" the smaller diameter CNT because of higher curvature surface. By systematically increasing the CO_2 concentration entering the reactor from 0.25% to 1.00%, the mean diameter shifted from 1.2 to 1.9 nm, respectively, thus changing the optical properties. Fig. 4–2 shows the absorbance spectra for the CNT for each of the CO_2 parameters as well as the CNT diameter histograms.

In order to illustrate the idea of void dimensions being critical to functionality, we provide an example of using CNT as a nano-reactor. Using nanocontainment for conducting a chemical reaction is known as *chemical nanoscience*. Chemical nanoscience has the very similar goals to that of traditional chemistry (Khlobystov, 2011): (1) studying structure, (2) harnessing functional physical properties, and (3) controlling the chemical reactions of molecules. However, these must be done on a nanometerscale. The central void

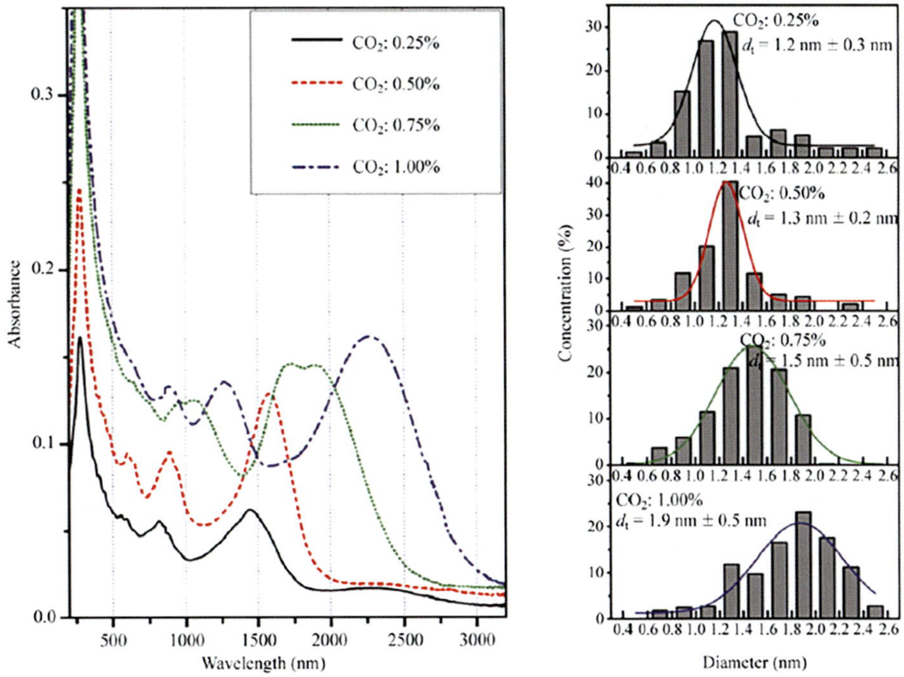

FIGURE 4–2 Absorbance spectra and CNT diameter histograms as a function of CNT diameter. By increasing CO_2 concentration in the reactor, the mean diameter of the CNTs increased. Source: *Tian, Y., Timmermans, M.Y., Kivistö, S., Nasibulin, A.G., Zhu, Z., Jiang, H., ... Kauppinen, E.I. (2011). Tailoring the diameter of single walled carbon nanotubes for optical applications.* Nano Research, *4(8), 807.*

dimension of the CNT is the most important parameter in determining whether a molecule can be contained within it. As a rule, internal diameter of a nanotube must be at least 0.6 nm wider than the diameter of the guest molecule (Khlobystov, 2011).

For SWCNTs, the cavity diameter ranges from 0.5 to 2.0 nm, whereas for MWCNTs it ranges between 2.0 and 60 nm. Once reactants are confined within the CNT, the guest species in the CNT nanocavities can have a considerable impact on the host CNT, including changes in their structure, size distribution, surface area, and dynamics (Axet & Serp, 2016). This leads to many interesting properties that are different from those in bulk chemical reactions and leads to new potential applications, such as quantum molecular sieving, nanofiltration, drug delivery, and catalysis. The environment within a CNT is very unique not only from a spatial standpoint but also from chemical standpoint, which can contribute to the reaction, for example, the electronic interaction due to the curved graphene walls (Axet & Serp, 2016). Khlobystov (2011) identified other effects that are unique when using the confined porosity space within CNTs as nanoreactors, see Table 4–1.

Because of the environment within a nanotube and the void dimensions, some novel reactions and subsequent structures can be synthesized. Fig. 4–3 highlights some of the reactions that can take place within nanotubes using fullerenes, C_{60}, as the reactant (Khlobystov, 2011).

Table 4–1 Interactions unique to the carbon nanotube nano-reactor environment.

Interaction between CNT and catalyst	Interaction between CNT and reactants	Interaction between CNT and products
Enhanced stability of catalyst inside nanotube	Attractive interactions between nanotube interior and reactant molecules leading to a higher local concentration and an effective pressure inside nanoreactors	Restriction of reaction space inside nanotube favoring formation of one product (e.g., linear isomer) over another (e.g., branched isomer)
Electron transfer between catalyst and interior surface of nanotube altering catalytic activity	Alignment of reactant molecules within nanotube facilitating reaction	Efficient transport of product molecules from nanoreactor to bulk phase
Chemical reactions between metal catalyst and nanotube interior leading to transformations of CNT sidewall	van der Waals or electron transfer interactions between reactants and nanotube interior lowering activation energy	

CNT, carbon nanotube.

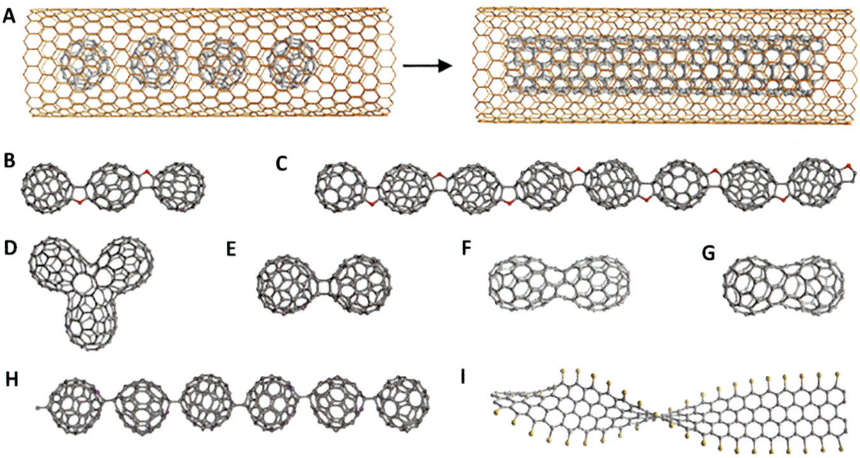

FIGURE 4–3 Using the unique environment within the central void of a CNT as a nano-reactor, fullerenes (C_{60}) can react to form (A) tubes, (B) C_{60} short chain polymers, (C) longer chain polymers, (D) trefoils, (E–G) various dimer molecules, (H) $C_{59}N$ polymers, and (I) graphene ribbons. Source: *Khlobystov, A.N. (2011). Carbon nanotubes: From nano test tube to nano-reactor. ACS Nano, 5(12), 9306–9312.*

4.3 Zeolites

Zeolites are aluminosilicates that have porosities in the nanometer range and are classified as microporous material, that is, the pores are less than 2 nm in size. The pores have a precisely defined system of microchannels (Feliczak-Guzik, 2018). These occur in nature and are also important synthetic materials. Water, alkali, and alkaline earth cations are contained within the pores of the zeolite framework (Moshoeshoe, Nadiye-Tabbiruka, & Obuseng, 2017). The presence of cations makes them well suited as molecular sieves and adsorbents in separation and purification processes. These are frequently used for environmental cleanup, such as adsorbents for the removal of heavy metal ions and other wastes. For a zeolite there is a general chemical formula (Wang & Peng, 2010):

$$M_{x/n}\left[Al_xSi_yO_{2(x+y)}\right]\cdot pH_2O$$

where

M is Na, K, Li and/or Ca, Mg, Ba, Sr
n is the cation charge
$y/x = 1-6$
$p/x = 1-4$

Table 4–2 lists the most common naturally occurring zeolite minerals with their chemical composition (Wang & Peng, 2010). Naturally occurring zeolites, depending

Table 4–2 Name and chemical formula of the most abundant zeolites found in nature.

Zeolite	Chemical formula
Clinoptilolite	$(K_2, Na_2, Ca)_3Al_6Si_{30}O_{72}.21H_2O$
Mordenite	$(Na_2, Ca)_4Al_8Si_{40}O_{96}.28H_20$
Chabazite	$(Ca, Na_2, K_2)_2Al_4Si_8O_{24}.12H_2O$
Phillipsite	$K_2(Ca, Na_2)_2Al_8Si_{10}O_{32}.12H_2O$
Scolecite	$Ca_4Al_8Si_{12}O_{40}.12H_2O$
Stilbite	$Na_2Ca_4Al_{10}Si_{26}O_{72}.30H_2O$
Analcime	$Na_{16}Al_{16}Si_{32}O_{96}.16H_20$
Laumontite	$Ca_4Al_8S_{16}O_{48}.16H_2O$
Erionite	$(Na_2K_2MgCa_{1.5})_4Al_8Si_{28}O_{72}.28H_2O$
Ferrierite	$(Na_2, K_2, Ca, Mg)_3Al_6Si_{30}O_{72}.20H_2O$

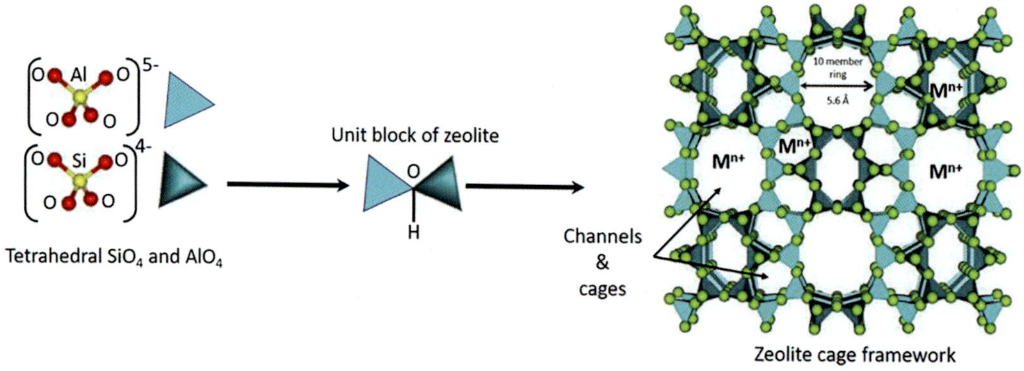

FIGURE 4–4 The basic formation of a zeolite. *Source: Courtesy of Geoffery Price - University of Tulsa.*

on the chemical formula, have a porosity from 17% to 47% by volume (Jha & Singh, 2011); analcimes have the smaller pores and chabazites have the larger pores. Another important characteristic of zeolites is their void volume, which can directly be correlated with the *cation exchange capacity* (CEC). CEC is defined as the amount of positive charge that can be exchanged per mass of zeolite.

Fig. 4–4 shows the structure of a zeolite from the *primary building units* (PBU) of tetrahedral SiO_4 and AlO_4. SiO_4 and AlO_4 combine, by sharing oxygen, into the complex zeolite with water molecules and cation in the pores and channels. The PBU assemble into a cage framework, that is, secondary block unit (SBU). The SBU is assembled into the complex zeolite containing pores and channels. The number of tetrahedra forming the pore and channel *rings* will determine the size of the opening. For example, a 10-member ring will have an opening of 0.56 nm while a 12-member ring will have an opening of 0.73 nm. Typical zeolites have 4, 5, 6, 8, 10, or 12 tetrahedrons in a ring (Moshoeshoe et al., 2017). The SBU zeolite structure further assembles into tertiary units and arranges into infinitely extended frameworks where pore sizes reach the single nanometer and sub-nanometer range. Synthetic zeolites are synthesized by chemical processes, which result in a more uniform and purer state as compared to the natural types in terms of their lattice structures, sizes of pores, and cages in their frameworks.

4.4 Nanoporous polymers

Typical porous cellular materials that we encounter in our daily life have micrometer sized pore structure. We refer to these as foams and they are also common in nature and in synthetic materials. When we reduce this micrometer scale voids to less than 200 nm, we refer to these foams as nanocellular polymers.

Notario, Pinto, and Rodriguez-Perez (2016) classify nanoporous polymer fabrication techniques as follows:

1. Phase separation
2. Imprinting or templating
   a. Molecular imprinting
   b. Colloid imprinting
   c. Self-organized templates
3. Foaming (gas dissolution)
   a. One step or batch foaming
   b. Two-step solid state foaming

Each of these have analogous techniques described in Chapters 6, Techniques for introducing intentional voids into balk materials, and 7, Techniques for introducing intentional voids into particles and fibers. In this section we provide an overview of the effects of reducing the cell size of foam into the nanometer range in polymers.

As shown throughout this book, the process by which voids are introduced into a material not only lead to geometrical differences in the voids, but changes in the material properties. Even if the same process is used, the smallest of perturbation to one of the process parameter can lead to drastic differences in material performance. There is a reality in materials science and engineering, and in many other disciplines (such as baking [Gladysz & Chawla, 2014]), that there is an interdependence of key factors that one must consider when designing a material. The material is the building block for a specific part and design criteria is imposed on that part. One must translate the part design criteria into material properties that can be tested and quantified. Fig. 4–5 is a schematic of the important factors that must be defined when designing a material to meet part-performance requirements (Liu, Chen, Spear, & Pollard, 2003). Defining these factors, raw materials

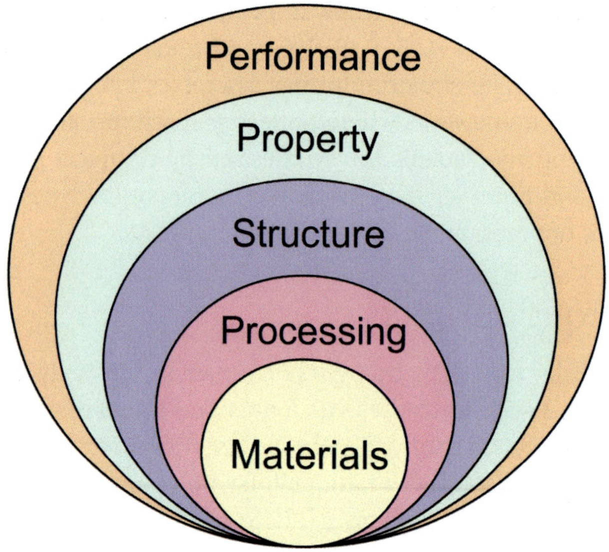

FIGURE 4–5 A schematic illustrating interdependence of key factors in materials science and engineering.

processing, structure, will yield a material that has certain materials properties and a component from that material will satisfy the performance requirements. Implicit to this interdependence is the fact that when one factor is changed, say for example *processing*, the result will be different material properties and part-performance. Applying Fig. 4–5, using the various processes results in changes in the structure, leading to variations in material properties and ultimately in different part performance.

As with the structure in micrometer-scale porosity, the structure in nanoporous polymers can be open or closed cell. The cell density N_p (pores/cm^3) can be calculated by (Kumar & Suh, 1990; Notario et al., 2016)

$$N_p = \frac{6\left(1 - \frac{\rho_{foam}}{\rho_{solid}}\right)}{\pi d^3},$$

where

ρ_{foam} is the density of the foam material
ρ_{solid} is the density of the solid phase
d is the average pore diameter (cm).

The quantity ρ_{foam}/ρ_{solid} is referred to as the relative density. Relative density provides a way of assessing/optimizing material performance, especially when comparing a foamed material to the nominally dense material from which it is made.

Miller and Kumar (2011) used relative density to illustrate the performance increase seen in nanocellular materials compared to microcellular materials made from polyethylenimine (PEI). The cell size of nanocellular and microcellular PEI ranged from 50–100 nm and 2–5 μm respectively. Fig. 4–6 shows several mechanical properties as a function of relative density, comparing nanocellular materials to microcellular materials. The implications of cell size on the mechanical properties are shown in Fig. 4–6. For example, the graph of strain at fracture versus relative density of 0.75; for the microcellular and nanocellular material the average strain at fracture is 0.30 and 1.03, respectively. This increase of 3.4 times in strain at fracture can be attributed to the reduction in cell size. Also note in Fig. 4–6 that the nanocellular PEI out performs microcellular PEI when we compare stress at fracture and toughness. Cell size has no effect on modulus of elasticity.

Thermal insulation properties can also be enhanced by reducing porosity size as well. When the pore size is below one micrometer, the thermal conductivity in

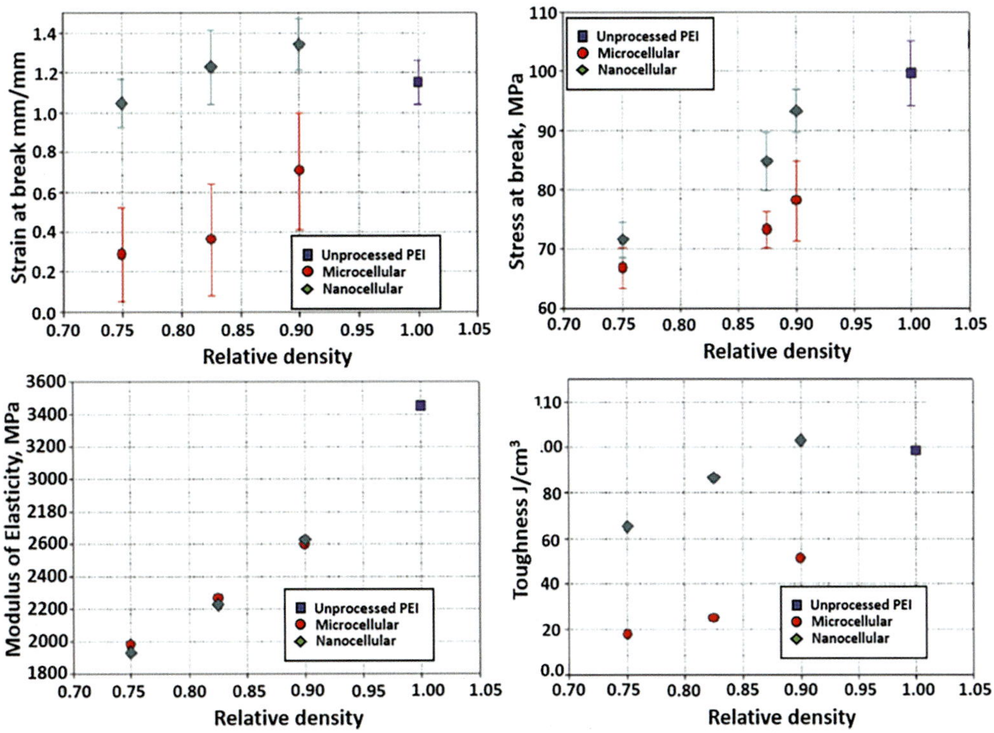

FIGURE 4–6 Mechanical properties of PEI foams; comparing microcellular to nanocellular performance. Source: *Miller, D., & Kumar, V. (2011). Microcellular and nanocellular solid-state polyetherimide (PEI) foams using sub-critical carbon dioxide II. tensile and impact properties https://doi.org/10.1016/j.polymer.2011.04.049.*

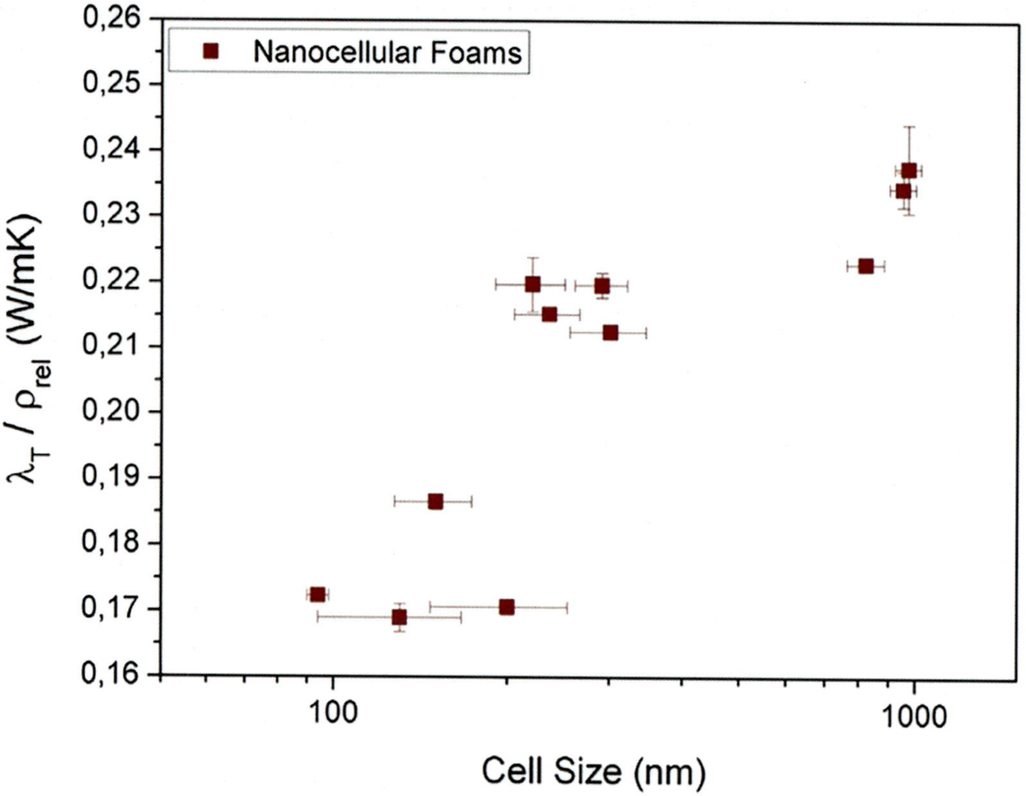

FIGURE 4–7 Thermal conductivity as a function of cell diameter for nanocellular PMMA. As the cell size is reduced there is a decrease in thermal conductivity. Source: *Notario, B., Pinto, J., Solorzano, E., de Saja, J.A., Dumon, M., & Rodríguez-Pérez, M.A. (2015). Experimental validation of the Knudsen effect in nanocellular polymeric foams https://doi.org/10.1016/j.polymer.2014.10.006*

nanoporous polymers shows a reduction in thermal conductivity. Phenomena behind this reduction are the confinement of both the gaseous phase and the solid phase (phonon scattering and increased tortuosity) (Notario et al., 2016), see, for example, the dependence of thermal conductivity on cell size using polymethylmethacrylate (PMMA) in Fig. 4–7. Note that the y-axis is normalized by dividing by the relative density. This was done to analyze the influence of the cell size independent of the density of the foams. The trend in the data shows that as the cell size is reduced, the thermal conductivity decreases.

4.5 Nanoporous organic networks

Nanoporous organic networks are a class of materials consisting solely of the lighter elements in the periodic table such B, C, N, and O. These materials differ from the nanocellular polymers presented earlier in this chapter, in that the void spaces in molecular network are intrinsic to the molecular structure. These materials do not require special techniques, such as templating, to introduce voids into the structure. These have potential uses in areas such as storage, separation, and catalysis (Dawson, Cooper, & Adams, 2012).

"Nanoporous' in this usage encompasses both microporous, pore size <2 nm, and/or mesoporous, pore size 2–50 nm materials. These nanoporous materials, consisting entirely of the lighter elements of the periodic table such B, C, N, and O, can have very low densities. To create a functional pore structure, such as hosting gases, an interconnected network of channels must be present. In addition, these channels must be stable, that is, they should not collapse. Typically, this stability is ensured by using rigid building blocks such as aromatic monomers. From these rigid building blocks the network is joined by:

1. Other aromatic monomers,
2. alkynes, and
3. alkenes.

Furthermore, to build a three dimensional network, at least one of the monomers must have a connectivity with two other groups forming a strut. Finally, the network will only be permanently porous after the removal of the solvent if the structure does not collapse and close the pores. There are four different classes of nanoporous polymer networks:

1. Covalent organic frameworks (COFs)
2. Covalent triazine frameworks
3. Polymers of intrinsic microporosity (PIMs)
4. Conjugated microporous polymers (CMP) and analogs

Porous polymer networks may be further divided into two subclasses: crystalline networks and amorphous networks. We will now discuss, briefly, each of the four classes of nanoporous organic networks.

4.5.1 Covalent organic frameworks

COFs are a class of crystalline porous polymers that allows the atomically precise integration of organic molecules into extended structures with periodic skeletons and ordered nanopores. The pores of COFs have clear polygonal shapes that can be designed to assume hexagonal, tetragonal, trigonal, rhombic, and kagome (star-shaped) structures. COFs are known for low density, good chemical stability, and high specific surface area, as well as regular pore structure, facile functional design, and some have luminous characteristics (Wang & Zhuang, 2019).

These are composed by different rigid monomers with symmetric reactive groups as vertices or edges in specific geometries. Three dimensional COFs have covalent bonding throughout the structure, whereas the two dimensional structures only have covalent bonding within layers and weak van der Waals ($\pi-\pi$ interactions) between layers. From a chemistry perspective, COFs are unique because their skeletons and pores are totally designable.

Fig. 4–8 shows different ways in which a hexagonal skeleton can be formed (Wang & Zhuang, 2019). Hexagonal skeletons can be assembled using just one monomer with two symmetric reactive groups (C_2). Another way is to use a $C_2 + C_3$ symmetric groups, see Fig. 4–8. An example of using a $C_2 + C_3$ symmetric groups reacting into hexagonal skeleton is shown in Fig. 4–9. The C_2 edges (with reactive MeO groups symmetric on the ring) react with the C_3 vertices (with 3 reactive NH_2 groups) to form a hexagonal skeleton. The structure extends and repeats, using all covalent bonding, in two dimensions from each of the vertices. These layers are stacked with weak van der Waals forces between each of the layers.

This building of porous polymers using diverse building blocks is called *reticular chemistry* (Cordova & Yaghi, 2017). One benefit of COF chemistry is the ability to predict a structure based on the choice of building units, or linkers. This predictability is based on the fact that given a set of linkers, there is often a just one or, at most a few crystalline structures that are likely to emerge (Lyle, Waller, & Yaghi, 2019). The porosity can be designed to have a wide range of sizes from small micropores to large mesopores. For example, the smallest hexagonal pore was synthesized using a C_2 +

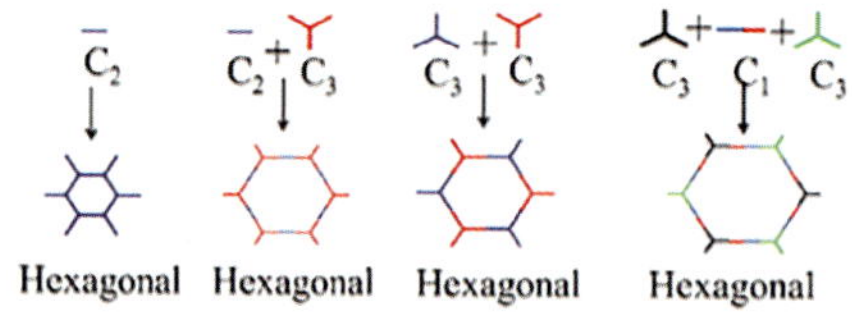

FIGURE 4–8 Various synthesis routes to obtain a hexagonal COF skeleton.

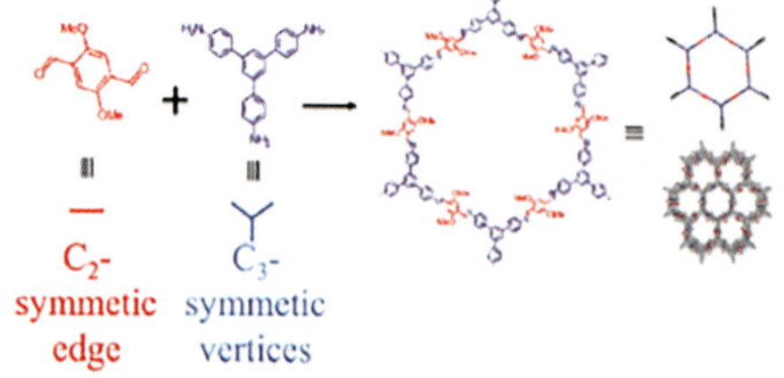

FIGURE 4–9 Using a C_2 and C_3 material, a two-dimensional layered hexagonal structure is synthesized. Within the hexagonal layer there is all covalent bonding while weak van der Waals forces bond layers together.

C_3 route yielding a 1.1 nm pore. Larger mesopores also used the $C_3 + C_2$ but with larger chemical units as edges and vertices yielded pores as large as 5.8 nm (Huang, Wang, & Jiang, 2016).

4.5.2 Covalent triazine frameworks

As we have discussed throughout this book, the control of void formation is reaching ever smaller scales. In covalent triazine frameworks, control of the void size can reach 1/10 of a nanometer. An example of the triazine framework is shown in Fig. 4–10. The chemical structure of triazine is similar to benzene. It is a six-member ring with alternating (delocalized) double bonds. However, in triazine the six-member ring contains three nitrogen and three carbon atoms. The reactant for making a triazine framework is 1,4-dicyanobenzene in the presence of zinc chloride ($ZnCl_2$). The temperature at which the reaction takes place and the quantity of $ZnCl_2$ present both have an effect on the pore diameter and surface area. The surface area of the structure in Fig. 4–10 is 791 m^2/g. Increasing the $ZnCl_2$ increases the surface to 1123 mm^2/g. Increasing the reaction temperature from 400°C to 700°C results in a pore diameter increase from 2.0 nm to 3.6 nm, respectively. Other reactants that can be used are thiophene and pyridines (Dawson et al., 2012).

4.5.3 Polymers of intrinsic microporosity (PIM)

PIMs are a class of nanoporous organic networks that result when the polymers pack ineffectively in the solid state. The term *intrinsic* in this usage means that the porosity is inherent, originating in the molecular structures because the large polymer chains do not pack efficiently. This packing leads to the nanoporosity. Most polymers like polyethylene and polypropylene are considered non-porous materials despite having free volume void between polymer chains, see Chapter 3, Intrinsic voids in polymer networks, PIM porosity is different in that it is not dependent on the thermal or processing history, or "aging" of the material as are *free volume holes*. Porosity in PIMs results from the rigid, bent monomers that contain a tetrahedral carbon atom referred to as a *site of contortion*. This site of contortion causes the polymer to fill space ineffectively, resulting in porosity between the polymer chains. The monomer 5,5',6,6'-tetrahydroxy-3,3,3',3'-tetramethyl-1,1'-spirobisindane in Fig. 4–11 has been used in a number of PIMs as it can effectively introduce a site of contortion into a polymer chain. Note the bent and contorted polymer structure in the spatial representation (Budd et al., 2004). Being a rigid structure PIM-1 has a stable porosity from 0.4 to 0.8 nm.

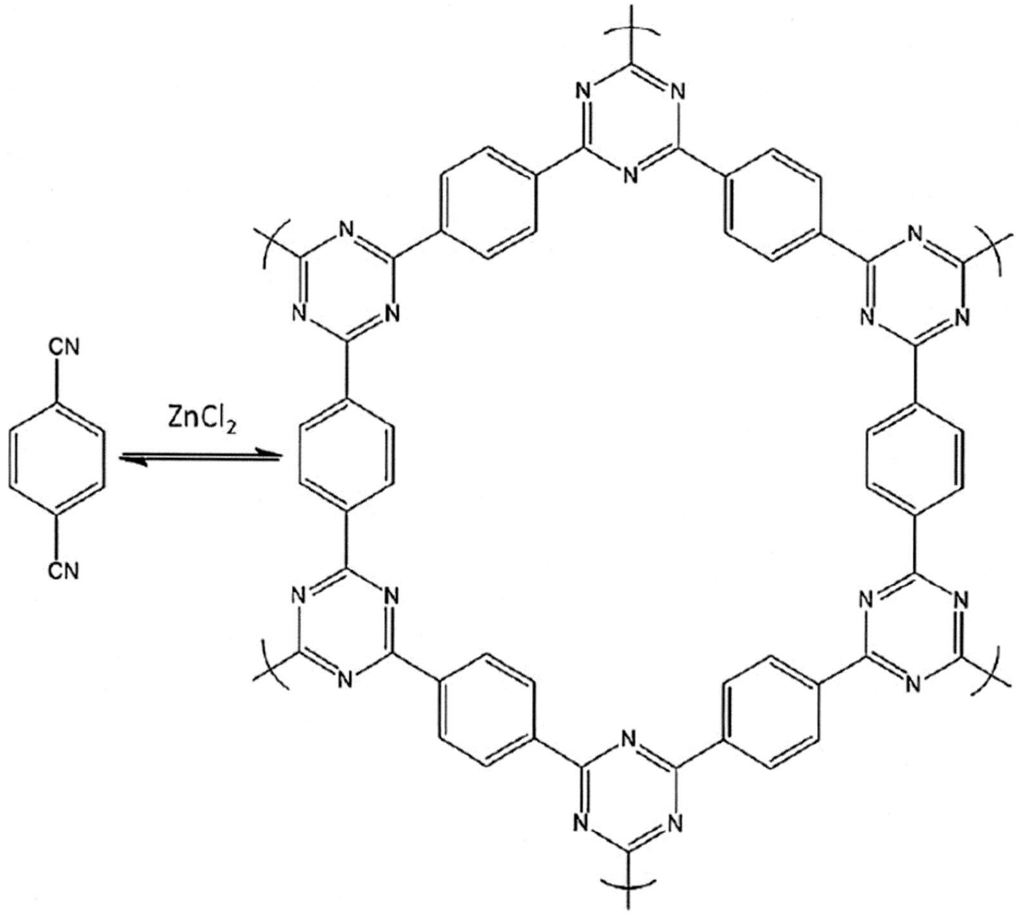

FIGURE 4–10 Nanoporous triazine network. Source: *Dawson, R., Cooper, A. I., & Adams, D. J. (2012). Nanoporous organic polymer networks. Progress in Polymer Science, 37(4), 530–563.*

4.5.4 Conjugated microporous polymers

A CMP is a three-dimensional semiconducting polymer in which rigid aromatic groups are linked together, either directly or via double or triple bonds. Conjugation in this sense means that there are alternating single and double- or triple-bonds throughout the extended network. This alternating bond formation leads to useful electronic properties (Lee & Cooper, 2020). Fig. 4–12 shows a chemical structure of a CMP known as CMP-1. CMP-1 is a poly(aryleneethynylene) network recognized as the first CMP molecule reported in 2007 (Jiang et al., 2007). It was synthesized by a palladium catalyzed cross-coupling of

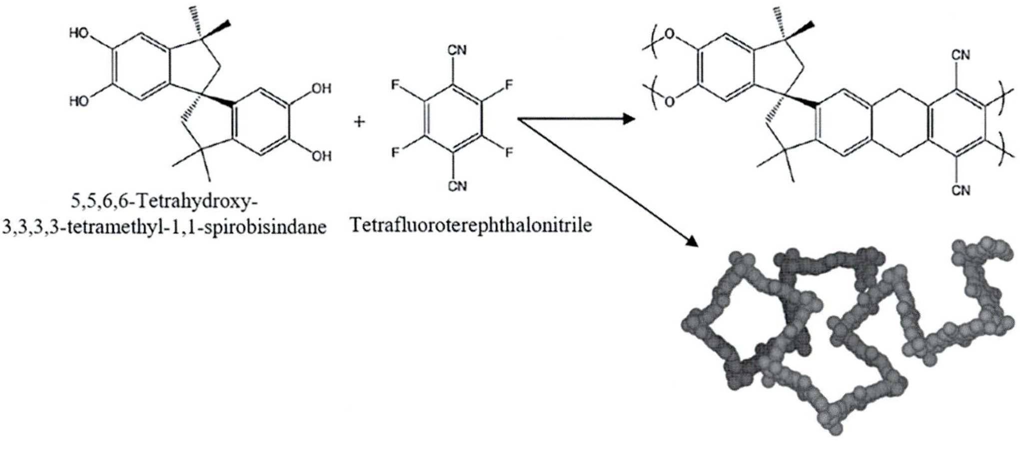

PIM-1

FIGURE 4–11 The reaction to form PIMs-1 (McKeown et al., 2005). The PIM-1 chemical structure is at the top and spatial representation (Budd et al., 2004) at the bottom right. Note the rigid and contorted polymer structure of PIM-1. Source: *McKeown N.B., Budd P.M., Msayib K.J., Ghanem B.S., Kingston H.J., Tattershall C.E. ... Fritsch D., Polymers of intrinsic microporosity (PIMs): Bridging the void between microporous and polymeric materials, Chemistry–European Journal 11 (9), 2005, 2610–262. Budd P. M., Ghanem B.S., Makhseed S., McKeown N.B., Msayib K.J. and Tattershall C.E., Polymers of intrinsic microporosity (PIMs): Robust, solution-processable, organic nanoporous materials, Chemical Communications 2, 2004, 230–231.*

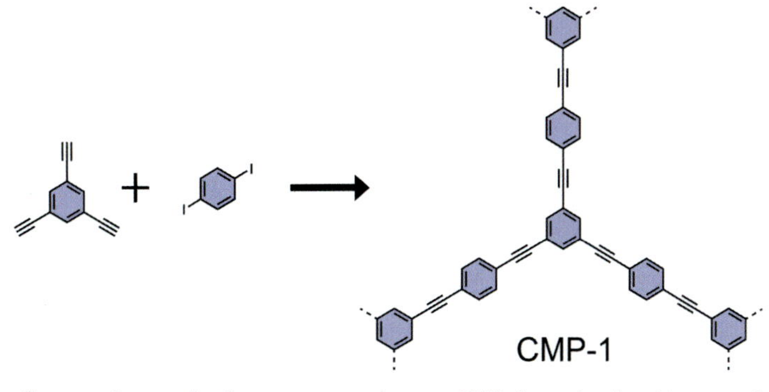

FIGURE 4–12 The first conjugated microporous polymer, CMP-1 synthesized by reacting1,3,5-triethynylbenzene with 1,4-diiodobenzene.

1,3,5-triethynylbenzene with 1,4-diiodobenzene. Note the defining features of CMPs: the aromatic rings and the alternating single bond and double- or triple-bonds throughout the network. The "dotted" bonds on the each of the three terminal 1,3,5 benzene rings in Fig. 4–12 are the points that the network extends.

This 1,3,5 benzene points of network extension is known as a *node* point. It can also be referred to as a *knot* (Xu, Jin, Xu, Nagai, & Jiang, 2013). Around the node points there are *struts* or *linkages* that can be tailored so the CMP has a certain surface area, pore volume and/or pore size, see Fig. 4–13. For example, the average pore size can be increased incrementally by increasing strut length.

Fig. 4–13 shows the various poly(aryleneethynylene) polymers (Xu et al., 2013). Note that these all have the same 1,3,5 benzene node, but the number of units that compose the struts are different. Length of the struts can be increased by the selective addition of ethyne (triple bonds unit) and benzene (ring structure

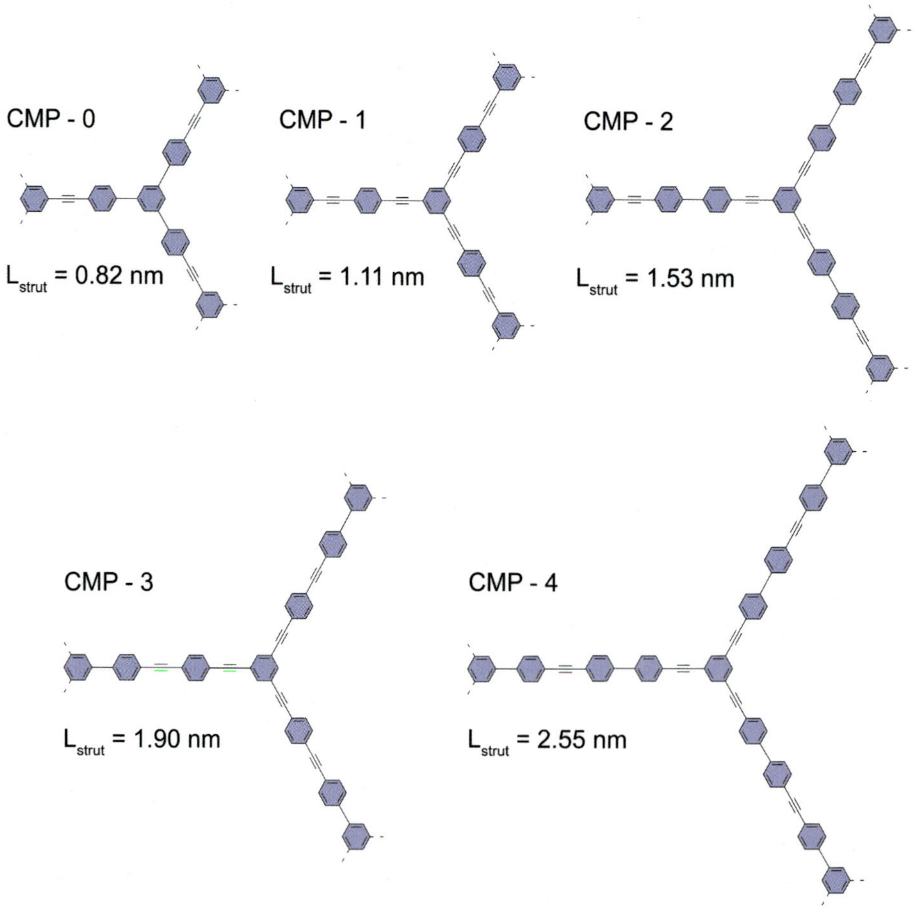

FIGURE 4–13 Chemical structures of poly(aryleneethynylene)-based CMPs with the strut length indicated. The strut length is an important parameter in tailoring the pore volume, surface area, and pore size in CMPs.

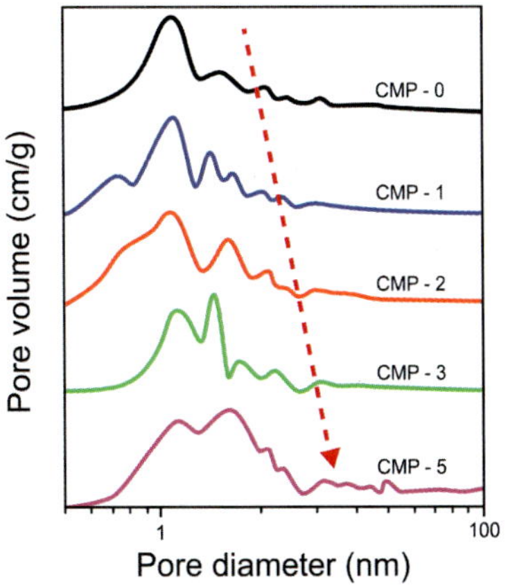

FIGURE 4–14 Pore size for poly(aryleneethynylene) -based CMPs. Note the decreasing trend of average pore size with pore volume (red arrow). This decreasing trend is ultimately a function of the strut length of each CMP. Source: *Adapted from Jiang J-X, Su F, Trewin A, Wood CD, Niu H, Jones JTA, Khimyak YZ,Cooper AI. Synthetic control of the pore dimension and surface area in conjugated microporous polymer and copolymer networks. J Am Chem Soc 2008;130:7710–20.*

unit). The strut lengths of CMP-0, CMP-1, CMP-2, CMP-3, and CMP-5 are 0.82, 1.11, 1.53, 1.90, and 2.55 nm, respectively. Pore volumes decrease as the strut length increases, decreasing from 0.38 to 0.33, 0.25, 0.18, and 0.16 cm^3/g, respectively (Jiang et al., 2008).

The pore size increases with the strut length. The pore sizes for CMP-0, CMP-1, CMP-2, CMP-3, and CMP-5 are shown in Fig. 4–14 with the blue arrow indicating the trend in average pore size. The shorter struts yield smaller voids, averaging less than 2 nm (microporous), and increase into the mesoporous range (2–50 nm) as the struts increase in length. Jiang et al. (2008) report 3% reproducibility on surface area and pore size of some CMP system.

4.6 Nanoporous noble metals

Nanoporous noble metals and noble metal-based alloys are important because of favorable properties such as, good temperature stability, corrosion resistance, high surface area, high conductivity, easy processing, tunable porosity, and good biocompatibility. Noble metals are elements that resist oxidation even at elevated temperatures. There is a general agreement on the elements that are classified as noble metals: silver (Ag), gold (Au), platinum (Pt), rhodium (Rh), iridium (Ir), palladium (Pd), ruthenium (Ru), and osmium (Os); however, few include rhenium (Re) also. Gold nanoparticles have been used as catalysts since the 1980s for CO oxidation (Haruta, Kobayashi, Sano, & Yamada, 1987; Kim, 2018). Research into the nanoporous noble metals was a natural progression from catalytic nanoparticles. Nanoparticles and nanoporous structures both have high surface areas; however, nanoparticles are difficult to recover from, for example, a water treatment process. In a nanoporous foamlike structure the metal is immobilized, which makes for easier recovery action (Kim, 2018).

There are many ways to make nanoporous noble metal-based structures. Many of the these have naming conventions unique to metals but have more generalized methods that we cover in Chapter 6, Techniques for introducing intentional voids into bulk materials, and Chapter 7, Techniques for introducing intentional voids into particles and fibers. These methods are as follows (Lu, 2019):

1. Dealloying
   a. Chemical dealloying
   b. Electrochemical dealloying
2. Template based
   a. Hard templating
   b. Soft templating
3. Assembly methods

The dealloying method relies on the corrosion or etching process that selectively attacks and removes the more active metal in an alloy; it is similar to the treatment known as *depletion gilding* (Erlebacher, Aziz, Karma, Dimitrov, & Sieradzki, 2001). *Chemical dealloying* is done using certain chemicals, such as an acid or base to remove one of the metals. For example, with a PdPtAl alloy, NaOH will selectively attack and remove Al leaving nanoporous PdPt alloy. *Electrochemical dealloying* is a corrosion process during which the more electrochemically active element is removed from a solid-electrolyte interface, resulting in the formation of a nanoporous microstructure of the more noble metal (Zeng et al., 2019). The *placeholder methods* covered in Chapter 6, Techniques for introducing intentional voids into bulk materials, and Chapter 7, Techniques for introducing intentional voids into particles and fibers are analogous to dealloying. The templating method has been mentioned in these and other chapters and the methodologies are similar. The assembly method has been referred to a self-assembly in Chapters 6, Techniques for introducing intentional voids into bulk materials, and 7, Techniques for introducing intentional voids into particles and fibers.

Fig. 4–15A and B are SEM images of a nanoporous gold film made by dealloying of Au and Ag. The original alloy was in the Au:Ag atomic ratio of 32:68. A chemical dealloying process, using nitric acid, was employed to selectively remove Ag. Fig. 4–15A is a cross-sectional view and (B) is the top view. Once Ag is removed, Ag foam structure with a void volume content of 68% remained.

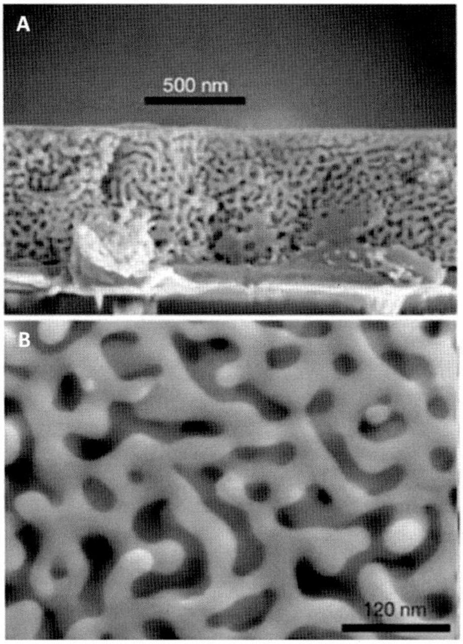

FIGURE 4–15 Nanoporous gold (A) cross-sectional view and (B) top view. The structure was made by etching silver with nitric acid from Au and Ag alloy. The alloy was 68 atomic at% Ag; therefore, the structures are 68 vol% void. Source: *Kim, S. H. (2018). Nanoporous gold: Preparation and applications to catalysis and sensors. Current Applied Physics, 18(7), 810–818.*

References

Ahmad, M., & Silva, S.R.P. (2020). Low temperature growth of carbon nanotubes—A review, *Carbon, 158*, 24–44. https://doi.org/10.1016/j.carbon.2019.11.061.

Axet, M. R., & Serp, P. (2016). *Carbon nanotube nanoreactors for chemical transformations. Organic nanoreactors* (pp. 111–157). Elsevier; Amsterdam.

Barhoum, A., Pal, K., Rahier, H., Uludag, H., Kim, I. S., & Bechelany, M. (2019). Nanofibers as new-generation materials: From spinning and nano-spinning fabrication techniques to emerging applications. *Applied Materials Today, 17*, 1–35. Available from https://doi.org/10.1016/j.apmt.2019.06.015.

Budd, P. M., Ghanem, B. S., Makhseed, S., McKeown, N. B., Msayib, K. J., & Tattershall, C. E. (2004). Polymers of intrinsic microporosity (PIMs): Robust, solution-processable, organic nanoporous materials. *Chemical Communications, 2*, 230–231.

Cadek, M., Coleman, J., Ryan, K., Nicolosi, V., Bister, G., Fonseca, A., ... Blau, W. (2004). Reinforcement of polymers with carbon nanotubes: The role of nanotube surface area. *Nano Letters, 4*(2), 353–356.

Cordova, K. E., & Yaghi, O. M. (2017). The 'folklore' and reality of reticular chemistry. *Materials Chemistry Frontiers, 1*(7), 1304–1309.

Dawson, R., Cooper, A. I., & Adams, D. J. (2012). Nanoporous organic polymer networks. *Progress in Polymer Science, 37*(4), 530–563. Available from https://doi.org/10.1016/j.progpolymsci.2011.09.002.

El Mel, A., Nakamura, R., & Bittencourt, C. (2015). The Kirkendall effect and nanoscience: Hollow nanospheres and nanotubes. *Beilstein Journal of Nanotechnology, 6*(1), 1348–1361.

Erlebacher, J., Aziz, M. J., Karma, A., Dimitrov, N., & Sieradzki, K. (2001). Evolution of nanoporosity in dealloying. *Nature, 410*(6827), 450–453.

Faucher, S., Aluru, N., Bazant, M. Z., Blankschtein, D., Brozena, A. H., Cumings, J., . . . Fourkas, J. T. (2019). Critical knowledge gaps in mass transport through single-digit nanopores: a review and perspective. *The Journal of Physical Chemistry C, 35*, 21309–21326.

Feliczak-Guzik, A. (2018). Hierarchical zeolites: Synthesis and catalytic properties. *Microporous and Mesoporous Materials, 259*, 33–45.

Gladysz, G.M., & Chawla, K.K. (November 26, 2014). *The space between: Reverse-engineering bread.* Retrieved from http://scitechconnect.elsevier.com/the-space-between-reversing-engineering-bread/

Haruta, M., Kobayashi, T., Sano, H., & Yamada, N. (1987). Novel gold catalysts for the oxidation of carbon monoxide at a temperature far below 0°C. *Chemistry Letters, 16*(2), 405–408.

Huang, N., Wang, P., & Jiang, D. (2016). Covalent organic frameworks: A materials platform for structural and functional designs. *Nature Reviews Materials, 1*(10), 1–19.

Jha, B., & Singh, D. (2011). A review on synthesis, characterization and industrial applications of flyash zeolites. *Journal of Materials Education, 33*(1), 65.

Jiang, J., Su, F., Trewin, A., Wood, C. D., Campbell, N. L., Niu, H., . . . Khimyak, Y. Z. (2007). Conjugated microporous poly (aryleneethynylene) networks. *Angewandte Chemie International Edition, 46*(45), 8574–8578.

Jiang, J., Su, F., Trewin, A., Wood, C. D., Niu, H., Jones, J. T., . . . Cooper, A. I. (2008). Synthetic control of the pore dimension and surface area in conjugated microporous polymer and copolymer networks. *Journal of the American Chemical Society, 130*(24), 7710–7720.

Khlobystov, A. N. (2011). Carbon nanotubes: From nano test tube to nano-reactor. *ACS Nano, 5*(12), 9306–9312.

Kim, S. H. (2018). Nanoporous gold: Preparation and applications to catalysis and sensors. *Current Applied Physics, 18*(7), 810–818.

Kuhn, P., Antonietti, M., & Thomas, A. (2008). Porous, covalent triazine-based frameworks prepared by ionothermal synthesis. *Angewandte Chemie International Edition, 47*(18), 3450–3453.

Kumar, V., & Suh, N. P. (1990). A process for making microcellular thermoplastic parts. *Polymer Engineering & Science, 30*(20), 1323–1329.

Lee, J. M., & Cooper, A. I. (2020). Advances in conjugated microporous polymers. *Chemical Reviews, 120*(3), 2171–2214.

Levshov, D. I., Tran, H. N., Paillet, M., Arenal, R., Than, X. T., Zahab, A. A., & Michel, T. (2017). Accurate determination of the chiral indices of individual carbon nanotubes by combining electron diffraction and Resonant Raman spectroscopy. *Carbon, 114*, 141–159. Available from https://doi.org/10.1016/j.carbon.2016.11.076.

Li, L., Peng, S., Lee, J. K. Y., Ji, D., Srinivasan, M., & Ramakrishna, S. (2017). Electrospun hollow nanofibers for advanced secondary batteries. *Nano Energy, 39*, 111–139. Available from https://doi.org/10.1016/j.nanoen.2017.06.050.

Liu, Z., Chen, L., Spear, K., & Pollard, C. (2003). An integrated education program on computational thermodynamics, kinetics, and materials design. *JOM-e, 12*.

Lu, L. (2019). Nanoporous noble metal-based alloys: A review on synthesis and applications to electrocatalysis and electrochemical sensing. *Microchimica Acta, 186*(9), 664.

Lyle, S. J., Waller, P. J., & Yaghi, O. M. (2019). Covalent Organic Frameworks: Organic Chemistry Extended into Two and Three Dimensions. *Trends in Chemistry, 1*(2), 172–184. Available from https://doi.org/10.1016/j.trechm.2019.03.001.

McKeown, N. B., Budd, P. M., Msayib, K. J., Ghanem, B. S., Kingston, H. J., Tattershall, C. E., ... Fritsch, D. (2005). Polymers of intrinsic microporosity (PIMs): Bridging the void between microporous and polymeric materials. *Chemistry–European Journal, 11*(9), 2610–2620.

Miller, D., & Kumar, V. (2011). Microcellular and nanocellular solid-state polyetherimide (PEI) foams using sub-critical carbon dioxide. *Polymer, 52*(13), 2910–2919. Available from https://doi.org/10.1016/j.polymer.2011.04.049.

Moshoeshoe, M., Nadiye-Tabbiruka, M., & Obuseng, V. (2017). A review of the chemistry, structure, properties and applications of zeolites. *American Journal of Materials Science, 7*(5), 196–221.

Ni, Z., Wang, H., Kasim, J., Fan, H., Yu, T., Wu, Y. H., ... Shen, Z. (2007). Graphene thickness determination using reflection and contrast spectroscopy. *Nano Letters, 7*(9), 2758–2763.

Notario, B., Pinto, J., & Rodriguez-Perez, M. A. (2016). Nanoporous polymeric materials: A new class of materials with enhanced properties. *Progress in Materials Science, 78-79*, 93–139. Available from https://doi.org/10.1016/j.pmatsci.2016.02.002.

Notario, B., Pinto, J., Solorzano, E., de Saja, J. A., Dumon, M., & Rodríguez-Pérez, M. A. (2015). Experimental validation of the Knudsen effect in nanocellular polymeric foams. *Polymer, 56*, 57–67. Available from https://doi.org/10.1016/j.polymer.2014.10.006.

Nunes, D., Pimentel, A., Santos, L., Barquinha, P., Pereira, L., Fortunato, E., & Martins, R. (2019). Introduction. In D. Nunes, A. Pimentel, L. Santos, P. Barquinha, L. Pereira, E. Fortunato, & R. Martins (Eds.), *Metal oxide nanostructures* (1st ed.). 1-19, Elsevier. Available from https://doi.org/10.1016/B978-0-12-811512-1.00001-1.

Pierotti, R., & Rouquerol, J. (1985). Reporting physisorption data for gas/solid systems with special reference to the determination of surface area and porosity. *Pure and Applied Chemistry. Chimie Pure et Appliquee, 57*(4), 603–619.

Ragab, T., & Basaran, C. (2011). Introduction for JEP special issue on carbon nanotubes. *Journal of Electronic Packaging, 133*(2).

Ratner, M. A., Ratner, D., & Waser, R. (2003). *Nanotechnology: A gentle introduction to the next big idea*. Prentice Hall Professional. New Jersey, USA.

Shanmugam, K., Manivannan, J., & Manjuladevi, M. (2019). Stupendous nanomaterials: Carbon nanotubes synthesis, characterization, and applications. *Nanomaterials-toxicity, human health and environment*. IntechOpen, London.

Taniguchi, N., Arakawa, C., & Kobayashi, T. (1974). On the basic concept of nano-technology. Paper presented at the *Proceedings of the International Conference on Production Engineering, 1974-8, 2* 18–23.

Tian, Y., Timmermans, M. Y., Kivistö, S., Nasibulin, A. G., Zhu, Z., Jiang, H., ... Kauppinen, E. I. (2011). Tailoring the diameter of single-walled carbon nanotubes for optical applications. *Nano Research, 4*(8), 807.

Wang, S., & Peng, Y. (2010). Natural zeolites as effective adsorbents in water and wastewater treatment. *Chemical Engineering Journal, 156*(1), 11–24.

Wang, J., & Zhuang, S. (2019). Covalent organic frameworks (COFs) for environmental applications, *Coordination Chemistry Reviews, 400*, 213046. https://doi.org/10.1016/j.ccr.2019.213046

Xu, Y., Jin, S., Xu, H., Nagai, A., & Jiang, D. (2013). Conjugated microporous polymers: Design, synthesis and application. *Chemical Society Reviews, 42*(20), 8012–8031.

Zeng, Y., Gaskey, B., Benn, E., McCue, I., Greenidge, G., Livi, K., & Elebacher, J. (2019). Electrochemical dealloying with simultaneous phase separation, *Acta Materialia, 171*, 8–17. Available from https://doi.org/10.1016/j.actamat.2019.03.039.

5 Hollow and porous structures utilizing the Kirkendall effect

5.1 Introduction

The Kirkendall effect can be a very useful processing tool to create hollow and porous structures. Therefore, we devote an entire chapter to this topic. This process for making hollow structures was first identified by Yin et al. (2004). Since then the number of publications each year has increased dramatically. We provide a background of the Kirkendall effect by discussing the original research. After the background, we discuss the work that led to the creation of hollow nanometer and micrometer scale tubes and particles as a well as porous structures.

The Kirkendall effect is a well known phenomenon especially in metallurgy. The mechanism was first reported in 1947 (Smigelskas & Kirkendall, 1947) which involved the observation of void formation in a Cu and Zn diffusion couple. The effect is a diffusional process that occurs at an interface of two materials that have unequal diffusivities. Kirkendall-type diffusion is a steady state process described by Fick's first law, the diffusion of mass and vacancies under the influence of a concentration gradient. Fig. 5–1 shows the effect; a larger flux of Zn, J_{Zn}, than that of Cu, J_{Cu}, across an interface results in the formation of voids at the original interface and a movement of the interface toward the faster-diffusing material.

Vacancy diffusion occurs to compensate for the inequality of the material flow across the interface. In terms of flux (J) (Tu & Gösele, 2005):

$$J_{void} = J_A - J_B,$$

where the subscripts A and B represent the materials that make the diffusion couple. Kirkendall et al. documented the movement of the initial interface as a direct consequence of this faster diffusion of zinc into the copper than that of copper into the zinc. The intrinsic diffusivity of Zn is $\sim$2.5 times that of Cu at 785°C (Fan, Gösele, & Zacharias, 2007). The void formation at the original interface is a result of vacancy diffusion from the side of the faster diffusing metal, Zn in Kirkendall's case, condensing and coalescing.

Voids in Materials. DOI: https://doi.org/10.1016/B978-0-12-819282-5.00005-3

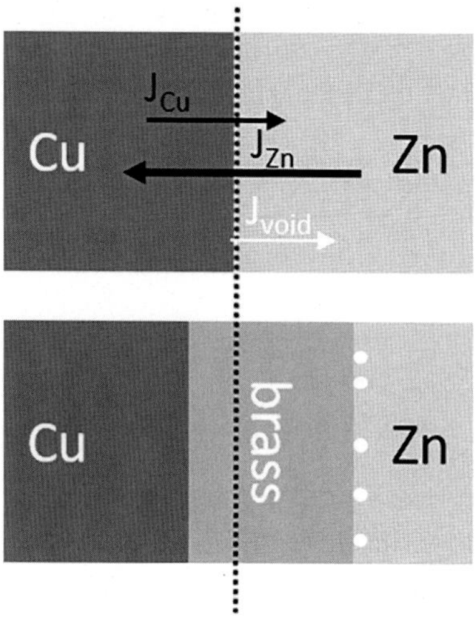

FIGURE 5–1 Schematic of the Kirkendall process in the Cu–Zn system. The Cu and Zn form a diffusion couple with the flux of the Zn, J_{Zn}, being greater than Cu, J_{Cu}, resulting in void flux, J_{void} and formation at the interface. Brass is a solid solution of copper and zinc.

5.2 Generalized Kirkendall mechanism for formation of hollow particles

The Kirkendall process for forming a hollow particle is very interesting in that it does not require a sacrificial core or porogen to form the central void. Because of this characteristic, it is referred to as template-free or sometimes as a self-templating process. We introduced the phenomenon of Kirkendall porosity in Chapter 2, Intrinsic voids in crystalline materials: ideal materials and real materials, as intrinsic voids that occur when two metals in contact have significantly different diffusivities. When the Kirkendall effect was first discovered, the appearance of Kirkendall voids was considered a detrimental process because the voids led to a weakening of the material and potential point of failure. In 2004, Yin and colleagues were thc first to bring the Kirkendall effect to light in a beneficial way to create hollow nanospheres. Yin et al. first observed this phenomenon in hollow cobalt sulfide (CoS) nanoparticles. Hollow CoS nanoparticles were formed by reacting a cobalt (Co) particle with a sulfur (S)-containing solution. The entire process of creating the hollow CoS particles occurred "within a few seconds" at 172°C. Yin et al. also reported the formation of hollow cobalt selenide (CoSe) and cobalt oxide (CoO) particles. The authors used a series of time sequenced TEM images showing the formation of hollow CoSe from a reaction with selenium suspended in o-dichlorobenzene (liquid) at 182°C from a solid Co nanoparticle to a hollow CoSe particle. Once a CoSe reaction layer forms on the Co nanoparticle, Kirkendall voids appear in and at the original interface.

The process of forming a hollow particle via the Kirkendall process starts with the particle as a reactant while the second reactant can be in the form of a solid coating, in a liquid phase (Yin et al., 2004) or a gas phase (Zheng et al., 2007). In all cases, whether the second reactant is solid, liquid, or gas, a diffusion couple must form a solid reaction layer on the starting particle. The reactant diffusing through the reaction layer to the surface (core material) of the particle must have a greater diffusivity than that of the reactant diffusing into the center of the particle

The process usually takes place at a high temperature. Once the solid phases are in contact forming the diffusion couple and with the appropriate energy input, the development of the hollow interior consists of two main steps. The first step is a bulk diffusion process characterized by the formation of the small Kirkendall voids near the original interface. The second step is a consequence of the surface diffusion of the core material (the faster diffusing species) along the pore surface.

Yu et al. (2019) demonstrated this process, making Mn_xFe_3-xO_4 while Zheng et al. (2007) made hollow nanoaluminum nitride (AlN) particles. The process due to Zheng et al. is an example of the nonparticle reactant being in the gaseous phase. It involved taking nanometer scale (solid) aluminum particles and reacting them with NH_3 (gas). The proposed mechanism of forming hollow particles is discussed in Chapter 7, Techniques for introducing void into particles and fibers. After an initial surface layer of AlN is formed, the diffusion of Al to the surface is faster than the diffusion of nitrogen toward the center. This difference in diffusion rates allows the aluminum to diffuse and react at the surface leaving a hollow core.

An experimental model of the formation of hollow nanostructures by the Kirkendall effect was developed by Jana, Chang, and Rioux (2013). The model is an isotropic diffusion model for nonreactive, homogeneous metal–metal hollow nanoparticle formation. The model utilizes Fick's first law of diffusion to describe the growth of the nanoparticles and subsequent void formation via the Kirkendall effect. *Nonreactive* in this case means that the metals are soluble in one another and form solid solutions as they diffuse. In other words, it is a diffusional process driven by concentration gradients without chemical reactions taking place.

Jana et al. (2013) used a nickel nanoparticle and Zn-containing organometallic liquid as their model system to make Ni–Zn hollow nanoparticles. The model included moving boundaries at the interfaces of void – solid and solid – bulk solutions. Apparent diffusion coefficients for both Zn and Ni and vacancy were evaluated from modeling the time-dependent growth of the central void of the nanoparticle. They compared simulations to experimental data for a Ni nanoparticle reacting with an organometallic Zn-precursor material, see Fig. 5–2.

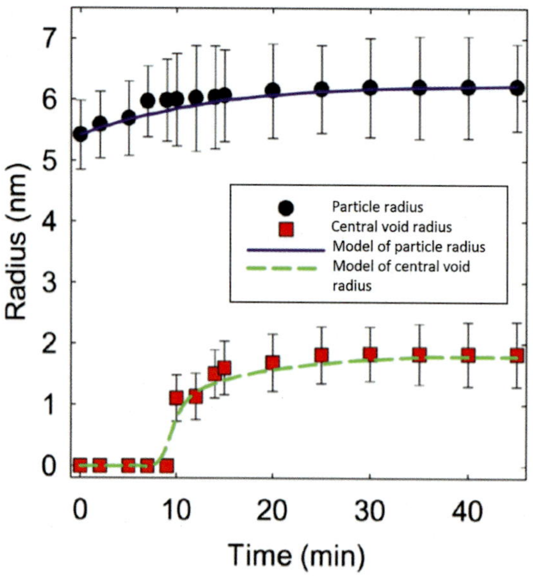

FIGURE 5–2 Experimental data compared to simulations for the formation of nickel–zinc hollow nanoparticles. *Source: Jana, S., Chang, J.W., & Rioux, R.M. (2013). Synthesis and modeling of hollow intermetallic Ni–Zn nanoparticles formed by the Kirkendall effect. Nano Letters, 13(8), 3618–3625.*

There have been many studies that have used the Kirkendall effect to fabricate hollow particles. Many of these have been tabulated in review articles (Fan et al., 2007; Vais & Heli, 2016).

When observing the transformation of a solid particle into a hollow one via the Kirkendall effect, two distinct conversion mechanisms have been identified; symmetric and asymmetric. These terms refer to the final geometry of hollow particle, one with a uniform shell thickness (symmetric) and the other with a nonuniform shell thickness (asymmetric). We discuss each of these and the processes that lead to them in the following sections.

5.2.1 Symmetric hollow particles

An example of a symmetric conversion mechanism of transforming a solid particle to a hollow particle is shown in Fig. 5–3 in the TEM images (Yin et al., 2004); this involves conversion of Co and Se into a hollow CoSe nanoparticle. The dynamic transformation process starts with a thin reaction layer of CoSe on the surface of the Co nanoparticle. Once this initial reaction layer is in place, the following mechanisms occur simultaneously (El Mel, Nakamura, & Bittencourt, 2015):

1. Co atoms diffuse outward until they reach the outer surface.
2. Se atoms diffuse inward to the Co core at a much slower rate than the Co diffusion rate.

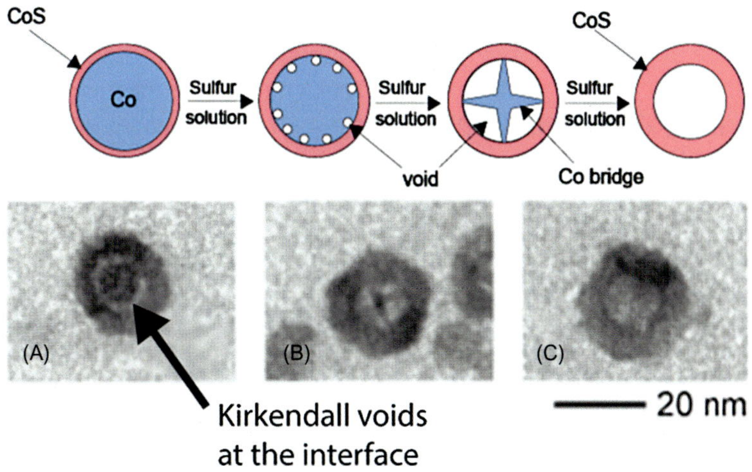

FIGURE 5–3 Top schematic shows the formation of a symmetrical hollow particle via the Kirkendall process. The key step to the uniform shell thickness is the formation of bridges that connect the core material to the reaction layer that will form the shell of the hollow particle. The micrographs in (A), (B) and (C) show corresponding TEM images (Yin et al., 2004) of the appearance of Kirkendall voids, formation of the Co bridge and the final CoS hollow nanoparticle, respectively.

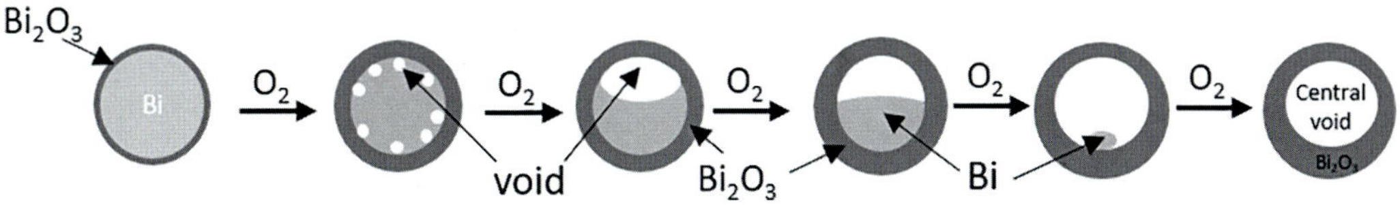

FIGURE 5–4 Schematic of the asymmetrical conversion process forming a hollow particle with a nonuniform shell thickness. *Source: Adapted from Niu, K., Park, J., Zheng, H., & Alivisatos, A.P. (2013). Revealing bismuth oxide hollow nanoparticle formation by the Kirkendall effect.* Nano Letters, 13 *(11), 5715–5719.*

3. Creation and ejection of vacancies at the Co/CoSe interface.
4. The formation of Kirkendall voids at the interface via the migration and agglomeration of vacancies.

In the case of symmetric transformation, Co bridges are formed and are shown schematically in Fig. 5–3. The bridges link the Co core to the CoSe reaction layer. Yin et al. (2004) reported that these bridges remain until the Co core is consumed by the reaction leaving a hollow particle.

5.2.2 Asymmetric hollow particles

In the early stage of asymmetric conversion, vacancies migrate and agglomerate into a single void. Fig. 5–4 is a schematic of the process involving oxidation of a bismuth

(Bi) particle. Instead of being distributed along the metal/metal oxide interface as observed in the symmetric conversion mechanism during the oxidation of Cu nanospheres, Niu, Park, Zheng, and Alivisatos (2013) showed that as the oxidation of Bi proceeds in time, the growth of Bi_2O_3 occurs preferentially opposite to the side of the single void that has agglomerated. In general, most authors report that the shell is thinner on the side where a large void is present during oxidation and thicker on the opposite side (El Mel et al., 2015).

When analyzing both conversion mechanisms from a surface energy standpoint, it is more favorable for vacancies and voids to agglomerate into a single large void compared to the formation of multiple smaller voids. The sum of surface energies of multiple small voids is greater than the surface energy of a single large void. However, the exact reason why some conversions are asymmetric and others are symmetric is not clearly understood. It is postulated that the asymmetric conversion occurs when the generation and the migration rate of vacancies injected at the metal/metal oxide interface are comparable. Therefore, the vacancies can migrate toward a position over a long range distance where they aggregate to form a large single void instead of small voids distributed at the interface. This asymmetric conversion occurs in oxidizing Ni nanoparticles but not with, for example, Cu. During Cu oxidation, the generation rate of vacancies in Cu was found to be much faster than the migration rate since the diffusion coefficient of Cu in Cu_2O is up to nine times higher at 100°C than the selfdiffusion coefficient of Cu.

In general, most authors report that during asymmetric conversion the shell is thinner on the side where a large void is present during oxidation and thicker on the opposite side as depicted in Fig. 5–3. However, not all hollow particles with the core containing a single large void form a nonuniform shell. Cabot et al. (2008) studied the formation of hollow cadmium sulfide (CdS) nanospheres. Despite the presence of an off-centered single void in the Cd core during the conversion process, the CdS shell grew isotropically and had a uniform thickness.

How one gets to an asymmetric or symmetric shell when a single large void is present in the core is unclear. In all cases, the term *asymmetric* does describe the geometry of the particle and the growth of the single void within the core material as the process proceeds.

5.3 Tubes

The same general mechanisms prevail in hollow tube formation as the ones we presented in the previous section on hollow nanoparticles. Note in this section we use the general term tube to describe the product of the reaction. The vast majority of the work on hollow tubes has been done on nanotubes. However, there has been

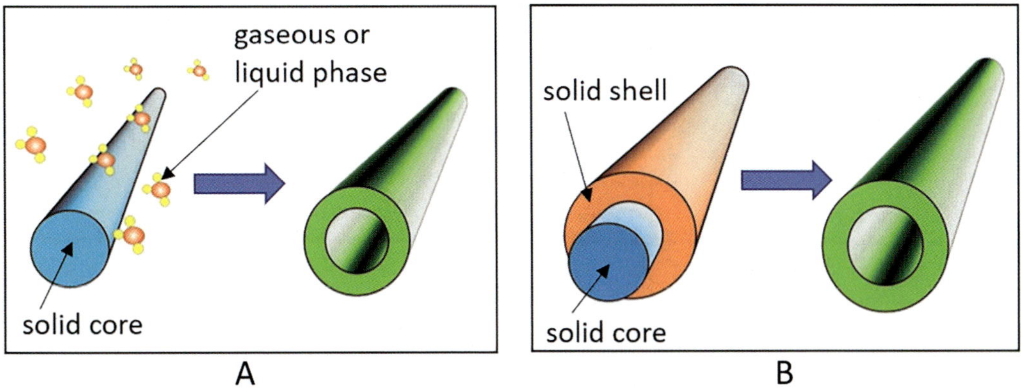

FIGURE 5–5 Schematic of the various reactant phases used to form hollow tubes using the Kirkendall effect (A) using a solid core material with liquid or gaseous phase or (B) a solid shell creating the diffusion couple with the solid core. *Source: Adapted from Fan, H.J., Gösele, U., & Zacharias, M. (2007). Formation of nanotubes and hollow nanoparticles based on Kirkendall and diffusion processes: A review.* Small, *3(10), 1660–1671.*

work on the micrometer scale also to produce microtubes (Paz y Puente & Dunand, 2018; Peng et al., 2009; Yang et al., 2019; Zheng, Lindquist, Yuan, Müllner, & Dunand, 2014). Fig. 5–5 is a general schematic showing the formation of a hollow tube via the Kirkendall effect whether nanometer or micrometer in scale. All routes to obtain tubes start out with a solid phase core material. The difference is the second reactant, which can be a gas (El Mel et al., 2013), liquid (Li et al., 2018), or solid (Jin Fan et al., 2006; Park et al., 2011; Zhou et al., 2007).

Peng et al. (2009) created micrometer scale coaxial tubes (tube-in-tube) of $ZrAl_2O_4$ (a spinel) using a series of coatings, $Al_2O_3/ZrO/Al_2O_3$ on a polyvinyl alcohol (PVA) fiber followed by heat treatments. Fig. 5–6 provides an axial and cross-sectional schematic view and annealing details of their process. Note that the central core of PVA is the sacrificial material that will create the central void. The PVA is removed in a subsequent annealing step by heating the structure to 450°C for 12 h. The Kirkendall process occurs during second annealing step at 700°C for 12 h. The specific coating arrangement is $Al_2O_3/ZrO/Al_2O_3$ and knowing vacancies will diffuse toward the materials with the faster diffusivity, ZrO in this case; there will be two separate Kirkendall processes occurring simultaneously. The result is that there will be two distinct Al_2O_3/ZrO interfaces, the vacancies diffuse in opposite directions but always toward the interlayer of ZrO. Vacancies move *inward* to ZrO from the outer Al_2O_3/ZrO interface. From the inner ZrO/Al_2O_3 interface, vacancies will move *outward* to the ZrO layer. This directionally opposing flow of vacancies results in single nanometer scale Kirkendall void space between the $ZrAl_2O_4$ microtubes. The Kirkendall void or gap could range from 6.4 to 12 nm. An SEM image of the coaxial microtubes is shown in Fig. 5–7 (Peng et al., 2009).

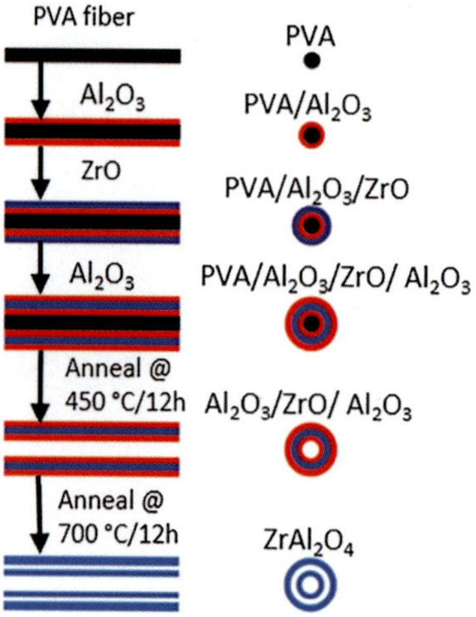

FIGURE 5–6 Process for making micrometer scale coaxial $ZrAl_2O_4$ microtubes. Note that the vacancy diffusion occurs toward the material with the faster diffusivity, ZrO, resulting in the void between the $ZrAl_2O_4$ tubes. *Source: Adapted from Peng, Q., Sun, X., Spagnola, J.C., Saquing, C., Khan, S.A., Spontak, R.J., & Parsons, G.N. (2009). Bi-directional Kirkendall effect in coaxial microtube nanolaminate assemblies fabricated by atomic layer deposition.* ACS Nano, *3(3), 546–554.*

A highly ordered array of nanotubes was constructed from Cu_2O hollow nanospheres by Xu et al. (2009). Two separate processes occur to make this hierarchical structure. In this Kirkendall process, $Cu(OH)_2$ nanorods are converted to Cu_2O tubes by immersing the nanorods in a 0.005–0.015 M ascorbic acid ($C_6H_8O_6$) solution at 55°C for 60 min. The reaction that occurs forming the shell of the tube is

$$2Cu(OH)_2(s) + C_6H_8O_6 \xrightarrow{55^\circ C} Cu_2O(s) + C_6H_6O_6 + 3H_2O$$

Initially a thin Cu_2O layer forms on the surface of the $Cu(OH)_2$ rod and acts as an interphase, setting up the diffusion couple that separates $Cu(OH)_2$ rod from the ascorbic acid molecules. The Cu^{+2} ions and ascorbic react to form Cu_2O crystallites, which aggregate into solid nanospheres forming the shell of the tube. With time and continued heating, the solid Cu_2O nanosphere can be converted to hollow Cu_2O spheres via the Ostwald ripening process (Ostwald, 1896). Ostwald ripening is a template free method of forming hollow structures. It is a thermodynamically driven process that minimizes the surface energy. When solids precipitate out of a solution, energy considerations will cause large precipitates to grow at the expense of smaller precipitates. Ostwald ripening occurs when we have agglomerates of small particles, and the smaller particles combine to form larger particles in order to minimize surface energy. As the particles grow larger, a void can form at the center of the particle cluster forming a hollow particle. This is shown in Fig. 5–8B–E where Cu_2O

FIGURE 5–7 TEM micrograph of the coaxial $ZrAl_2O_4$ microtubes. *Source: Adapted from Peng, Q., Sun, X., Spagnola, J.C., Saquing, C., Khan, S.A., Spontak, R.J., & Parsons, G.N. (2009). Bi-directional Kirkendall effect in coaxial microtube nanolaminate assemblies fabricated by atomic layer deposition.* ACS Nano, *3(3), 546–554.*

crystallites agglomerate into nanospheres and the Cu_2O nanospheres eventually transform into larger hollow Cu_2O nanospheres.

5.4 Porous and hollow structures

Utilizing the Kirkendall effect one can make hollow structures, as the previous examples illustrate, but also porous structures. The Kirkendall effect has become very popular in the synthesis of hollow micro- and nanoparticles and micro- and nanotubes covering a wide range of materials including sulfides, oxides, selenides, tellurides, fluorides, phosphides, and metals (El Mel et al., 2015). In this section we highlight the role the Kirkendall effect has in creating porous fibers, particles and films. By porous we mean the voids are dispersed throughout the bulk materials as opposed to a having single, centralized void.

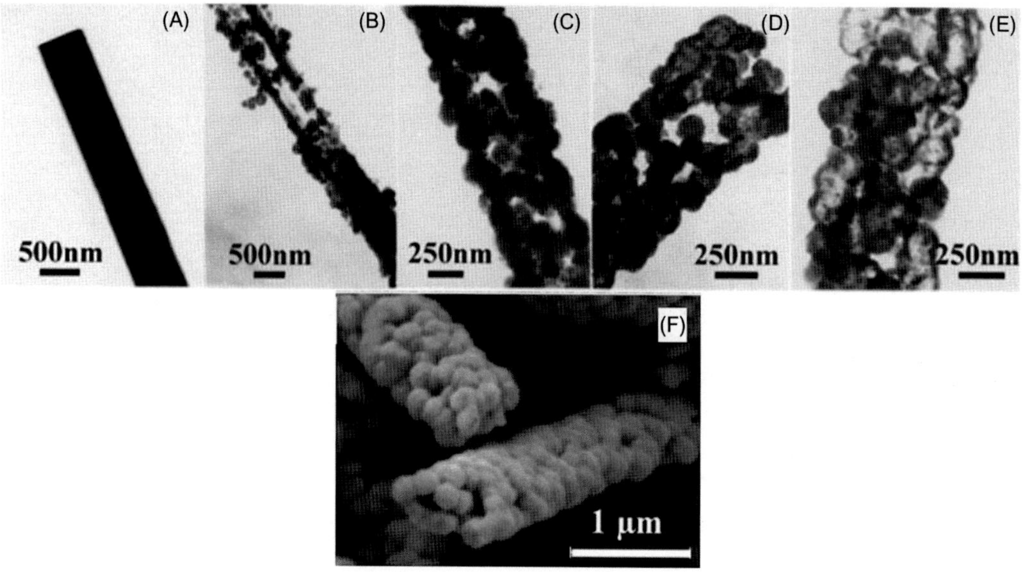

FIGURE 5–8 Formation of nanotubes with hierarchical porosity. TEM images show (A) $Cu(OH)_2$ nanorod, (B) $Cu(OH)_2$ nanorod immersed in ascorbic acid react to form a thin layer of Cu_2O crystallites in the nanorod surface, (C) crystallites form solid Cu_2O nanospheres on the forming the shell of the nanotube via the Kirkendall process, (D) and (E) solid nanospheres transform into hollow ones via Ostwald ripening, and (F) SEM image of the final Cu_2O tube with the shell made from hollow Cu_2O spheres. *Source: Adapted from Xu, J., Tang, Y., Zhang, W., Lee, C., Yang, Z., & Lee, S. (2009). Fabrication of architectures with dual hollow structures: Arrays of Cu2O nanotubes organized by hollow nanospheres.* Crystal Growth & Design, *9(10), 4524–4528.*

Yost, Erdeniz, Paz y Puente, and Dunand (2019) demonstrated that the final Kirkendall pore structure of a titanium (Ti) coated nickel (Ni) wire was different depending on the diffusion distance. Yost et al. started with nickel wires having diameters of 25, 50, and100 μm. The diffusion distance is shortest for the 25 μm diameter wire and longest for the 100 μm wire. The titanization was done via solid-state reaction by embedding the Ni wire in Ti powder bed at 925°C. The time of the titanization step varied, depending on the diameter of the Ni wire, such that the average wire composition had an approximate Ni:Ti atomic ratio of 1:1. The titanization time for the 25, 50, and 100 μm Ni wire was 1, 4, and 8 h, respectively. After titanization the wires were removed from the powder bed. This was followed by a homogenization step carried out at 925°C in an inert atmosphere. The times again varied, and homogenization was allowed to continue until the material (alloy) phases within the wire were constant. Fig. 5–9 shows, schematically, the development and final structure for each diameter wire as Ni and Ti interdiffused leaving Kirkendall voids.

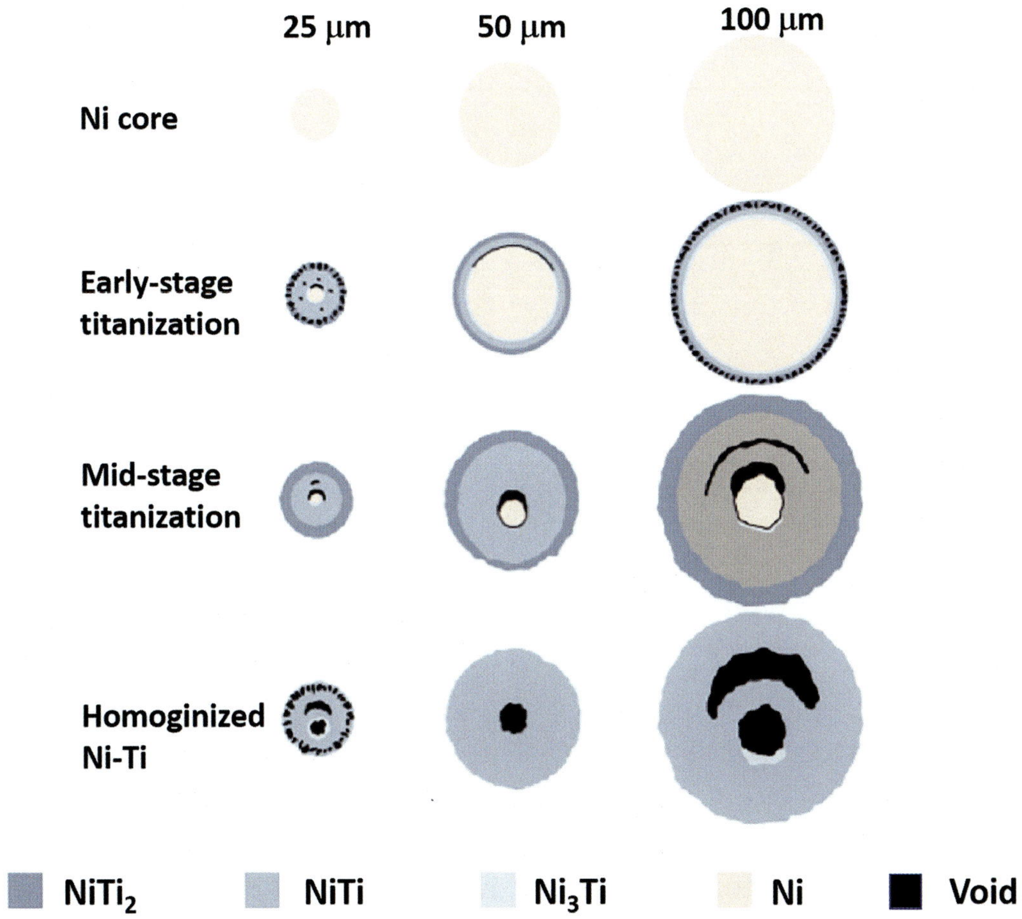

FIGURE 5–9 Schematic showing the development of NiTi microtubes created via the Kirkendall effect between Ni and Ti. The phase development and void structure within end product was found to be dependent on the diffusion distance. The diffusion distance was manipulated by changing the initial diameter of the Ni wire. *Source: Adapted from Yost, A.R., Erdeniz, D., Paz y Puente, A.E., & Dunand, D.C. (2019). Effect of diffusion distance on evolution of Kirkendall pores in titanium-coated nickel wires. https://doi.org/10.1016/j.intermet.2018.10.020*

Upon completion of the homogenization step at 925°C, the majority phase present in all microtubes, independent of initial Ni diameter, was NiTi with a small amount of Ni_3Ti near the core. Fig. 5–10 shows the development of the pore-channel structure during the homogenization of the 25 μm Ni wires. Vacancies migrated and agglomerated into nanometer scale Kirkendall voids near the exterior surface of the microtubes. Larger micrometer scale Kirkendall voids appeared near the core. The final structure of the 50 μm Ni wire was a single, ~20 μm diameter void near the center of the NiTi-wire, transforming the wires

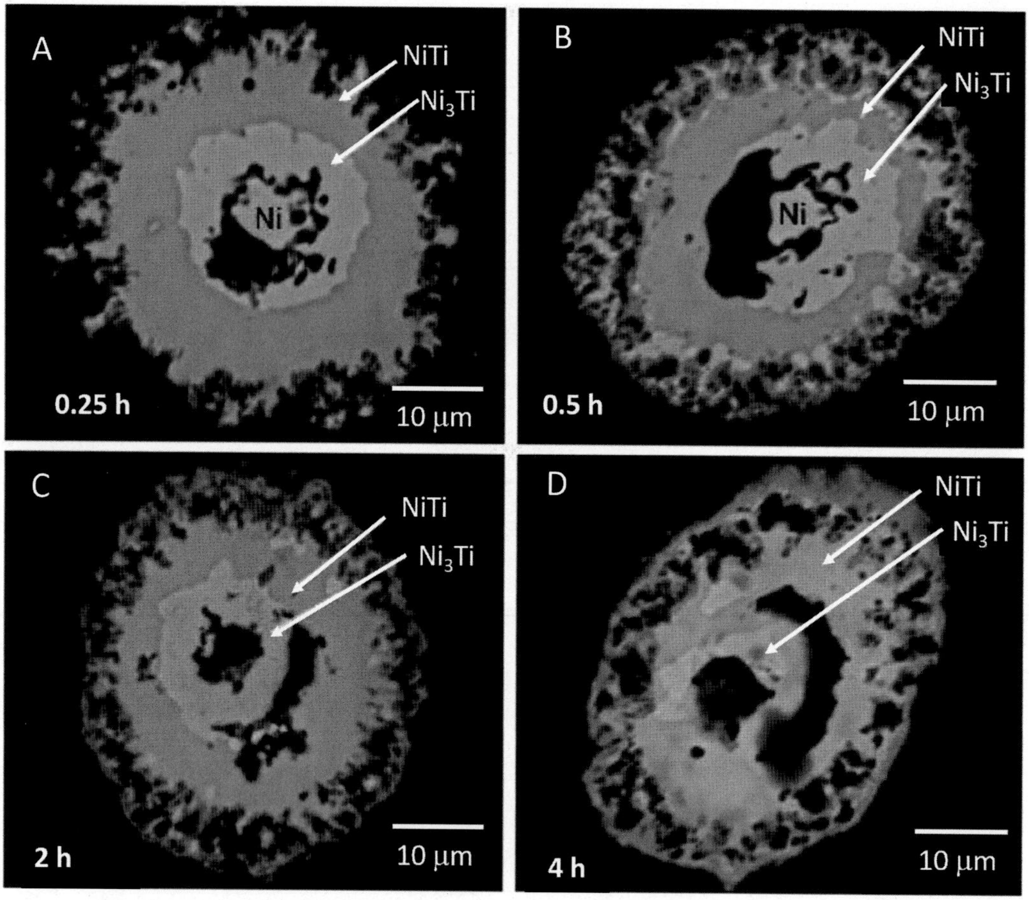

FIGURE 5–10 The Ni–Ti phase and Kirkendall void development in the wire with an initial Ni diameter of 25 μm. Once titanization resulted in a 1:1 atomic ratio of Ni:Ti, the structure was homogenized at 925°C for (A) 0.25 h, (B) 0 0.5 h, (C) 2 h, and (D) 4 h. *Source: Adapted from Yost, A. R., Erdeniz, D., Paz y Puente, A.E., & Dunand, D.C. (2019). Effect of diffusion distance on evolution of Kirkendall pores in titanium-coated nickel wires. https://doi.org/10.1016/j.intermet.2018.10.020.*

into an (non-porous) microtube, see Fig. 5–11. The final structure of the 100 μm Ni wire was quite different compared to the 25 and 50 μm wires. In the 100 μm Ni, an approximately 50 μm diameter irregular pore was formed near the NiTi-wire center, along with an eccentric crescent-shaped pore of similar cross sectional area. There was no nanometer scale porosity as was seen in the 25 mm wire. Yost et al. attributed the crescent shaped pore in the 25 and 100 μm Ni wire to the interruption of a single diffusion path, due to the longer diffusion distances and time.

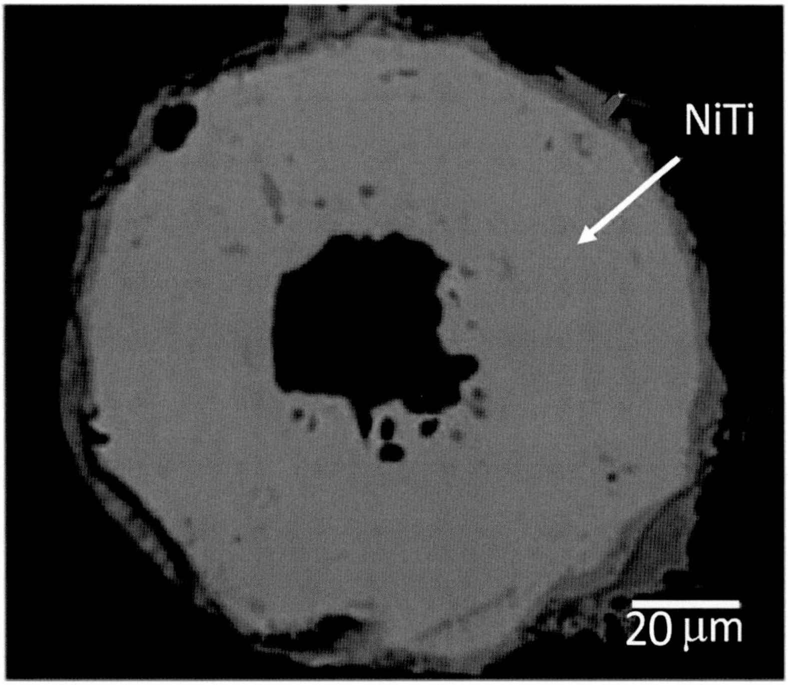

FIGURE 5–11 The final NiTi tube structure when starting with a 50 μm Ni wire. The Kirkendall effect transformed the Ni wire into a nonporous microtube via solid-state diffusion with Ti. This microtube contained a single, ~20 μm diameter void near the center. *Source: Adapted from Yost, A.R., Erdeniz, D., Paz y Puente, A.E., & Dunand, D.C. (2019). Effect of diffusion distance on evolution of Kirkendall pores in titanium-coated nickel wires. https://doi.org/10.1016/j.intermet.2018.10.020.*

There is considerable amount of work on creating porous structures for battery applications using the Kirkendall effect to create the porosity. Some of the material systems being investigated are CuO (Dong, Jiang, Mo, Zhou, & Zhou, 2020; Hu et al., 2010), SnO_2 (Park, Oh, Kim, & Kang, 2020; Shi & Lu, 2014), NiO (Park, Hong, Park, & Kang, 2019), $Mn_xFe_{3-x}O_4$ (Yu et al., 2019), metal sulfides (Liu et al., 2020), etc. Many of these are in the form of films. In general, increasing accessible void space is required (i.e., high surface area) for applications such as energy storage, catalysis, sensor systems, and environmental engineering increases performance. The Kirkendall effect is an effective tool to create these highly porous functional materials with high accessible surface area.

The benefits of highly porous structures for energy storage applications are that they provide buffer space against electrode volume changes during repetitive charge and discharge processes, increase the electrolyte/electrode contact area, and reduce the lithium-ion diffusion path (Ko, Park, & Kang, 2014; B. Park et al., 2020; Xu et al., 2014; Yao et al., 2011). Xu et al. (2019) fabricated hierarchical porosity in AgO films by oxidizing Ag films in an environment of reactive magnetron sputtering deposition of NiO. The method is capable of creating hierarchically structured AgO films with nanometer scale porosity. The formation of AgO nanorods and the creation of hierarchical porosity result from the Kirkendall effect between the oxidized Ag films and the deposited NiO. Xu et al. divided the process of creating the hierarchically porous AgO films into three stages:

1. Oxidation of Ag films.
2. Formation of AgO nanorods.
3. Creation of hierarchical porosity.

The different stages can be seen in Fig. 5–12, which are SEM images taken during the 60 min process of creating the porous AgO film. During the first stage, oxidation

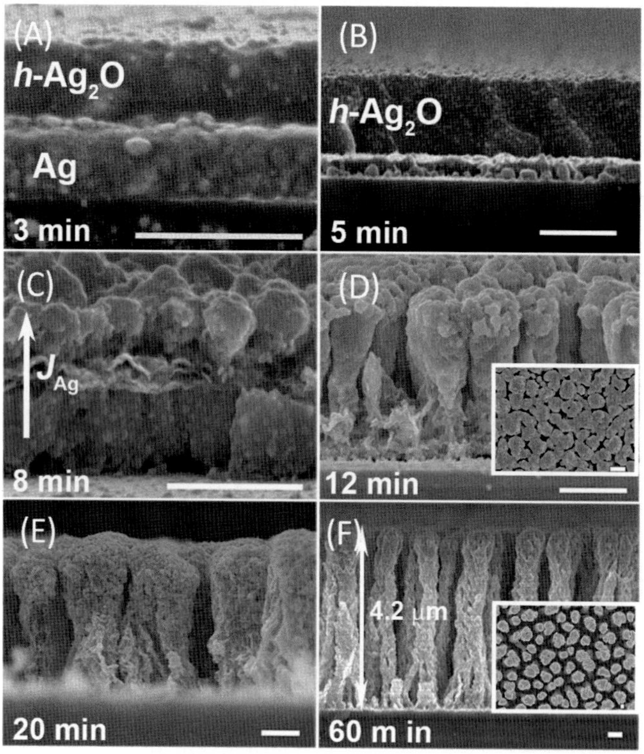

FIGURE 5–12 SEM micrographs illustrating the three-stage process to form AgO hierarchical porous film via the Kirkendall effect. The first stage, (A) and (B), is the layer-by-layer oxidization forming Ag_2O with hexagonal structure (h-Ag_2O). The second stage is the transform from a bulk solid film to a nanorod-like structure accompanied by a transition from h-Ag_2O to cubic AgO (c-AgO). The third and final stage is characterized by the appearance of nanometer scale porosity within the nanometer scale rods. All scale bars are 400 nm (Xu et al., 2019). The mesoporosity (2–50 nm) is present between the nanorods and <2 nm voids within each nanorod. *Source: Xu, W., Wang, S. Q., Zhang, Q. Y., Ma, C. Y., Wang, Q., Wen, D. H., & Li, X. N. (2019). Hierarchically structured AgO films with nano-porosity for photocatalyst and all solid-state thin film battery. Journal of Alloys and Compounds, 802, 210–216.*

occurs from 0–5 min of reactive sputtering deposition of NiO. Fig. 5–12A and B illustrate the layer-by-layer oxidization of Ag films forming Ag_2O with hexagonal structure (h-Ag_2O). In the second stage, from 5–20 min, the oxidized Ag films transform from a bulk solid film to a nanorod-like structure accompanied by a transition from h-Ag_2O to cubic AgO (c-AgO,), see Fig. 5–12C and D. The third stage occurs at reactive sputtering deposition of NiO for time greater than 20 min. During this final stage, porosity developed within AgO nanorods, see Fig. 5–12E and F. In the final structure thc porosity between the nanorods is mesoporous (2–50 nm), whereas the porosity contained in the nanorods is less than 2 nm.

References

Cabot, A., Smith, R. K., Yin, Y., Zheng, H., Reinhard, B. M., Liu, H., & Alivisatos, A. P. (2008). Sulfidation of cadmium at the nanoscale. *ACS Nano*, *2*(7), 1452–1458.

Dong, Y., Jiang, X., Mo, J., Zhou, Y., & Zhou, J. (2020). *Hollow CuO nanoparticles in carbon microspheres prepared from cellulose-cuprammonium solution as anode materials for Li-ion batteries* Chemical Engineering Journal, *381*, 122614. Available from https://doi.org/10.1016/j.cej.2019.122614.

El Mel, A., Buffière, M., Tessier, P., Konstantinidis, S., Xu, W., Du, K., ... Snyders, R. (2013). Highly ordered hollow oxide nanostructures: The Kirkendall effect at the nanoscale. *Small*, *9* (17), 2838–2843.

El Mel, A., Nakamura, R., & Bittencourt, C. (2015). The Kirkendall effect and nanoscience: Hollow nanospheres and nanotubes. *Beilstein Journal of Nanotechnology*, *6*(1), 1348–1361.

Fan, H. J., Gösele, U., & Zacharias, M. (2007). Formation of nanotubes and hollow nanoparticles based on Kirkendall and diffusion processes: A review. *Small*, *3*(10), 1660–1671.

Hu, Y., Huang, X., Wang, K., Liu, J., Jiang, J., Ding, R., & Li, X. (2010). *Kirkendall-effect-based growth of dendrite-shaped CuO hollow micro/nanostructures for lithium-ion battery anodes*, Journal of Solid State Chemistry, *183*(3), 662–667. Available from https://doi.org/10.1016/j.jssc.2010.01.013.

Jana, S., Chang, J. W., & Rioux, R. M. (2013). Synthesis and modeling of hollow intermetallic Ni–Zn nanoparticles formed by the Kirkendall effect. *Nano Letters*, *13*(8), 3618–3625.

Jin Fan, H., Knez, M., Scholz, R., Nielsch, K., Pippel, E., Hesse, D., ... Gosele, U. (2006). Monocrystalline spinel nanotube fabrication based on the Kirkendall effect. *Nature Materials*, *5*, 627. Available from https://doi.org/10.1038/nmat1673.

Ko, Y. N., Park, S. B., & Kang, Y. C. (2014). Design and fabrication of new nanostructured SnO_2-Carbon composite microspheres for fast and stable lithium storage performance. *Small*, *10* (16), 3240–3245.

Li, W., Xiong, Y., Wang, Z., Bao, M., Liu, J., He, D., & Mu, S. (2018). Seed-mediated synthesis of large-diameter ternary TePtCo nanotubes for enhanced oxygen reduction reaction. *Applied Catalysis B: Environmental*, *231*, 277–282.

Liu, Q., Hong, X., You, X., Zhang, X., Zhao, X., Chen, X., & Liu, X. (2020). *Designing heterostructured metal sulfide core-shell nanoneedle films as battery-type electrodes for hybrid supercapacitors*, Energy Storage Materials, *24*, 541–549. Available from https://doi.org/10.1016/j.ensm.2019.07.001.

Niu, K., Park, J., Zheng, H., & Alivisatos, A. P. (2013). Revealing bismuth oxide hollow nanoparticle formation by the Kirkendall effect. *Nano Letters, 13*(11), 5715–5719.

Ostwald, W. (1896). *Lehrbruck der allgemeinen chemie, vol. 2, part 1.* Leipzig, Germany: Engelmann.

Park, B., Park, J., Yu, S., Cho, S., Byun, J. Y., Oh, J., & Lee, S. Y. (2020). *Hollow/porous-walled SnO_2 via nanoscale Kirkendall diffusion with irregular particles,* Acta Materialia, *186*, 20–28. Available from https://doi.org/10.1016/j.actamat.2019.12.039.

Park, G. D., Hong, J. H., Park, S., & Kang, Y. C. (2019). *Strategy for synthesizing mesoporous NiO polyhedra with empty nanovoids via oxidation of NiSe polyhedra by nanoscale Kirkendall diffusion and their superior lithium-ion storage performance,* Applied Surface Science, *464*, 597–605. Available from https://doi.org/10.1016/j.apsusc.2018.09.122.

Park, J., Oh, Y. J., Kim, J. H., & Kang, Y. C. (2020). *Porous nanofibers comprised of hollow SnO2 nanoplate building blocks for high-performance lithium ion battery anode,* Materials Characterization, *161*, 110099. Available from https://doi.org/10.1016/j.matchar.2019.110099.

Park, M., Cho, Y., Kim, K., Kim, J., Liu, M., & Cho, J. (2011). Germanium nanotubes prepared by using the Kirkendall effect as anodes for high-rate lithium batteries. *Angewandte Chemie International Edition, 50*(41), 9647–9650.

Paz y Puente, A. E., & Dunand, D. C. (2018). Synthesis of NiTi microtubes via the Kirkendall effect during interdiffusion of Ti-coated Ni wires. *Intermetallics, 92*, 42–48.

Peng, Q., Sun, X., Spagnola, J. C., Saquing, C., Khan, S. A., Spontak, R. J., & Parsons, G. N. (2009). Bi-directional Kirkendall effect in coaxial microtube nanolaminate assemblies fabricated by atomic layer deposition. *ACS Nano, 3*(3), 546–554.

Shi, W., & Lu, B. (2014). *Nanoscale Kirkendall Effect Synthesis of Echinus-like SnO_2@SnS_2 Nanospheres as High Performance Anode Material for Lithium Ion Batteries,* Electrochimica Acta, *133*, 247–253. Available from https://doi.org/10.1016/j.electacta.2014.04.013.

Smigelskas, A. D., & Kirkendall, K. O. (1947). Zinc diffusion in alpha brass. *Transactions of AIME, 171*, 130.

Tu, K., & Gösele, U. (2005). Hollow nanostructures based on the Kirkendall effect: Design and stability considerations. *Applied Physics Letters, 86*(9), 093111.

Vais, R.D., & Heli, H. (2016). The Kirkendall effect: Its efficacy in the formation of hollow nanostructures, *Journal of Biology and Today's World,* **5**(8), 2016, 137–149.

Xu, J., Tang, Y., Zhang, W., Lee, C., Yang, Z., & Lee, S. (2009). Fabrication of architectures with dual hollow structures: Arrays of Cu_2O nanotubes organized by hollow nanospheres. *Crystal Growth & Design, 9*(10), 4524–4528.

Xu, Y., Jian, G., Liu, Y., Zhu, Y., Zachariah, M. R., & Wang, C. (2014). Superior electrochemical performance and structure evolution of mesoporous Fe_2O_3 anodes for lithium-ion batteries. *Nano Energy, 3*, 26–35.

Xu, W., Wang, S. Q., Zhang, Q. Y., Ma, C. Y., Wang, Q., Wen, D. H., & Li, X. N. (2019). *Hierarchically structured AgO films with nano-porosity for photocatalyst and all solid-state thin film battery,* Journal of Alloys and Compounds, *802*, 210–216. Available from https://doi.org/10.1016/j.jallcom.2019.06.188.

Yang, P., Wang, X., Ge, C., Fu, X., Liu, X. Y., Chai, H., ... Chen, K. (2019). *Fabrication of CuO nanosheets-built microtubes via Kirkendall effect for non-enzymatic glucose sensor,* Applied Surface Science, *494*, 484–491. Available from https://doi.org/10.1016/j.apsusc.2019.07.197.

Yao, Y., McDowell, M. T., Ryu, I., Wu, H., Liu, N., Hu, L., ... Cui, Y. (2011). Interconnected silicon hollow nanospheres for lithium-ion battery anodes with long cycle life. *Nano Letters, 11*(7), 2949–2954.

Yin, Y., Rioux, R. M., Erdonmez, C. K., Hughes, S., Somorjai, G. A., & Alivisatos, A. P. (2004). Formation of hollow nanocrystals through the nanoscale Kirkendall effect. *Science, 304*(5671), 711–714. Available from https://doi.org/10.1126/science.1096566.

Yost, A. R., Erdeniz, D., Paz y Puente, A. E., & Dunand, D. C. (2019). *Effect of diffusion distance on evolution of Kirkendall pores in titanium-coated nickel wires,* Intermetallics, *104,* 124–132. Available from https://doi.org/10.1016/j.intermet.2018.10.020.

Yu, X., Zhang, C., Luo, Z., Zhang, T., Liu, J., Li, J., ... Cabot, A. (2019). *A low temperature solid state reaction to produce hollow MnxFe3-xO4 nanoparticles as anode for lithium-ion batteries,* Nano Energy, *66,* 104199. Available from https://doi.org/10.1016/j.nanoen.2019.104199.

Zheng, J., Song, X., Zhang, Y., Li, Y., Li, X., & Pu, Y. (2007). Nanosized aluminum nitride hollow spheres formed through a self-templating solid–gas interface reaction. *Journal of Solid State Chemistry, 180*(1), 276–283. Available from https://doi.org/10.1016/j.jssc.2006.10.011.

Zheng, P., Lindquist, P., Yuan, B., Müllner, P., & Dunand, D. C. (2014). Fabricating Ni–Mn–Ga microtubes by diffusion of Mn and Ga into Ni tubes. *Intermetallics, 49,* 70–80.

Zhou, J., Liu, J., Wang, X., Song, J., Tummala, R., Xu, N. S., & Wang, Z. L. (2007). Vertically aligned Zn_2SiO_4 nanotube/ZnO nanowire heterojunction arrays. *Small, 3*(4), 622–626.

6

Techniques for introducing intentional voids into materials

6.1 Introduction

In this chapter, we describe the intentional introduction of voids into a material. A major part of this will cover traditional cell-forming methods. Most of these techniques are used in the industry to manufacture foams. We will also describe some newer methods. In Section 6.9, we introduce the topic of hierarchical design of materials using voids. A basic characteristic of this hierarchical design is that the voids occur simultaneously at multiple length scales.

The most widely studied foams are polymer based. However, in principle, one can introduce voids intentionally in any type of material. In this chapter, we examine the commonalities of the different approaches of making foams, that is, how the voids are introduced into different types of materials: polymers, metals, ceramics, and composites. We will then categorize them into common processing techniques followed by a description and examples of each technique.

Foams, in general, and composite foams, in particular, can provide some very attractive combinations of physical and mechanical properties. Among these are the following:

1. Very low density: the density is significantly decreased by the presence of empty cells or cells containing gas.
2. High specific stiffness and compressive strength, that is, stiffness and strength divided by density: these can be further modified via sandwich construction (two dense faces and a foam core).
3. Excellent energy absorption characteristics.
4. Noise and vibration control.

The specific method one chooses to introduce voids into any material depends on many factors including the type of material, processing route, and targeted void characteristics. The size and distribution of the voids introduced will vary depending on the process chosen, which, in turn, will control the properties of the foam.

Voids in Materials. DOI: https://doi.org/10.1016/B978-0-12-819282-5.00006-5

6.2 Commonalities of foam formation processes

In Sections 6.2–6.7, we review and classify the various methods for creating cellular materials, independent of material type. There have been reviews and even books written on methods of introducing voids into all types of materials: metals, ceramics, polymers, and composites. Several reviews on the methods for manufacturing metal based foams have been done (Banhart, 2001; Wadley, 2002), while Smith, Szyniszewski, Hajjar, Schafer, and Arwade (2012) focused on steel-based foams. Colombo (2008) reviewed the methods for introducing voids into ceramics for the manufacture of cellular ceramics. Polymer foams have been extensively studied and there are many references that can be consulted. In a general overview, Gibson and Ashby (1999) discuss methods of making polymer foams. Mills (2007) gives a more detailed discussion relating methodologies for making particular classes of polymer foams, that is, open/closed cell, reticulated, flexible/rigid, and rubber. Gladysz and Chawla (2002) review methods for polymer based composite foams. However, when doing a detailed comparison of methods across different material types, it becomes obvious that there are many similarities among the methods having different names.

It should be further pointed out that many of these material-specific methods are classified by the processing methods used after the voids have been introduced. For example, a ceramic foam can be made by simply mixing air into a ceramic slip (a slip is a suspension with some additives to make it stable), followed by drying, binder burnout, and sintering. These processing methods are typical methods for ceramic type of material. The way the *void* is introduced, by simply mixing and entraining gas, is a process that can be employed in any kind of liquid phase material: polymers, ceramics, metals, and composites.

In this chapter, we choose to classify all foams by the way the voids are introduced into the material. We highlight these common instances and propose a generalized category that will incorporate all of the related methods, independent of the material type. Thus we classify the intentional void introduction processes under the following broad categories:

1. Introduction of a gas by one of the following methods:
   a. Mixing,
   b. Physical blowing agent,
   c. Chemical blowing agent;
2. Templating and sacrificial pore former;
3. Bonding together of spheres, powders, fibers, etc.;
4. Additive manufacturing;
5. Mechanical stretching;
6. Exploiting chemically selective weakness in solids.

6.3 Introduction of a gas

We can introduce gas in a material by three general methods. The first method involves vigorous mixing, whereby a gaseous phase is introduced into the material (generally, a liquid) thus entraining gas, generally air, creating a voided structure. The other two methods involve the use of physical or chemical blowing agents. The gas introduction occurs by the activation of a *latent-gas* agent that, when certain processing conditions are reached, will release a gas and form voids in the material. A *physical blowing agent* undergoes a simple phase change such as vaporization during processing. A *chemical blowing* involves (1) decomposition of a chemical additive or (2) a reaction between starting materials and one of the reaction products is the gaseous phase.

6.3.1 Mixing

Creating voids in a material by mixing is relatively easy to achieve. The procedure involves vigorous stirring of a gas into the liquid phase of the material or precursor material, thus causing entrainment of that gas. Although it is easy to create foams by this method, it is very difficult to control the type, size, distribution, and volume fraction of voids in the finished cellular structure.

Solidification is important in any liquid processing method. Whether it is by the degree of cure or exothermal reaction in a thermoset polymer/prepolymer, or in the case of thermoplastic polymers and molten metals, there must be careful control of the temperature during the foaming process. The processing parameters such as mixing speed and viscosity must be controlled to get a reproducible material with consistent properties.

Mixing is commonly used in the food industry; mixing air into food is called *aeration.* In the early 2000s, the food industry started adopting the processes and characterization methods that had been established earlier for polymer based foams. As in engineered foams, the microstructure is important in controlling the mechanical properties, texture, and esthetics in the foods (Lim & Barigou, 2004). The chocolate bar shown in Fig. 6–1 illustrates a wide cell size distribution created during aeration.

Void structure, size, shape, and distribution are important to ensure consistency of foods such as ice cream, soufflé, sponge cakes, chocolate bars, marshmallows, biscuits, and breads. Lim and Barigou (2004) used X-ray microcomputed tomography, a technique that is commonly used in engineered foams, to characterize the cell structure of many cellular foods. Fig. 6–2 illustrates the wide cell size and shape distributions caused via the aeration process in cellular foods. There are differences in size, shape, size distribution, open/closed cell, density gradients, etc.

In contrast to the above examples from the food industry, Mao, Wang, and Shimai (2008) were able to achieve a regular cell structure with controlled cell

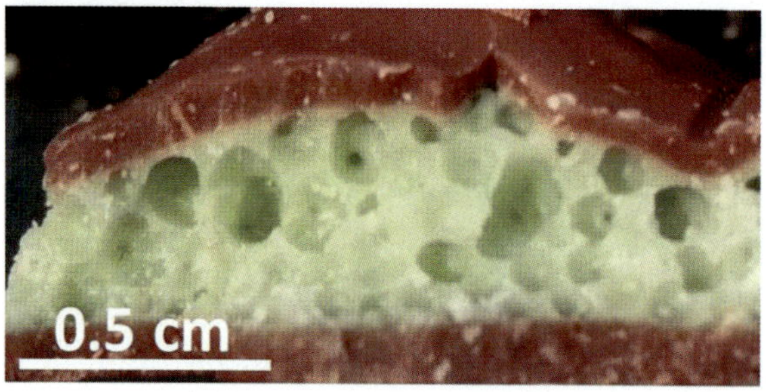

FIGURE 6–1 A chocolate bar in which the center filling has been formed by aeration. Note the large distribution of cell sizes.

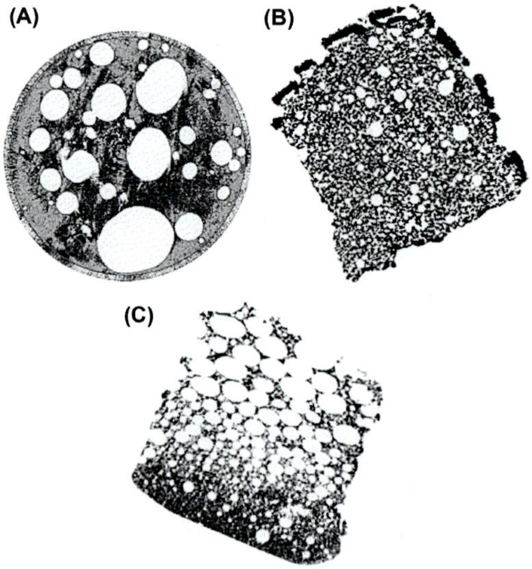

FIGURE 6–2 X-ray microcomputer tomography sections showing the pore structure of various aerated foods. (A) Mousse, (B) marshmallow, and (C) chocolate (Lim & Barigou, 2004). Note the size and shape difference of voids within a single material.

size and size distributions in a ceramic by mixing/aeration. The process of forming the ceramic foam follows a slurry processing route. The control of the cell structure relies on the use of a surfactant to maintain stable bubble formation through the drainage and drying process. The final structure is a porous SiO_2, as shown in Fig. 6–3, with three different pore sizes of 50, 10, and 1 μm. As shown in the figure, the cell structure is regular, which is achieved by careful control of slurry additives to make the suspension stable after the air is introduced.

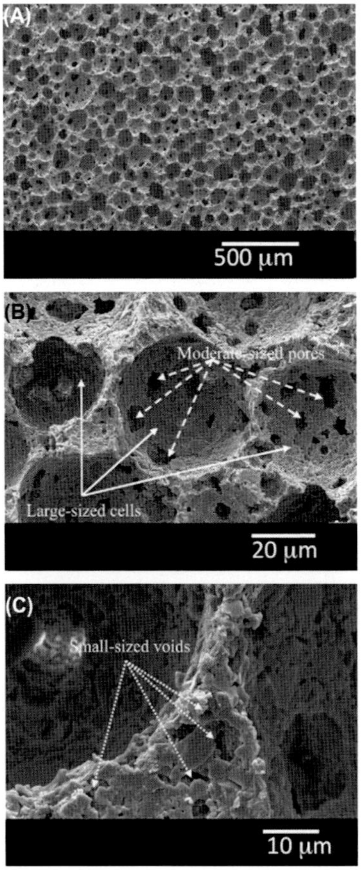

FIGURE 6–3 Micrographs of an SiO_2 foam made by simple mixing of air into a ceramic slurry followed by sintering. (A) Low, (B) medium, and (C) high magnification (Mao et al., 2008). Note the consistent pore structure achieved by the addition of a surfactant during the process. SEM.

6.3.2 Physical blowing agent

In this process, a latent gas agent is introduced into a material that subsequently undergoes vaporization. The generated gas creates the void and cellular structure in the material. Examples of latent-gas agents are hydrochlorofluorocarbons and pentane, which are commonly used when making polymer based foams. Also included in this category are gases such as hydrogen, argon, nitrogen, and carbon dioxide, which can dissolve into a molten material such as borosilicate glass (Wang, Matsumaru, Yang, Fu, & Ishizaki, 2012), metal, or polymers under high pressure. When the pressure is released, the dissolved gas comes out of solution, leaving a void.

Formation of lotus shaped foams in porous metal is carried out with a physical blowing process. It is referred to as "lotus shaped" because the structure resembles lotus roots (Babcsán, Banhart, & Leitlmeier, 2003; Hyun & Nakajima, 2002). The process involves dissolution of a gas in a molten metal. During solidification, the gas becomes insoluble and comes out of solution forming a void, see Fig. 6–4. A metal foam formed in this way is called a *gasar metallic foam.* This process, patented by Shapovalov and Serdyuk (1980), gets its name from combining the word "gas" with an "ar" for the abbreviation of the Russian verb to reinforce, *armirovat* (армировать). Loosely translated, it means "reinforced by air" (Shapovalov, 1998).

The general process for making a gasar foam has two stages. The first stage is creating a molten metal that is saturated with gas. Sievert's law governs the equilibrium concentration, C_g, of a gas in a melt as a function of temperature. The following expression gives C_g as a function of temperature:

$$C_g = \frac{\sqrt{p_g}}{p_0}\exp\left[\frac{-\Delta G_m^0}{RT}\right], \tag{6.1}$$

where ΔG_m^0 (J/mol) is the change in molar free energy of the gas during solution, p_g (Pa) is the partial pressure, p_0 (Pa) is the standard pressure of the gas, R is the molar gas constant (8.314 J/mol K), and T (K) is the temperature of the molten metal.

The second stage of the process is unidirectional solidification of the melt. As the melt cools and eventually solidifies, the dissolved gas becomes less soluble in the

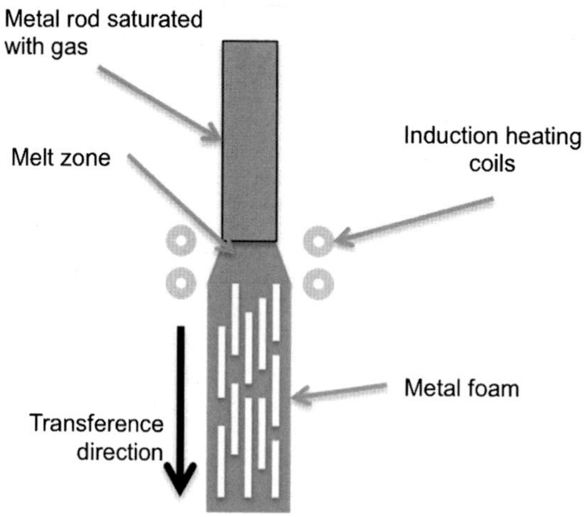

FIGURE 6–4 Schematic of the general method for fabricating lotus-type metal foam by the continuous zone melting method. The directional melt solidification front yields a columnar or cylindrical void geometry. *After Kashihara, M., Hyun, S.K., Yonetani H., Kobi, T., & Nakajima, H. (2006). Fabrication of lotus-type porous carbon steel by unidirectional solidification in nitrogen atmosphere. Scripta Materialia, 54(4), 509–512.*

molten metal and comes out of solution as a series of bubbles at the solid/liquid interface. With careful control of the temperature, the directional solidification will proceed and the voids take on a columnar or cylindrical geometry. Such lotus-type foams have been made using steel (Kashihara, Hyun, Yonetani, Kobi, & Nakajima, 2006; Kujime, Tane, Hyun, & Nakajima, 2007), copper (Du et al., 2013; Liu, Liu, & Xie, 2012), magnesium (Liu, Li, Jiang, & Xie, 2012), iron (Hyun, Ikeda, & Nakajima, 2004), and Ni-Al (Lee et al., 2012). Fig. 6–5 shows an example of a lotus-type foam of copper. Note the columnar pore structure in Fig. 6–5B.

6.3.3 Chemical blowing agent

Introducing porosity into a polymer using a *chemical blowing agent* involves one of the two mechanisms: (1) addition of a chemical additive that will decompose to form a gas and (2) a reaction occurs during material processing in which one of the products has the gaseous form. An example of this is a "hydrogen-blown" polydimethylsiloxane (PDMS); the simplified chemistry of which is shown in Fig. 6–6.

When a chemical blowing agent evolves from a cross-linking reaction, such as the hydrogen blown PDMS, careful control is required of the rheological properties of the polymer and kinetics of the reaction generating the blowing agent. Often times, fumed silica or some other nucleating agents are added to control the rheology as the cross-linking and gas evolution occur. If the gas evolves and the polymer does not have a high enough viscosity, the foam will overblow and collapse. The interplay among the reinforcement, viscosity, and gas generation also has a direct effect on the cell size in the finished structure.

Examples of chemical blowing agents via decomposition into a gas phase are carbonates, sulfites, nitrates, bicarbonates, and azodicarbonamide. These chemical agents are usually activated by temperature, which causes chemical decomposition and release of a latent gas. However, UV-activated decomposition is also possible (Schlögl, Reischl, Ribitsch, & Kern, 2012).

In porous ceramics, the combustion synthesis route of pore formation also involves a chemical blowing agent. The initial mixture consists of the chemical

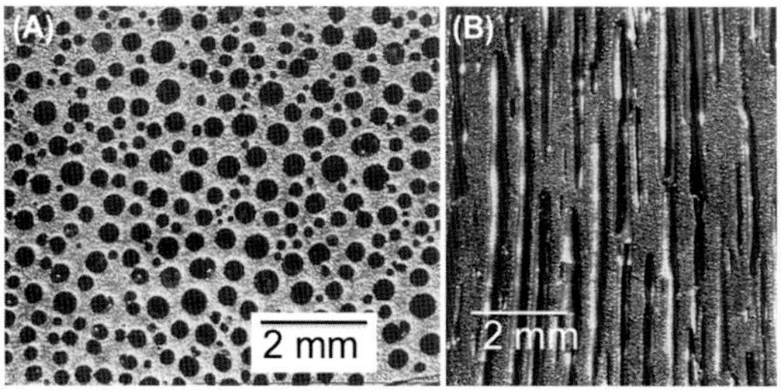

FIGURE 6–5 The structure of a lotus-type copper foam. (A) Transverse and (B) longitudinal cross sections (Nakajima, Hyun, Ohashi, Ota, & Murakami, 2001).

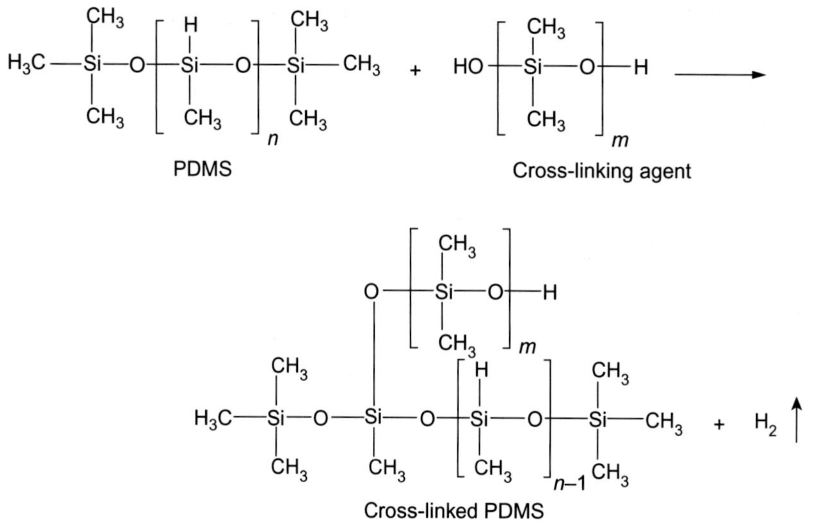

FIGURE 6–6 The chemical reaction of a polydimethylsiloxane (PDMS) and a cross-linking agent resulting in both cross-linked PDMS and hydrogen gas generation. The gas is responsible for forming the cells in the foam; a chemical-blown foam.

blowing agent dispersed with a ceramic powder and a binder material. This powder mixture undergoes typical ceramic processing: compaction of green body, binder burnout, followed by sintering. During the sintering process, there is a thermally activated reaction involving the chemical blowing agent. The reaction causes the pore forming material to vaporize, which expands to create a void or pore.

Maca, Dobsak, and Boccaccini (2001) used charcoal mixed with alumina to create both evenly distributed and functionally graded voids or pore structures. The sintering of alumina was done in air and vacuum. Sintering in air leads to the following gas forming reaction between oxygen and charcoal:

$$C(s) + O_2(g) \stackrel{\Delta}{\Longrightarrow} CO_2(g)$$

The delta symbol in the reaction above indicates the addition of heat for the reaction to occur. Under these conditions, the pores were uniformly distributed throughout the material. Sintering in vacuum led to more complex reactions and a graded final structure. In the absence of a significant partial pressure of oxygen, the following initial reaction leads to pore formation.

$$Al_2O_3(s) + 2C(s) \stackrel{\Delta}{\Longrightarrow} Al_2O(g) + 2CO(g)$$

Near the surface where there is higher oxygen partial pressure, Al_2O_3 recondenses and oxidizes by the following equation:

$$Al_2O(g) + O_2 \Longrightarrow Al_2O_3(s)$$

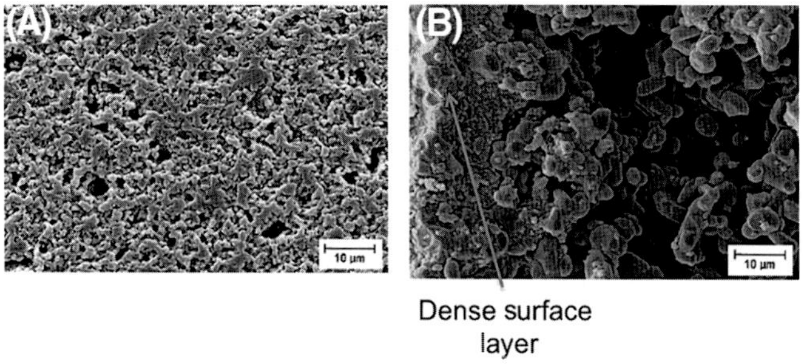

FIGURE 6–7 Cross section of a sintered Al_2O_3 foam where charcoal was used as a pore forming agent. The sample in (A) was sintered in air and has a uniformly distributed porosity and (B) sintered in vacuum and has a nonuniform distribution of porosity and dense surface layer (Maca et al., 2001). SEM.

Microstructurally, this reduction–evaporation–condensation–oxidation mechanism leads to a dense surface layer and necks between the particles, see Fig. 6–7. Interestingly, the vacuum sintered samples exhibited a higher strength structure at an equivalent density compared to the strength of the structure obtained by sintering in air. This feature of similar density but higher strength is an important finding. The design of the entire structure, including the voids, is extremely important in the performance of the final material.

Self-sustaining combustion synthesis reaction is a rapid method of creating a transition metal foam with porosity on the nanometer and micrometer scale. Researchers at Los Alamos National Laboratory (LANL) invented the nanoFOAM process. Depending on the selection of the precursor materials, the process can produce iron, cobalt, copper, and silver foams.

Fig. 6–8 shows a series of events in the nanoFOAM process. The initial frames show a slanted U-shape object, which is a resistively heated wire, resting on the top of a pellet. The pellet is a cold-pressed metal precursor material made from high nitrogen content transition-metal complexes. The nitrogen will become the blowing agent in this chemically blown metal foam.

When an electrical current passes through the wire, it heats up and ignites the pellet at the point of contact. This initial input of heat energy (ΔH) from the hot wire starts the decomposition of the precursor material into metal and nitrogen gas. The decomposition reaction is exothermic and releases enough heat to create a self-sustaining reaction that propagates through the entire volume of the pellet. During the reaction, there is a dramatic increase in volume caused by the release of nitrogen gas. The nitrogen release creates nanometer and micrometer scale porosity and the transition metal particles then form the solid struts, see Fig. 6–9.

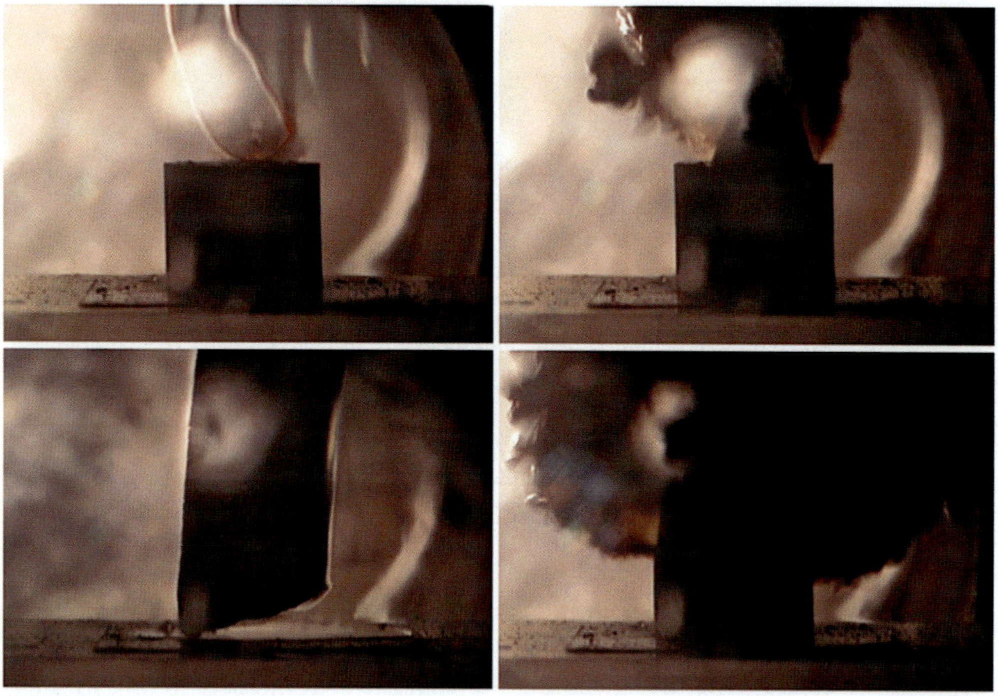

FIGURE 6–8 A series of photographs illustrating the chemical reaction to create a transition metal foam referred to as nanoFOAM by the inventors at Los Alamos National Laboratory. In the top right frame, the U-shaped wire is resistively heated and ignites the pressed pellet of transition metal complex. The foam blowing reaction is exothermic, thus self-propagating and starts at the point where the wire is in contact and spreads through the entire volume. The reaction releases nitrogen leading to nanometer and micrometer scale porosity in the transition metal that remains and a dramatic increase in volume. *Link to the video (used with permission): https://booksite.elsevier.com/9780444563675/metal_foam_full_high.php.*

The general chemical structure of a high-nitrogen transition-metal complexes reactant is

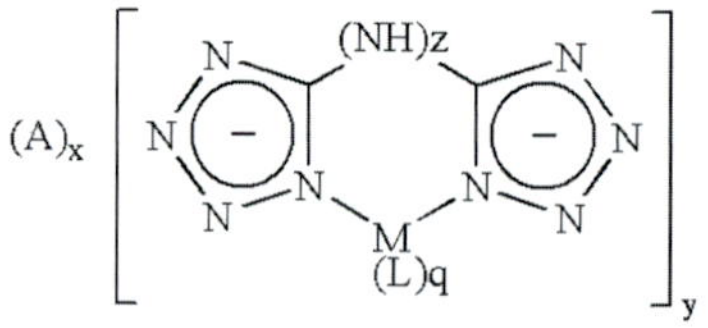

where A is selected from ammonium, hydrazinium, guanidinium, aminoguanidinium, diaminoguanidinium, and triaminoguanidinium; x is zero or an integer from 1 to 3, y is an integer from 1 to 3; z is 0 or 1, L is amine; q is 0 or 2; and M is a transition metal.

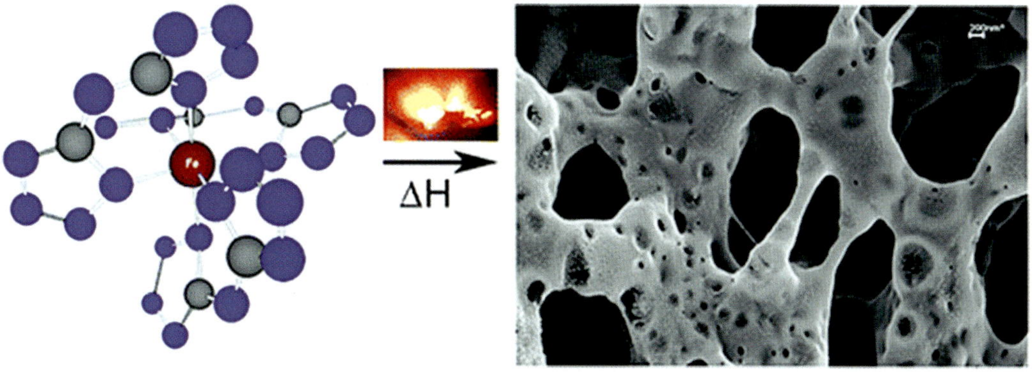

FIGURE 6–9 The general reaction synthesis of nanoFOAM technology is shown above. It is a chemically blown foam. The reactant, a transition metal complex, contains significant amounts of nitrogen. The reaction to produce the foam is exothermic. After an initial input of energy (Δ*H*), there is enough heat released to create a self-sustaining reaction. As the reaction proceeds through the material, nitrogen gas (the blowing agent) is liberated; forming the nanometer scale porosity and creating the final structure that is a reticulated metal foam (Tappan et al., 2006). *Reprinted with permission from the American Chemical Society.*

As shown in Fig. 6–9, this produces a highly porous, reticulated metal foam. The nanoFOAMs have a pore diameter between 20 nm and 1 μm, surface areas as high as 258 m^2/g, and densities as low as 0.01 g/cm^3. These values are comparable to aerogels, the lightest known solids. The comparison to aerogels is significant because, with only very few exceptions, aerogels are metal oxides.

LANL's nanoFOAM technology can be used for

1. the catalytic production of ammonia, sulfuric acid, fuels, plastics, and other chemicals, and products;
2. oil-refining processes and electrical generation from fuel cells that run on hydrocarbons;
3. silver biocidal filters that destroy liquid or airborne germs on contact;
4. improving the strength and heat-transfer properties of jet-turbine blades while decreasing their weight;
5. reducing emissions of nitrogen oxides from internal combustion engines and coal-fired power plants;
6. remediation chlorohydrocarbons in the environment; and
7. enhancing the sensitivity of biomedical detectors.

6.4 Templating or sacrificial pore former

Removal of a sacrificial material is a method of ensuring cell size uniformity; however, in order for the method to be suitable for large scale manufacturing, a very high volume of sacrificial material should be present so that we get close to 100% open cell porosity. This allows for removal of the sacrificial material in a reasonable time period. A common foam prepared by this method is silica reinforced PDMS. The PDMS in this case starts as a gummy (high viscosity, uncross-linked polymer) material, and urea particles with dimensions of the desired cell size are introduced by roll milling. Once the desired loading of urea particles is added to ensure an open cell structure, the material is molded into the desirable geometry and cured. At this point, the composite foam contains the cross-linked PDMS polymer matrix, silica reinforcement, and the sacrificial urea particles. Urea, being soluble in water, is then leached out by placing the component in a water bath. The leaching process can be accelerated if the bath is agitated. A drawback of this method is that the leaching process becomes less efficient for thick cross sections.

When making metal foams using this sacrificial pore forming method, it is usually referred to as the *placeholder method*. Furthermore, in many instances, a pore-forming material is the same material that is used as a chemical blowing agent, which is activated by thermal decomposition. For example, carbonates, sulfites, nitrates, and bicarbonates can act as both blowing agents and pore formers. Even though the material chemistry and removal process is the same, there is a distinct difference between the two. During chemical blowing, there is an expansion of the cell size, making it larger than the initial latent gas precursor particle size. In the simpler placeholder method, there is no additional expansion when the placeholder material is vaporized.

In metal and ceramic foam processing, a sacrificial material is usually removed during processing rather than afterward as is done in polymer processing. During sintering or consolidation of metal or ceramic, the temperature is high enough to vaporize the sacrificial material. Examples of placeholders are magnesium (Mg) (Aydogmus & Bor, 2012), sodium chloride (NaCl) (Jha et al., 2013; Minami, Kobayashi, & Okubo, 2005; Sun & Zhao, 2005; Torres, Pavón, & Rodríguez, 2012; Ye & Dunand, 2010; Zhao, Han, & Fung, 2004), BaF_2 (Brothers & Dunand, 2005, 2006; Brothers, Scheunemann, Defouw, & Dunand, 2005; Xie, Fukuhara, Louzguine-Luzgin, & Inoue, 2010), poly(methyl methacrylate) (Irausquín, Pérez-Castellanos, Miranda, & Teixeira-Dias, 2013; Manonukul, Muenya, Léaux, & Amaranan, 2010), and urea/carbamide (Bekoz & Oktay, 2012).

6.4.1 Aerogels

An aerogel is a highly porous material that has the lowest density, thermal conductivity, refractive index, sound velocity, and dielectric constant of any solid (Fricke & Tillotson, 1997). The process of making an aerogel involves the removal of a liquid phase placeholder (porogen). Common aerogel materials are silica, carbon, titania, and alumina. The density of a silica aerogel can be as low as 0.003 g/cm^3 and the surface area as high as 1000 m^2/g. The silica particles are on the order of 2–5 μm with a mean interstitial pore size on the order of 40 nm. Silica aerogels are made by the sol–gel process. The sol–gel process for making aerogels consists of

1. Gel preparation;
2. Aging of the gel;
3. Drying of the gel.

The process starts by reacting a silicon alkoxide, such as tetraethoxysilane (TEOS), with water, catalyst, and solvent. Typical catalyst and solvent used with TEOS are hydrochloric acid and ethanol, respectively. The critical step in forming an aerogel is drying of the gel, which involves removal of the porogen liquid. It must be done without damaging the aerogel. The important parameters in the drying process are the pore volume and surface area. The pores should not be too large or too small. If the pores are too large, the aerogel structure will be fragile and break from shrinkage that accompanies the drying process. If the pores are very small, fracture can occur from capillary forces that develop within a pore. Examples of drying methods are supercritical drying, ambient temperature drying, and freeze drying (Soleimani Dorcheh & Abbasi, 2008).

6.5 Bonding together of spheres, fibers, powders, or particles

Bonding together of spheres, powders, and fibers is a broad category that encompasses many different types of materials including syntactic foams as well as frit type material. Syntactic foams are a composite type foam consisting of hollow spheres distributed in a matrix and have become quite important. The word *syntactic* comes from Greek, meaning "to arrange parts together in a unit." In syntactic foams, the matrix can be a polymer, metal, or ceramic. Hollow spheres can range in size from nanometers to a few millimeters. They may consist of glass, carbon, metal, polymer, or ceramic materials, although most commonly hollow glass spheres (HGSs, 20–200 μm in diameter) are used. The fabrication methods for microballoons will be covered in Chapter 5, Hollow and Porous Structures Utilizing the Kirkendall Effect and Chapter 7, Techniques for Introducing Intentional Voids into Particles and Fibers. These microballoons can be mixed with a binder or matrix material or simply sintered together connecting at the contact points. When a binder or matrix is used it can be in a liquid or solid powder form. By far, the most common syntactic foam is the glass microballoon (GMB)/epoxy matrix material. When the HGSs are in the micrometer-size range, they are commonly referred to as hollow glass microspheres, *HMGSs*, or glass microballoons, *GMBs*.

Syntactic foams can be further subdivided into two, three, or multiphase foams. The microstructure of a two phase foam has only the binder and hollow microspheres (see Fig. 6–10A). In contrast, in a three-phase material there is not sufficient binder volume to fill the spaces between hollow microspheres (Fig. 6–10B). The voids in the binder constitute the third phase. Of course, fibers can be added to make the fourth constituent in a multiphase syntactic foam (Fig. 6–10C).

The bonding or entangling of fibers, especially nanofibers, is important for biomaterial, energy absorption, and superhydrophobic surfaces (Wang, Ding, Yu, & Wang, 2011). Although not typically thought of as foam, the void structure can lend interesting functionality to the bulk materials. These materials are referred to as webs, felts, and *sintered fibrous media*. This type of porous material can be described as having three-dimensional internal pore structures. These nonwoven fibrous structures are quite common. They can consist of fibers and/or nanofibers, see Fig. 6–11.

The processing of these materials involves electrospinning the fibers and collecting them on a substrate where they become randomly oriented, see Fig. 6–11A. This random orientation leads to the nonwoven or weblike structure. Many times these nonwoven structures are used as sacrificial templates, especially for porous ceramic structures.

Micrometer scale metal fibers have commonly been assembled to form nonwoven structures like "steel wool" and coarse filter media. Similar structures have been gaining attention for both structural and functional applications. Some of the applications include biomedical materials, catalyst carriers, fuel cells, acoustic applications, and metal electrodes (Tang et al., 2011).

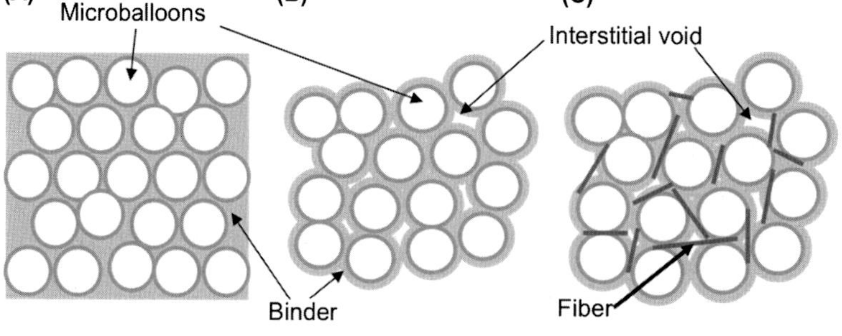

FIGURE 6–10 A schematic of a (A) two-phase, (B) three-phase, and (C) multiphase syntactic foam. Note that in the three-phase foam, there are both reinforced (microballoon) and unreinforced (interstitial) voids present.

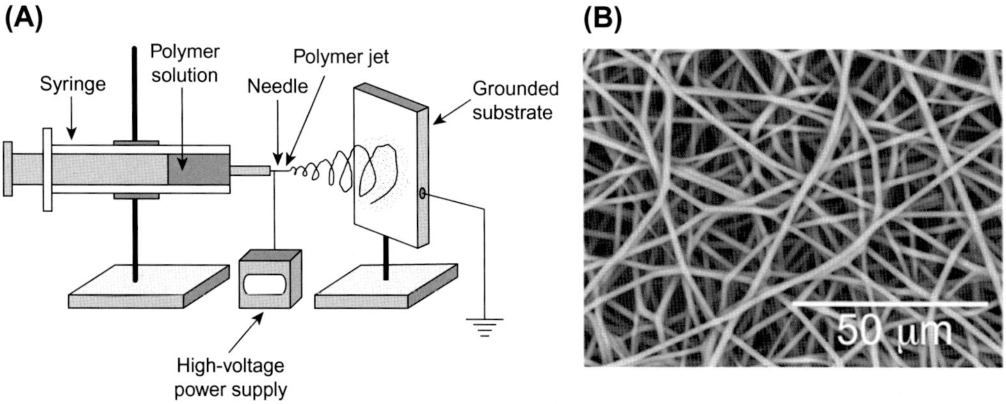

FIGURE 6–11 (A) A schematic of the electrospinning process and (B) a scanning electron micrograph of an electrospun polyurethane web (Beachley & Wen, 2010).

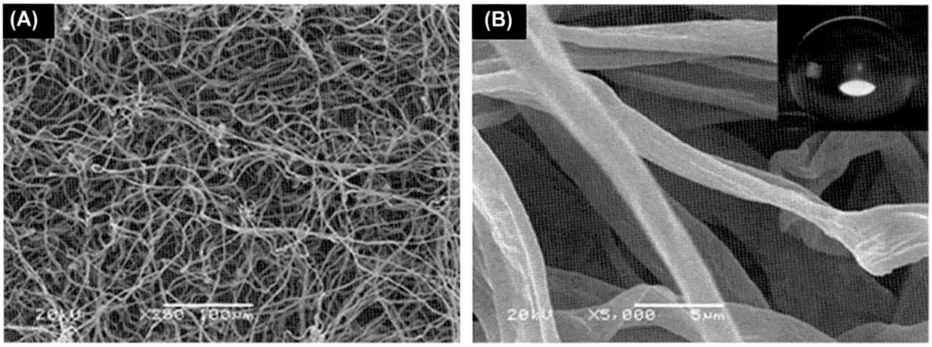

FIGURE 6–12 The surface of the superhydrophobic ramie leaf in (A) low and (B) high magnification. The nonwoven fiber arrangement as well as the smooth fiber surface texture contributes to the (B) (inset) 164 degrees surface angle with water (Guo & Liu, 2007). SEM.

Many examples of nonwoven structures can be found in nature, particularly in nonwetting or superhydrophobic surfaces. Fig. 6−12 is an SEM image of a ramie leaf that creates a contact angle with water of 164 degrees. The surface of the leaf has a three-dimensional pore structure created by the entanglement of 1−2 μm diameter fibers. Additionally, these fibers have a very smooth surface texture. The combination of the fiber texture and the pore structure makes the leaf superhydrophobic.

6.6 Additive manufacturing of cellular structures

Additive manufacturing (AM), 3D printing, and solid freeform fabrication are synonymous terms describing various techniques of building finished three-dimensional parts, layer-by-layer, based on computer models. Interest in this technology as a research tool started in the mid-1990s. As the names suggest, solid prototype parts and more recently, functional parts are manufactured by building up materials into a three-dimensional geometry as opposed to removing material from a larger piece. Conventionally, a component with a complex geometry is manufactured by the removal of material, for example, by drilling, grinding, and machining. AM builds a part from smaller units, layer-by-layer, into a finished component. Ideally, there is no need for a machining process once a build is complete. AM has also allowed for new design concepts. It is possible to create and features of parts and combinations of parts via AM that are not possible in traditional subtractive manufacturing methods, see Fig. 6–13.

Standards organizations such as ASTM and ISO (ASTM International, 2015) have categorized AM into seven distinct types of processes or technologies. These are categories are outlined below along with examples of intentional porosity incorporated into parts by each method.

1. *Vat polymerization* is an AM process in which a vessel containing liquid photopolymer is selectively cured by light-activated polymerization. To build a

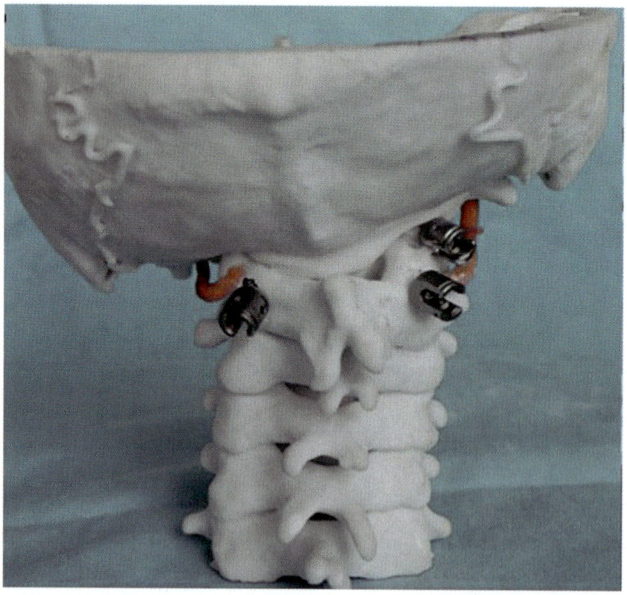

FIGURE 6–13 An example of the complex geometry that 3D printing can create in a single part. Creating this as a multiple part assemble of a skull and vertebrae by traditional subtractive manufacturing techniques would be impossible (Du et al., 2020). Using AM this assembly can be completed in a single build saving time and material.

part by vat polymerization, a build platform is first positioned in the tank of liquid photopolymer at a distance that is one-layer height from the surface of the liquid for *top-down process* or the bottom of the vat for a *bottom-up process*, see Fig. 6–14. In a top-down process, the light source that initiates the curing of the photopolymer is located above the vat and the part is submerged in the polymer as additional layers are added. In the bottom-up process, the light source is below the vat and the build plate moves upward out of the polymer as layers are added. The first layer of the part is made by focusing the light source/laser in a predetermined path across the resin bath; selectively curing and solidifying the photopolymer resin. The laser is focused and moves along a predetermined path using a set of mirrors as determined by the model until the entire layer of the part is fully solidified. When a layer is finished, the platform moves and a new layer of uncured photopolymer resin coats the surface. The process then repeats until all the layers in part are complete. Alternative to the point curing using a laser, one can project an entire "slice" of a UV light source onto the vat of resin to cure a complete layer at once. This method allows for both fine detail and fast printing (Molitch-Hou, 2018). After printing, the part is in a green, partially cured state and requires postprocessing under UV light to achieve optimum properties. A well known example of this type of AM is the first 3D printed athletic shoe, see Fig. 6–15. The adidas Futurecraft 4D midsole is printed in a vat process using a proprietary resin blend of UV curable resin and polyurethane. The potential of this technology lies in designing the midsole foam based on the analysis of the weight distribution as an individual walks

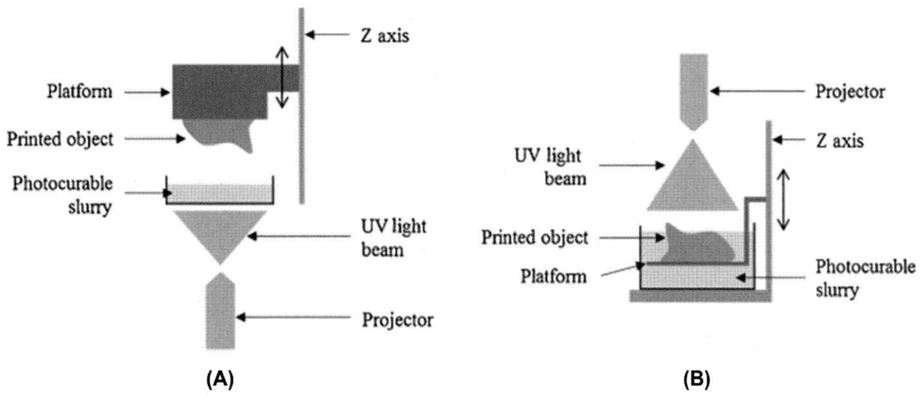

FIGURE 6–14 Schematic of the (A) bottom-up and (B) top-down of a digital light process (Santoliquido, Colombo, & Ortona, 2019), examples of vat additive manufacturing techniques.

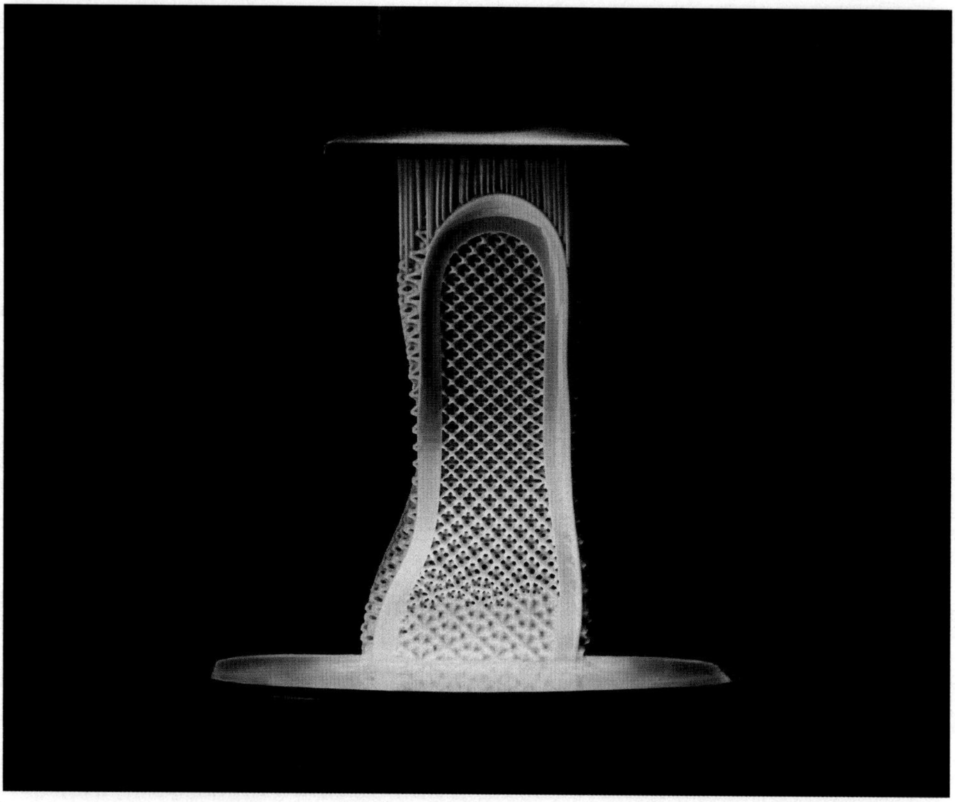

FIGURE 6–15 The finished sole of the adidas Futurecraft 4D midsole printed by a bottom-up digital light, vat polymerization. *Source: Courtesy of adidas*

or runs. No two persons are alike and no two persons will have the same weight distribution as they walk/run; therefore, each foamed midsole will have a different strut configuration. Each midsole strut configuration will be tuned specifically for an individual; this is the concept of *mass customization.*

2. *Material jetting* is an AM process in which droplets of build material are selectively deposited on a build platform, see Fig. 6–16. The first step is to heat the liquid resin to achieve optimal viscosity for printing. The process temperature depends on the print material chosen. Material is deposited as the printhead travels over the build platform and hundreds of tiny photopolymer droplets are jetted to the desired locations per a CAD model. A UV light source is attached to the printhead and cures the deposited material, creating the first layer of the part. The build platform then moves downward one-layer height and the process repeats until the whole part is complete.

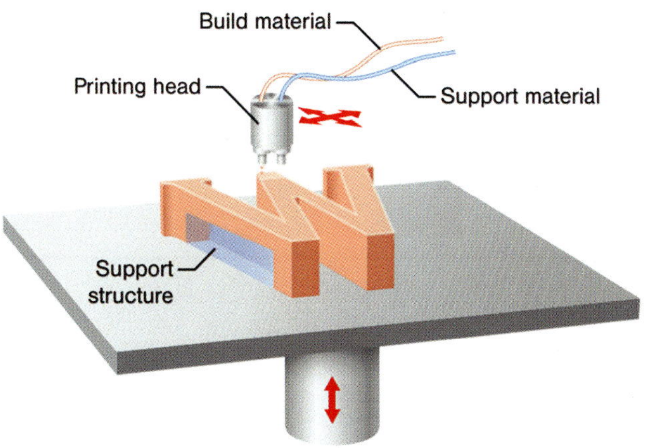

FIGURE 6–16 Schematic of the binder jet process printing both build material (orange) and support structure (blue). Support material is only printed where features are not self-supporting and is designed to be removed in a postprinting process (Rietzel, Friedrich, & Osswald, 2017).

Velasco, Lancheros, and Garzón-Alvarado (2016) used reaction-diffusion model (Turing, 1952) to design and build a porous bioscaffold using material jetting technology. Fig. 6–17 shows the design of pore and channel structure that is the output from the reaction-diffusion model. The reaction-diffusion model is the most widely used mathematical theory to describe pattern formation that occurs in biology (Kondo & Miura, 2010; Velasco et al., 2016). A discussion of the model is beyond the scope of this book. Fig. 6–16 shows a structure created using the reaction-diffusion model (Velasco et al., 2016) and printed via material jetting. The printed structure, shown in Fig. 6–18, compares designed feature dimensions in Fig. 6–18A and B with the printed features in Fig. 6–18C, D, and E.

3. *Binder jetting* is an AM process in which a liquid bonding agent is selectively deposited to join loose powder materials located on a build platform. In the binder jet process, see Fig. 6–19, a layer of powder is spread evenly onto a build platform. A carriage with nozzles passes over the bed, selectively deposits droplets of a binding agent, bonding the powder together. When the layer is complete, the build platform moves downward and another layer of powder is spread over the build platform. The process then repeats, applying enough binder to bond the current layer of powder together in addition to bonding the current layer to the previous layer. This process repeats until the part is complete. After printing, the part is within the build envelope, surrounded by loose powder. Depending on the binder, the part may be allowed to postcure to gain strength. The part can then be removed from the

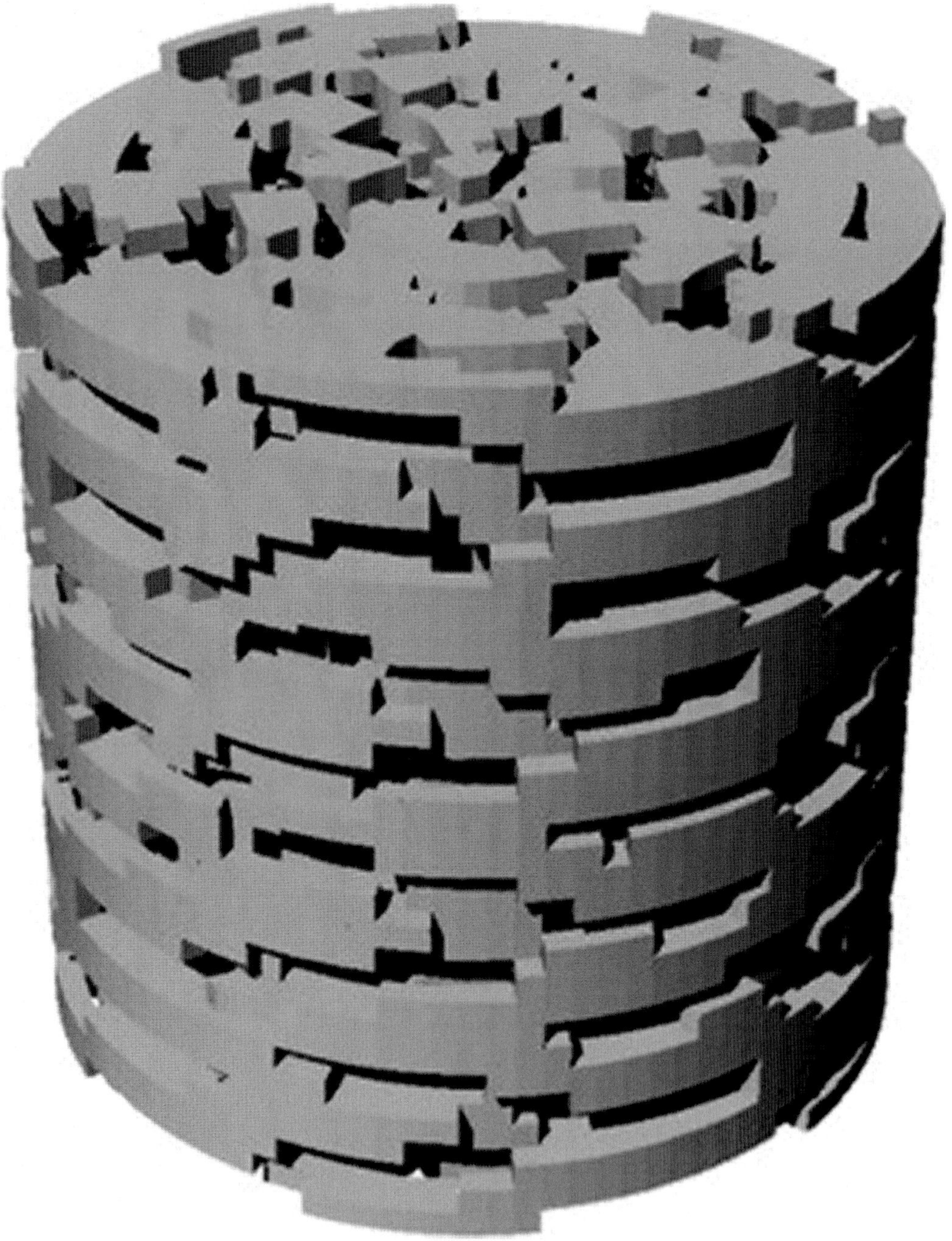

FIGURE 6–17 Computer model showing the porous cylindrical form to be printed via material jetting (Velasco et al., 2016).

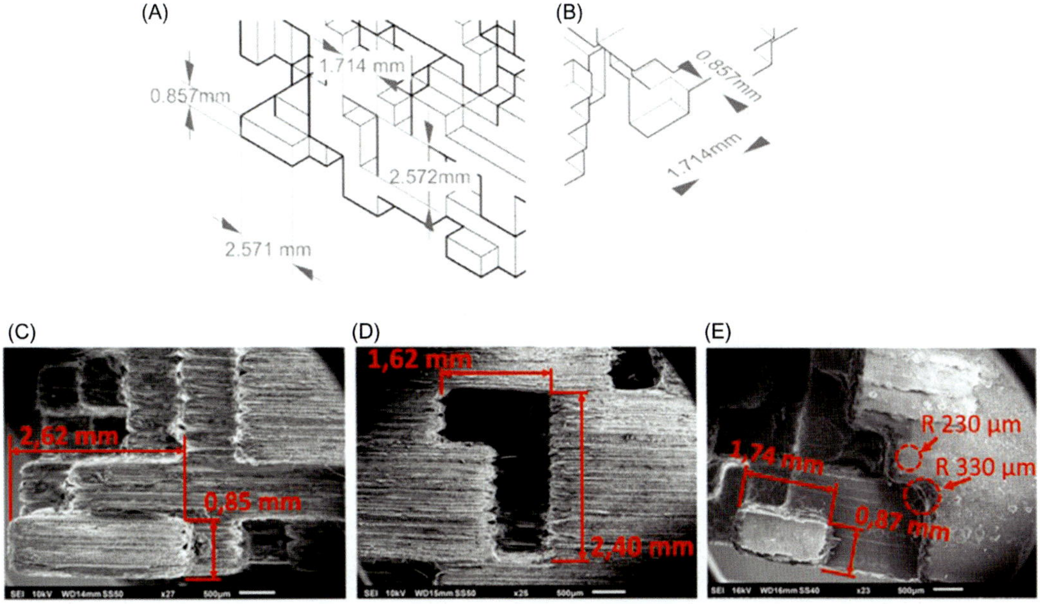

FIGURE 6–18 Dimensional comparison of the features based on computer models (A) and (B), to those structures made via material jetting (C) trabecular detail, (D) L shape pore, and (E) trabecular measurements in a vertex half (Velasco et al., 2016).

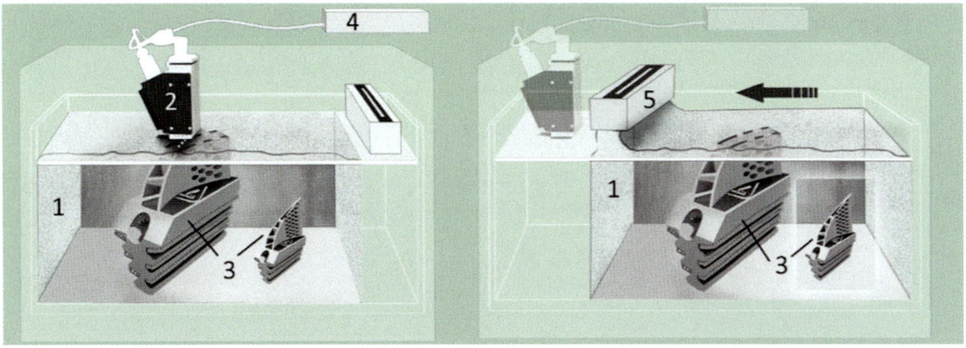

FIGURE 6–19 A schematic of the binder jet process (Fleisher et al., 2019). (A) 1 indicates the unbonded region powder bed, 2 is the binder nozzle and printhead being fed by 4-the binder through a pumping system, 3 is the printed sample. The hopper and powder delivery system 5, are shown on the right side. Following binding of a layer, this powder delivery system deposits a thin layer of powder on top of the previous layer.

powder bed and the excess powder is removed. Depending on the application, the porous part could be the finished part but typically it is thought of as a preform and the porosity will be infiltrated with a matrix material.

Myers et al. (2015) used the binder jet process to create a syntactic foam having a metal matrix with hollow ceramic microspheres. The composition of the foam was hollow mullite spheres with an aluminum binder or matrix phase. As it relates to Fig. 6–19, the hollow mullite microspheres are the powder bed component and polymer binder is sprayed through a nozzle onto selected regions of that powder bed binding the mullite together in a layer. Another layer of loose mullite is delivered from the hopper to the powder bed. The binder is then sprayed onto this new layer, bonding the mullite together as well as bonding it to the previous layer. Fig. 6–20 shows the product of the binder jet process, a three-phase syntactic foam of hollow mullite held together by the polymer binder phase at the contact points between the microballoons. The final step is infiltration of molten aluminum into the interstitial porosity between the mullite microballoons, and once solidified, it yields a two-phase metal matrix syntactic foam, see Fig. 6–21.

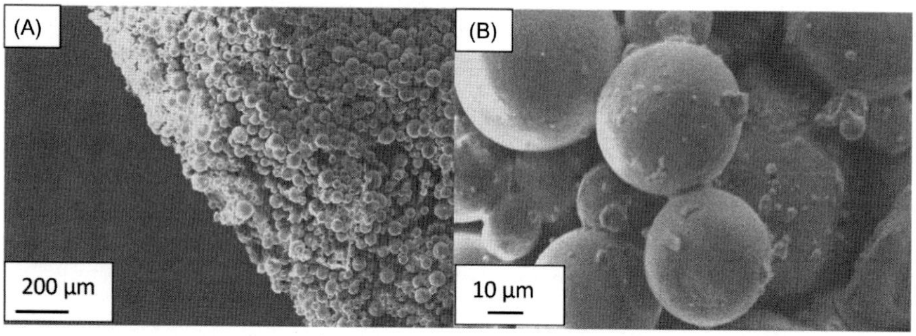

FIGURE 6–20 A three-phase syntactic foam made by the binder jet AM process: (A) low magnification and (B) high magnification.

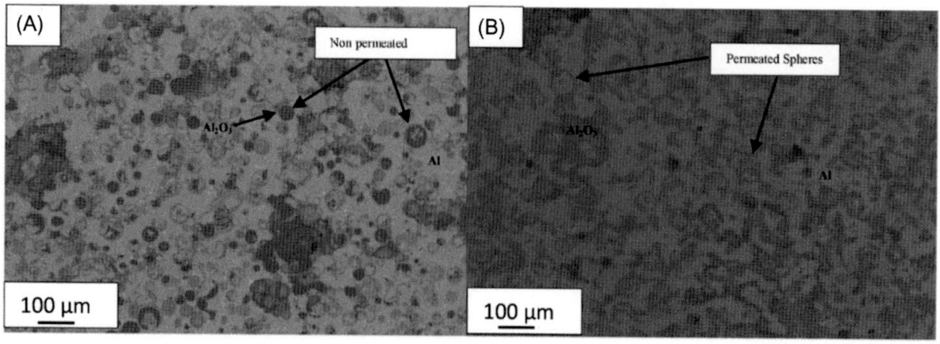

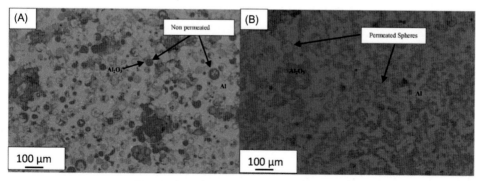

FIGURE 6–21 A two-phase syntactic foam made by aluminum infiltration into a three-phase syntactic foam and heat treated at 1200°C for (A) 4 h and (B) 16 h.

4. *Material extrusion* is an AM process in which material is selectively extruded through a nozzle or orifice, see Fig. 6–22. The temperature of the printer nozzle is material dependent and enough heat needs to be transferred to melt the filament being fed. The original AM material extrusion process used spools of thermoplastic filament but later, thermoplastic pellets, similar to feed pellets in traditional thermoplastic extruders, were used. The extrusion head is attached to a three-axis system that allows it to move in the x, y, and z directions. The material is melted and extruded into thin strands through a printer nozzle and is deposited layer-by-layer in predetermined locations as defined by CAD models. Once extruded, as the material begins to cool and the viscosity increases. There are competing forces at work when it comes to having an optimum temperature of the layer. If the layer and part bed are too warm, the viscosity and strength of the layer(s) will be too low and unable to support the weight of subsequent layers. In this case, the structure will collapse. If the temperature is allowed to cool too much, there will be insufficient entanglement of polymer chains between layers leading to lower interlaminar strength, that is, low strength in the z direction. To maintain the printed material at the optimum temperature, sometimes a fan is attached to the printhead to accelerate cooling while other times, on larger parts, heaters are needed on the part bed. When a layer is complete, the build platform moves down or the extrusion head moves up and a new layer is deposited. This process is repeated until the part is complete.

   Although material extrusion AM initially focused on thermoplastic extrudates, thermosetting polymer systems are becoming more common. Silicone (Elsayed, Colombo, & Bernardo, 2017), epoxy (Lindahl et al., 2019), and vinyl esters are

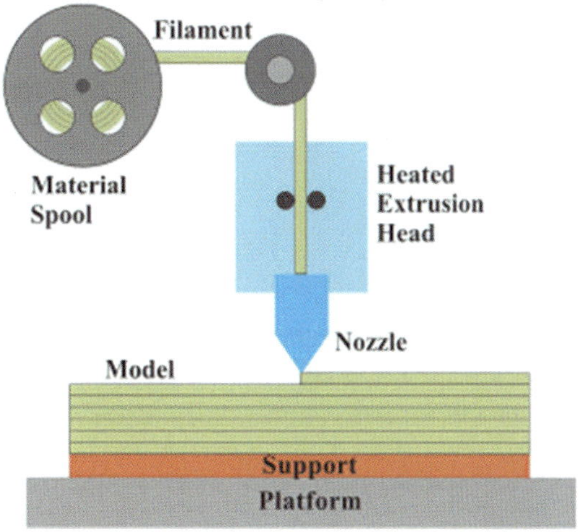

FIGURE 6–22 Schematic of the melt extrusion process (Zou et al., 2016).

examples of thermosetting materials that have been successfully used in AM via material extrusion. Elsayed et al. (2017) investigated the fabrication of wollastonite-diopside based bioceramic scaffolds using extrusion deposition. The extrudate was a silicone based preceramic polymer ink. The silicone was diluted in isopropyl alcohol loaded with powders to create a preceramic formulation. The silicone not only is the base liquid material to allow the powders to be extruded, it also serves as a source of silica (SiO_2). The silicone decomposes into SiO_2 during the heat treatment at 1100°C, transforming the extrudate into a finished fired bioceramic. Elsayed et al. investigated many bioceramic formulations but a common feature for bioactive materials is the needed hierarchical porosity. Fig. 6–23 illustrates this concept of hierarchical porosity showing the cross-hatched design creating a macroscale nonstochastic porosity and in a cross sectioned view, micrometer scale stochastic voids in the extruded filament. We would like to stress the point that the voids on both length scales are by design and have functionality in the bioscaffold (Holzapfel et al., 2013). On the macroscale, voids are an integral part of the bioscaffold and are part of the *digital design* of the component. On the micrometer scale, the origin of the voids lies in the careful formulation of the extrudate and firing, that is, the *material design* of the part. Both are equally important in the bioactivity of the finished bioscaffold.

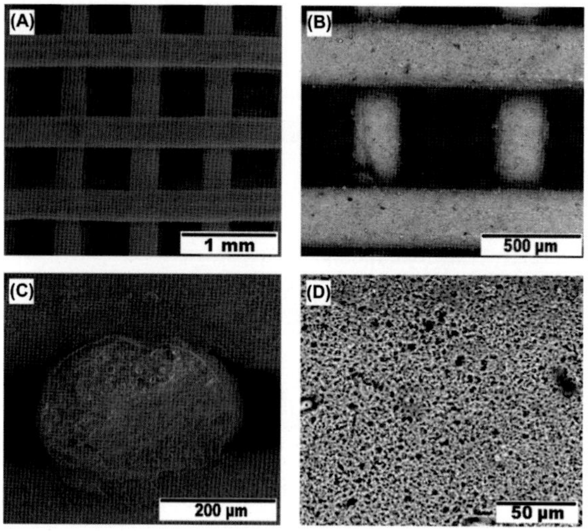

. SEM images of 3D printed W-D scaffolds +10 wt% W-D glass after ceramization: a, b) top view (XY) and; c) cross section through the Z direction and d) microstructural is of the filament.

FIGURE 6–23 SEM images showing the details of a nonstochastic composite scaffold made from 10 wt.% wollastonite and bioglass in a matrix: (A) and (B) are a top view down the z-axis, (C) is a cross section of an individual filament, and (D) is a higher magnification of the stochastic pore structure of the filament. *Source: Elsayed, H., Colombo, P., & Bernardo, E. (2017). Direct ink writing of wollastonite-diopside glass-ceramic scaffolds from a silicone resin and engineered fillers. Journal of the European Ceramic Society, 37(13), 41877–4195.*

Powder bed fusion is an AM process in which thermal energy selectively fuses regions of a powder bed, see Fig. 6–24. A layer of approximately 0.1 mm thick of material is spread evenly over the build platform. A laser rasters across the surface of the powder bed fusing the first layer or first cross section of the model. The fusion is a result of heating and melting of the powder. A new layer of powder is then spread above the fused layer using a roller. The process of adding a new powder followed by fusing the powder layer is repeated until the part is built. When the build is complete, the final part is surrounded by loose, unfused powder. Once cooled, the part can be removed and can undergo any required postprocessing.

Heinl, Müller, Körner, Singer, and Müller (2008) used microcomputer tomography to image three-dimensional structures with an interconnected macroscale porosity. Samples of two different cellular titanium alloy structures with a nominal layer thickness of 100 μm were fabricated via powder bed fusion using an electron beam as an energy source to selectively melt Ti-6Al-4V powder. The diamond cellular structure with a void volume content of 80.5% is shown in Fig. 6–25A, C, and E and hatched structure with a 61.3% void volume content is shown in Fig. 6–25B, D, and F.

5. *Sheet lamination* is an AM process in which sheets of material are bonded to form a three-dimensional object, see Fig. 6–26. There are two slightly different process procedures for sheet lamination. The first is when the individual laminae are cut prior to stacking and bonding to the previous layer. The second is when a lamina is first stacked, bonded to the previous layer, and then cut. The required shape for each layer is cut from the sheet by laser or knife.

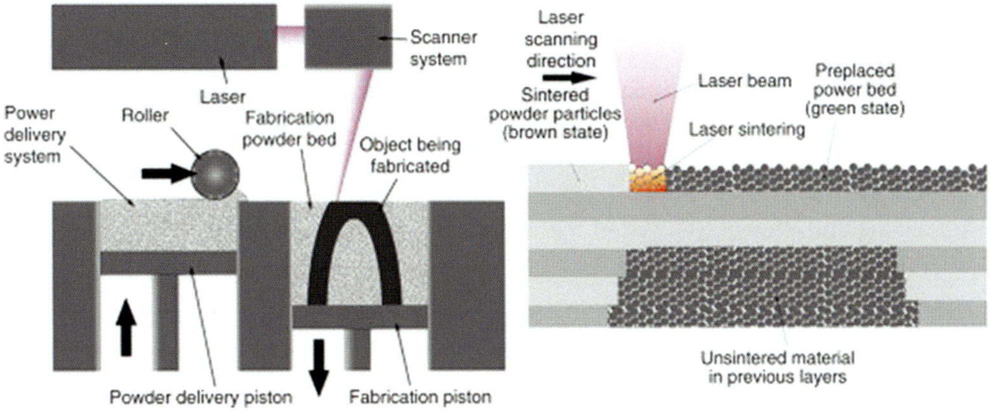

FIGURE 6–24 A schematic of the powder bed fusion process. Once a laser fuses a layer, the piston drops one build layer, the powder delivery system raises to provide enough powder for the roller to deliver another layer of loose powder. *Source: Molitch-Hou, M. (2018). Overview of additive manufacturing process. In Additive Manufacturing (pp. 1–38). Butterworth-Heinemann.*

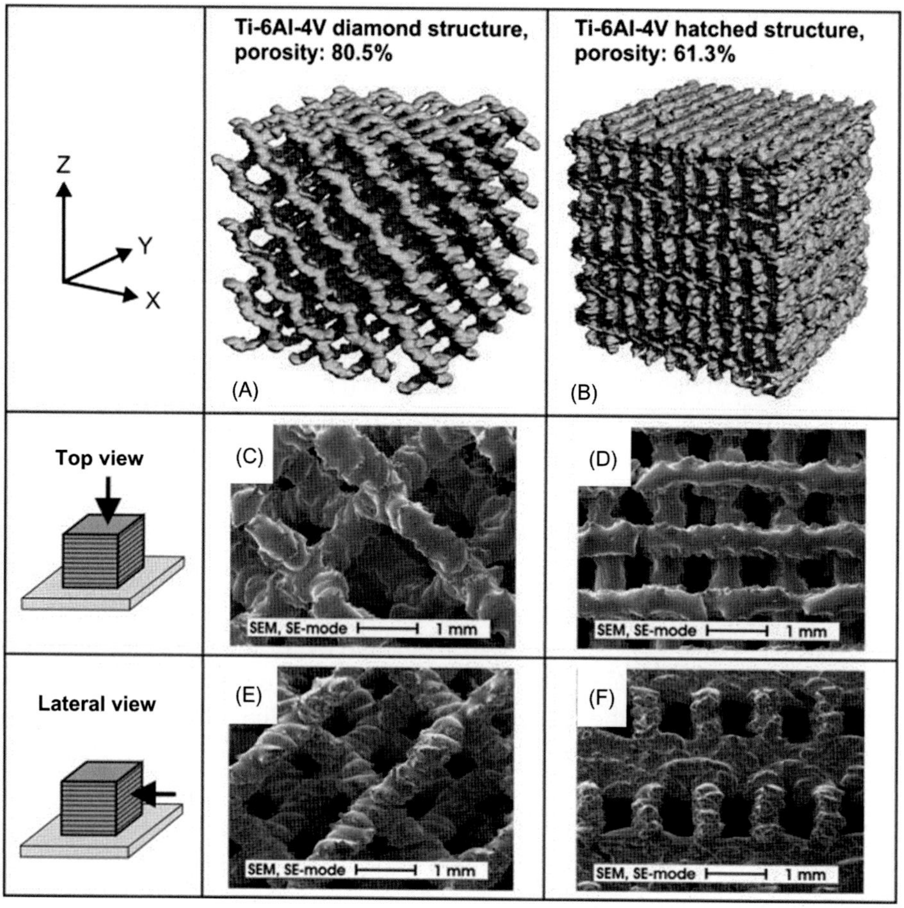

FIGURE 6–25 Computer rendering from a micro-CT scan of a (A) diamond and (B) hatched nonstochastic foam structure. Top view down the z-axis of a powder bed fusion processed (C) diamond and (D) hatched nonstochastic foam structure. Lateral view down the x-axis of a powder bed fusion processed (E) diamond and (F) hatched nonstochastic foam structure (Heinl et al., 2008).

Zhang and Wang (2017) used the sheet lamination AM process to make eight different porous polylactic acid (PLA) structures, including the configurations of open versus closed cell, round versus square void, and constant versus graded porosity. Individual PLA layers were fabricated via material extrusion, in a separate AM process, see Fig. 6–27. Arranging layers systematically, they were thermally bonded together at 140°C. The structures made from the individual layers are shown in Fig. 6–28.

6. *Directed energy deposition* is an AM process in which focused thermal energy is used to fuse materials by melting as they are being deposited, see Fig. 6–29. A four- or five-axis arm with nozzle moves around a fixed object depositing material. The material is

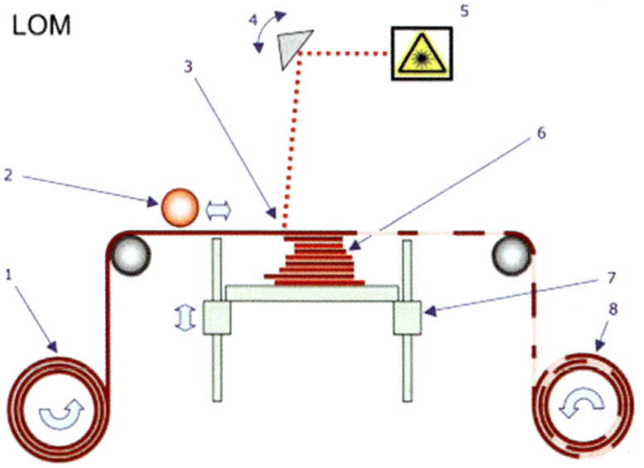

FIGURE 6–26 Diagram of the sheet lamination process. A roll of material fed into the build area and heat pressed with a heated roller onto the previous layers, at which point a laser cuts out a predetermined shape (Molitch-Hou, 2018).

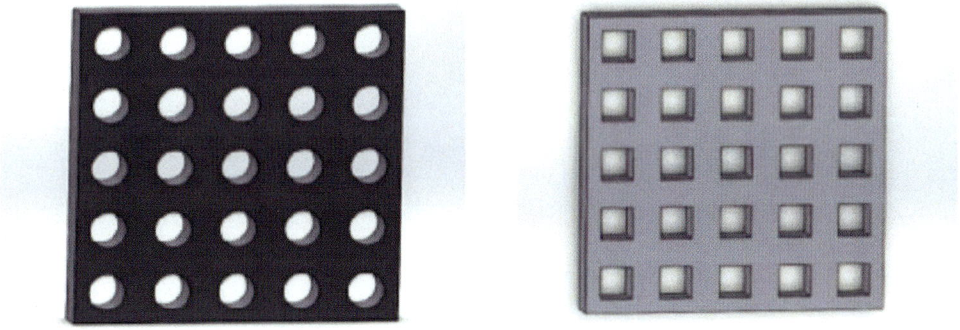

FIGURE 6–27 CAD generated in a single layer with (A) circular and (B) square void cross sections (Zhang & Wang, 2017).

melted using a laser, electron beam, or plasma arc upon deposition. Further material is added, layer-by-layer, and solidifies until the part is complete.

Both Seo and Shim (2018) and Koike et al. (2018) used direct energy deposition AM to make a metal foam. They blended Ti6-Al4-V powder and titanium hydride (TiH_2) (20, 30, 40, and 50 wt.%), a physical blowing agent, to create voids on the submillimeter scale. Thermal decomposition of TiH_2 releasing hydrogen gas occurs above 500°C by the following reaction (Kennedy & Lopez, 2003; Rasooli, Boutorabi, Divandari, & Azarniya, 2013):

$$TiH_2(s) \underset{\Delta}{\longrightarrow} Ti(s) + H_2(g)$$

Macroscale voids were created by depositing the Ti alloy into a cross-hatched pattern with track spacings ranging from 0.5 mm up to 1.75 mm.

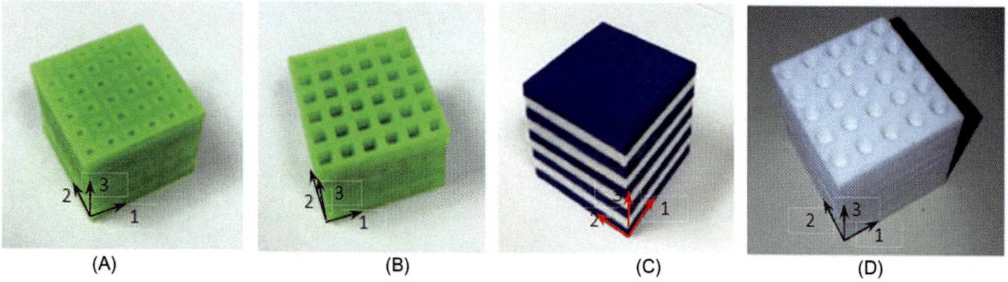

FIGURE 6–28 Porous PLA structure made by the sheet lamination AM process: (A) open cell-top view of cuboid voids, (B) open cell-bottom view with cuboid voids, (C) closed cell functionally graded PLA, and (D) functionally graded PLA with cylindrical voids (Zhang & Wang, 2017).

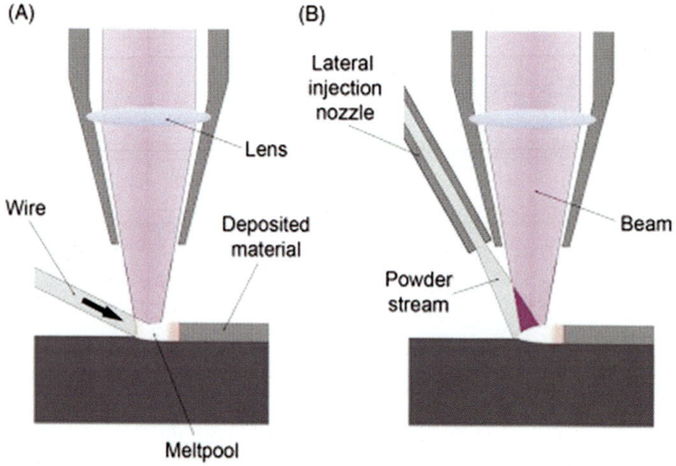

FIGURE 6–29 Schematic of an electron beam melting (EBM) apparatus, a type of directed energy deposition. The energy source can be an electron beam, plasma, or laser. The feedstock can be in the form of a (A) wire or (B) powder (Molitch-Hou, 2018).

The thermal decomposition and liberation of hydrogen gas occurs in the meltpool, see Fig. 6—29, initiating the blowing of the Ti alloy, see Fig. 6—30. Careful timing of the solidification of molten Ti alloy is needed to optimize the pore structure. To enhance the pore structure, it is important to cool the molten metal before the voids coalesce into larger, nonspherical cells or totally escape from the molten metal. Fig. 6—30C and F compares materials, each made with 30 wt.% TiH_2, but at 1100 and 900 W electron beam power, respectively. The sample process at 1100 W had a void volume content of 11% while the sample processed at 900 W had a higher void volume content at 22%. The difference in

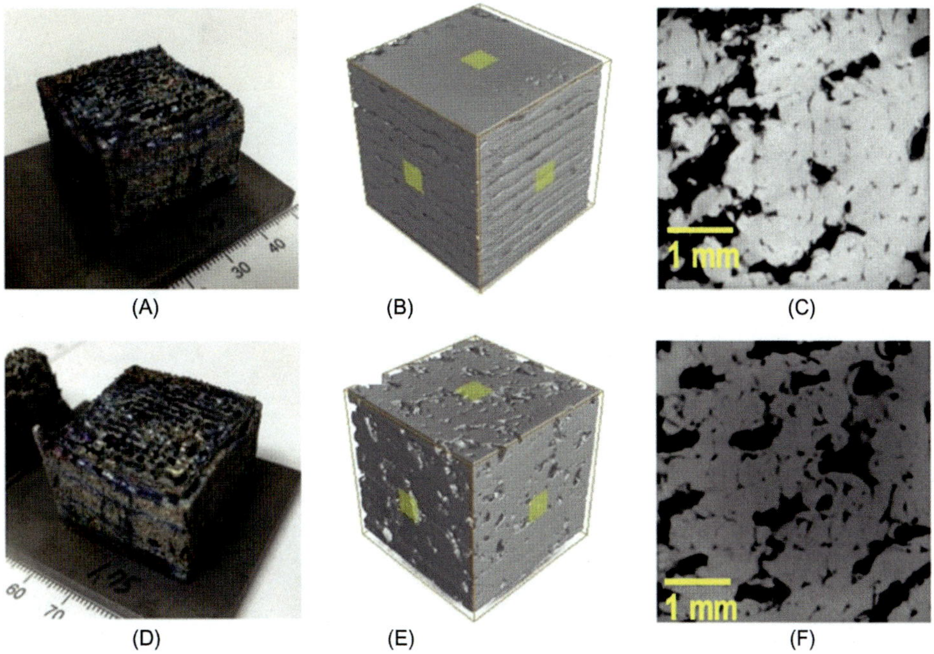

FIGURE 6–30 Ti6Al4V alloy foam fabricated via electron beam melting with a laser power of 1100 W with 30 wt.% TiH_2 foaming agent: (A) photograph of sample, (B) CT scan, and (C) micrograph. Ti6Al4V alloy foam fabricated via electron beam melting with a laser power of 900 W with 30 wt.% TiH_2 foaming agent: (D) photograph of sample, (E) CT scan, and (F) micrograph (Seo & Shim, 2018). Analysis of the images finds that the void volume of C and F is 11% and 22%, respectively, concluding that the lower the power the larger the void volume.

content was attributed to the longer time to solidify the Ti alloy processed at 1100 W leading to loss of the hydrogen gas from the molten metal.

6.7 Mechanical stretching

Mechanical stretching can be used to produce a unique pore structure. We provide an example from the field of lithium ion batteries. A separator in a lithium ion battery is a porous membrane that separates the anode and the cathode. It allows the flow of ionic charge carriers but prevents electrical contact between the electrodes (Arora & Zhang, 2004). All separators contain pores or voids. The process of making a porous separator involves melting a polyolefin resin, extruding it into a film, annealing, and subjecting to controlled tensile stretching to form submicrometer voids (~40 vol.%). An example of the microstructure of such separators made by Celgard is shown in Fig. 6–31. Note the slitlike form of voids. The mechanical properties of the separator are obviously anisotropic. Expanded metal sheet is made by a similar mechanical stretching process.

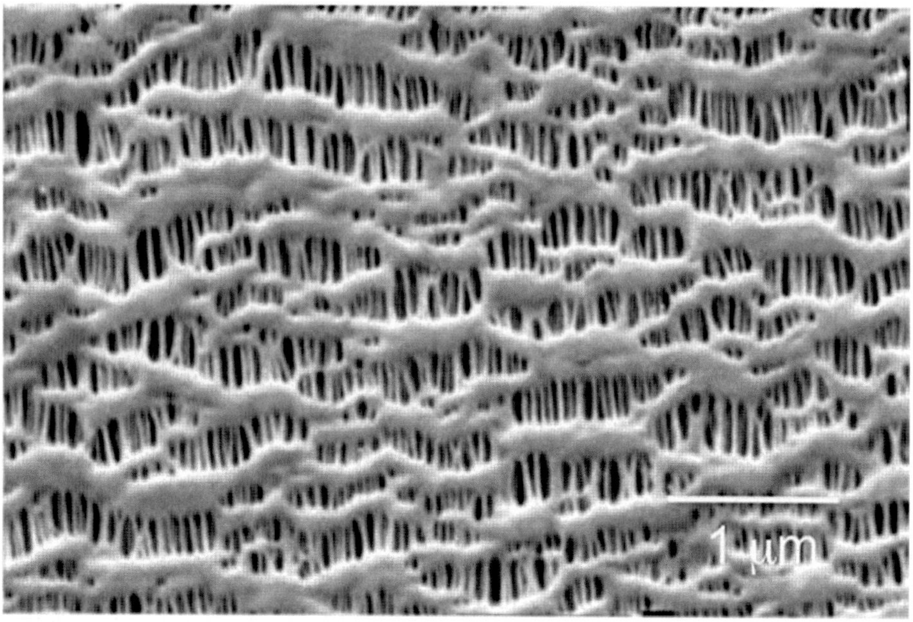

FIGURE 6–31 An example of the slitlike microstructure obtained by controlled tensile forces on a polyolefin polymer. This is a common membrane material used to separate the anode and the cathode in a lithium ion battery. SEM. *Courtesy of Celgard.*

6.8 Exploiting chemically selective weakness in solids

Just as this book views voids neutrally, not judging them as a positive or negative attribute, so does this method of introducing voids via exploiting weaknesses in the material. Morris and Čejka (2015) studied the inherent weakness in zeolite materials with the intent of exploiting the weakness to create new structures with tailored porosity. By not viewing traditional material weaknesses as a detriment, they developed ways to regioselectively target portions of the molecule to be removed through chemical, mechanical, or physical stimuli. These perceived weaknesses have now become pathways to create new structures with more functionality.

The first example is transforming a well-known zeolite into zeolites that have never been synthesized. Because zeolites have inherent porosity and this porosity is important to functionality, these methods of engineering voids are therefore of interest. There are many different structures and chemistries of zeolites both natural and synthetic, we will limit discussion in this section to synthetic. Zeolites are aluminosilicate and have a general chemical formula of

$$M_{2/n}OAl_2O_3 \cdot xSiO_2 \cdot yH_2O,$$

where

M is charge-balancing cation the nonframework;
n is the valence charge of the cation, *M*;
x is 2.0 or more; and
y is the moles of water in the voids.

The Al and Si tetrahedral atoms or T-atoms form a three dimensional framework of AlO_4 and SiO_4 tetrahedra linked together by shared oxygen ions.

The synthetic zeolites that are the starting point for the process, in and of themselves, are a very important class of aluminosilicate materials with porosity in the nanometer range that find applications in catalysts and adsorbents. Although computer simulations predict over 3 million hypothetical zeolite structures (Walker & Guiang, 2007), approximately 200 have been synthesized. The well known zeolite conundrum asks the question "why have scientist only been able to produce a small subset of predicted zeolite structures?" This structure modification via exploiting selective weaknesses is one approach to address the zeolite conundrum.

Typical zeolite pore sizes using oxygen packing models are shown in Fig. 6–32. They include small pore zeolites with eight-ring pores with free diameters of 0.30–0.45 nm, medium pore zeolites formed by a 10- ring, 0.45–0.60 nm in free

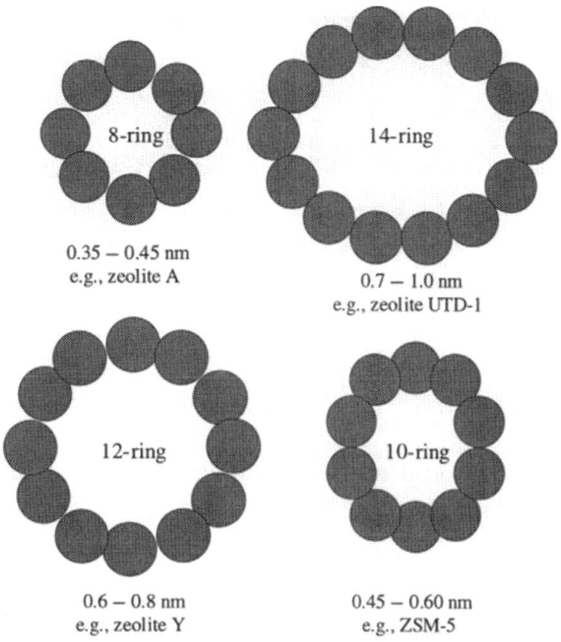

FIGURE 6–32 Typical zeolite pore size as a function of oxygen packing (Flanigen, 2001).

diameter; large pore zeolites with 12-ring pores, 0.6–0.8 nm; and extralarge pore zeolites with 14-ring pores (Flanigen, 2001).

One general method of creating tunable porosities by targeting regioselective weakness specifically in germanium (Ge) containing zeolites is called the assemble-disassemble-organization-reassembly (ADOR) method.

The naming convention and structures in zeolites are very complicated. The example we provide is simplified to highlight the general ADOR methodology. Interested readers should refer to the references for detailed information. The starting structure assembled in Fig. 6–33 (Eliášová et al., 2015) is the germanosilicate zeolite UTL. It was reported in 2004 as the first extralarge pore zeolite with intersecting 14- and 12-ring channels with pore diameters of 0.95 nm × 0.85 nm × 0.55 nm, respectively. However, it is

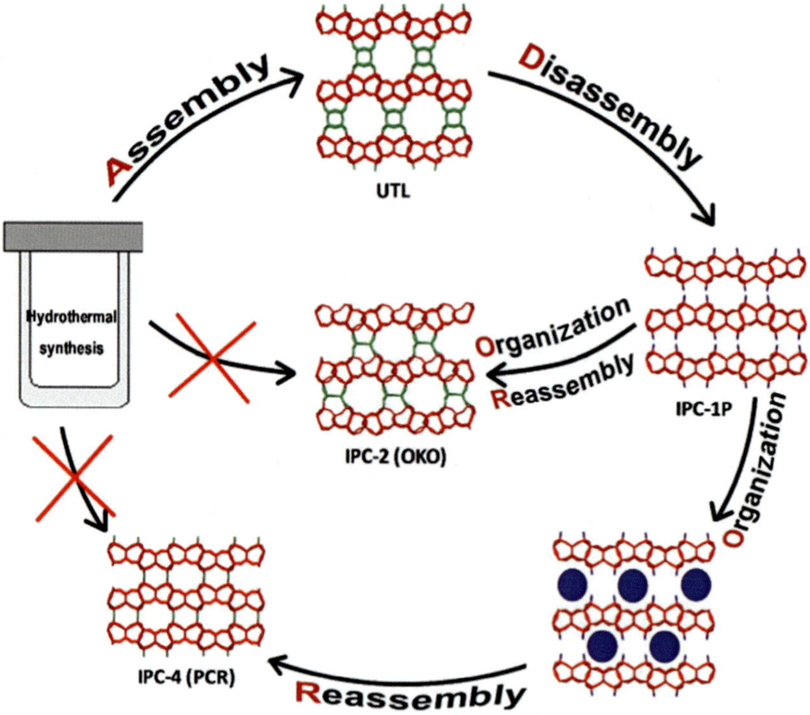

FIGURE 6–33 Schematic of the ADOR process highlighting the pathways to tuning pore size for creataing zeolite structures cannot be prepared via traditional synthesis methods. The assembly of UTL, known zeolite structure, is the first step, "A" in the ADOR process. UTL is dissembled by removing the interlayer linkages, shown in green, forming a structure called IPC-1P with silanol interlayer groups. The processes branch at the organization point depending on the size of porosity desired in the final structure (Morris & Čejka, 2015). The size of the porosity is related to the concentration of the acid used in the process.

unstable under certain reaction conditions, especially in the presence of water, where the framework will be irreversibly changed. The instability in the structure is the interlayer linkages highlighted in green in Fig. 6–33. These instabilities or weaknesses in the structure are the key to creating new zeolite structures the ability to tailor porosity characteristics.

During the disassembly step, these linkages are hydrolyzed, indicated in blue and the structure "unzips" in a dilute acid solution leaving the IPC-P1 structure. The process branches in the organization step of the ADOR process and the path taken depends on the porosity desired. Table 6–1 (Morris & Čejka, 2015) shows all of the various pore

Table 6–1 Overview of pore dimensions created when applying the ADOR method on the parent UTL zeolite.

	Channel size					
Structure type	**14-ring (nm)**	**12-ring (nm)**	**12-ring (nm)**	**10-ring (nm)**	**10-ring (nm)**	**8-ring (nm)**
UTL 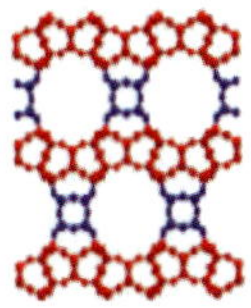	0.95 × 0.71	0.85 × 0.55	–	–	–	–
IPC-7 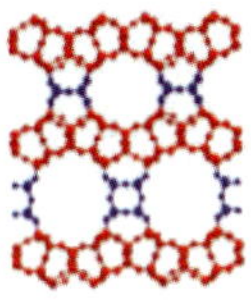	0.95 × 0.71	0.85 × 0.55	0.66 × 0.62	0.54 × 0.53	–	–
IPC-2 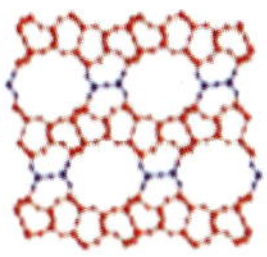	–	–	0.66 × 0.62	0.54 × 0.53	–	–
IPC-6 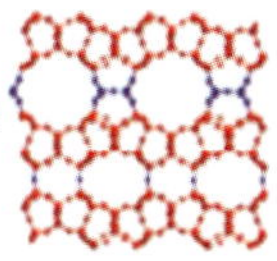	–	–	0.66 × 0.62	0.54 × 0.53	0.58 × 0.38	0.45 × 0.36
IPC-4 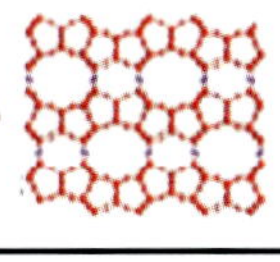	–	–	–	–	0.58 × 0.38	0.45 × 0.36

structures that have been produced from IPC-1P via the ADOR method. The two outcomes illustrated in Fig. 6–33 are achieved by different conditions in the organization and reassembly steps of the ADOR process. A low acidity condition favors the removal of residual species remaining between the IPC-1P layers (deintercalation route) and final reassembly into IPC-4 (PCR) with smaller pore sizes. High acidity favors the rearrangement and building of silica within the layers yielding the larger porosity of IPC-2 (OKO) on reassembly.

The first example illustrates the ADOR process, which uses chemical methods to attack selective weaknesses and to tailor porosities in Ge containing zeolites. The second example of using mechanical means to manipulate selective weakness uses hydrostatic pressure to engineer porosity. Typically, when applying hydrostatic pressure on a material, one anticipates reducing porosity as the material densifies. Lapidus, Halder, Chupas, and Chapman (2013) started with a nonporous double interpenetrating network of zinc cyanide, $Zn(CN)_2$ that is similar to a metal-organic framework. Structurally, it appears as two independent diamondoid frameworks interpenetrating in the overall cubic system (Pn-3m, $a = 0.59$ Å, "diac" topology).

Although the material is nominally dense, a specific gravity of 1.9, under hydrostatic pressure the $Zn(CN)_2$ framework will accommodate a pressure transmitting fluid. The density changes and the appearance of porosity are driven by an increase in overall atomic packing density leading to the creation of new Zn $(CN)_2$ polymorphs. These transitions are driven not only by the magnitude but also by the identity of the type of fluid molecules infiltrating into the pores of a metastable open framework, see Fig. 6–34. The porosity generated under pressure expands the framework instead of densifying as might be expected under pressure. Table 6–2 (Trousselet, Boutin, & Coudert, 2015) outlines the new polymorphs eventified by Lapedus et al. and Trousselet et al. used theoretical studies

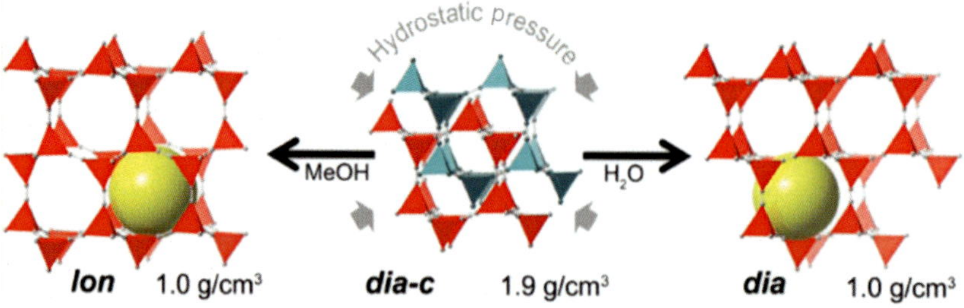

FIGURE 6–34 Schematic of the new metastable phases containing porosity identified by Lapedus et al. (Lapidus, Halder, Chupas, & Chapman, 2013). The initial non-porous dia-c zinc cyanide phase was altered using hydrostatic pressure. With methanol (MeOH) and water as the pressure transmitting fluid the zinc cyanide transformed to the porous lon and dia polymorphs, respectively.

Table 6–2 Theoretical density and porosity calculations (Trousselet et al., 2015) of the two polymorphs identified by Lapidus et al. (2013).

	$Zn(CN)_2$ polymorph		
	dia-c	**dia**	**Lon**
Pressure transmitting fluid		Methanol	Water
Nominal density, g/cm^3	1.2987	0.6495	0.6494
Framework density, g/cm^3	1.829	0.905	0.904
Porosity volume fraction	0.0	0.352	0.349

to quantify the amount of porosity of these new phases. Trousselet et al. (2015) showed the initial nonporous $Zn(CN)_2$ dia-c phase transformed to lio and dia phases with porosity volume fraction of 0.349 and 0.352, respectively. They concluded that these inefficiencies in space filling are likely to be a general characteristic of interpenetrated framework systems, and consequently, it may be possible to access a broad range of new porous framework materials in this way (Lapidus et al., 2013).

6.9 Hierarchical design with voids

It is fitting to end this chapter on intentional voids by discussing the concept of hierarchical design of materials using voids. The term hierarchical design, as it relates to materials science, is the control of material features simultaneously on multiple length scales. The hierarchical design using voids is often referred to as *hierarchical porosity*. These features impart particular functionalities at each length scale.

We focus this section on the control of voids and functionality at multiple length scales. We recognize that the field of hierarchical design can encompass other features as well. There have been many reviews on the characterization of natural materials and structure–functionality relationships; a review of all of these is beyond the scope of this book.

Nature has many examples of hierarchical materials, from lotus and rice leaves to bones. Inspired by the widespread examples in nature, researchers are studying the processes that are needed to create hierarchical materials as well as functionality–structure relationships. The lotus leaf shown in Fig. 6–35, which is found in swamps and shallow waters, has the ability to keep dirt and dust off of it. Dust particles are suspended by micrometer scale surface features, similar to inverted cones,

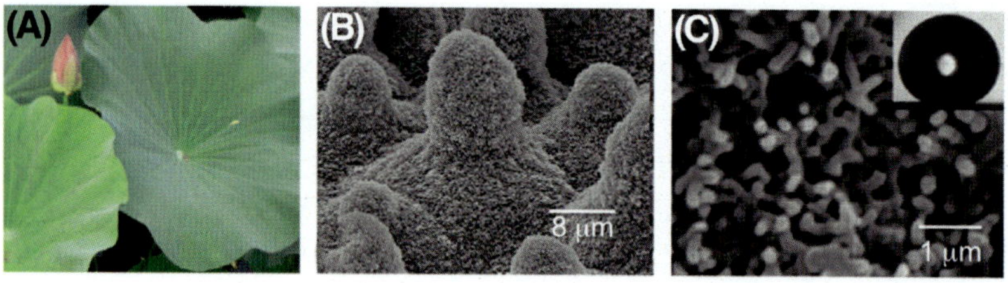

FIGURE 6-35 The macro- (Wang et al., 2011), (B) micro- (Yan, Gao, & Barthlott, 2011), and (C) nanometer scale (Guo & Liu, 2007) images of a lotus leaf. The structure on each level gives it the ability to clean dirt and dust from its surface. At the nanometer scale, the lotus leaf has a three-dimensional pore structure leading to a wetting angle of 161 degrees with water.

see Fig. 6–35B. Nanometer scale surface features prevent water from wetting the surface. This surface is superhydrophobic creating a high contact angle of 161 degrees with water. The water simply rolls down the surface of the leaf, carrying away any dust particle in its path. The nanometer scale structure of the lotus leaf in Fig. 6–35C is similar to a fibrous web discussed in Section 6.5, creating the superhydrophobic surface.

Synthetic bone and bone scaffolding are widely studied in materials. We would, however, like to highlight the role of voids as it relates to the design of bone scaffolds to illustrate the importance of multiscale material design at the interface of synthetic biomaterials and the regenerative nature of bone and tissue.

Bone is an example for the need of hierarchical pore structure. We try and mimic this pore structure in biomaterial bone scaffold on the macro-, micro-, and nanometer scale to facilitate bone regeneration. Each scale of porosity has essential interactions with the body that are required for successful tissue regeneration (Holzapfel et al., 2013). Extracellular matrix (ECM) is a term used to describe all of the supporting materials required to allow the cells to grow and function. Functions of ECM can be, for example, structural and biochemical. It is important to have interconnected porosity (open cell) larger than 100 μm (Hulbert et al., 1970) and, in certain cases, as large as 300 μm (Harvey et al., 1999). The maximum cell size on this level is usually controlled by mechanical properties. If cells size gets too large, the strength and fracture toughness are degraded to a point where the material is no longer suitable for scaffolding applications. Pores of <10 μm are important for intensifying adsorption of cell differentiation inducing factors and ion exchange (Holzapfel et al., 2013). In addition, an increase in surface area is needed for the proliferation and differentiation of anchorage-dependent cells for tissue regeneration. Nanoscale texture and surface features, such as porosity, facilitate interactions between host cells and the biomaterial. Surface features and properties determine the organization of adsorbed protein layers, which in turn determine specific cellular responses. Fig. 6–36 illustrates the functionality that the porosity on macro-, micro-, and nanometer scale facilitates.

In the previous sections of this chapter, we have described methods of sub-micrometer and micrometer void into materials. However, much of the work in hierarchical porosity field involves control of void structure at a nanometer scale. This multiscale control of pore structure has important applications in catalysis, biosensing, tailored mechanical properties, bioactivity, and even nanomedicine. Table 6–3 lists some of the important materials and functionality of materials.

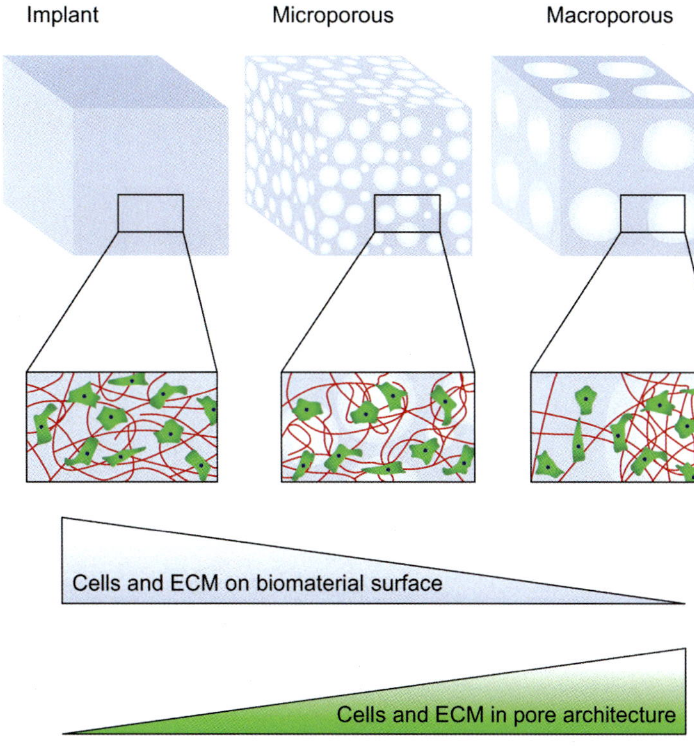

FIGURE 6-36 A schematic highlighting the importance of porosity to host cells and extracellular matrix (ECM) on three length scales. The interconnectivity on a macroporous level has a relatively low surface area but allows for many cells/ECM within the pore and ingrowth between pores. On the micrometer scale, there is a higher surface area which allows for a high number of cell attachment points. On the implant level, where there are nanometer scale texturing and surface features, there are relatively few cells and ECM within a surface feature but a very large surface area for providing sites for attachment (Holzapfel et al., 2013).

Table 6–3 Examples of materials that have been used for hierarchical porosity. Also listed are the size of controlled porosity and potential application.

Material	Void size			Application	Reference
	Small	Medium	Large		
SiO_2	<1 μm	10 μm	50 μm	Biomaterials, catalyst support	Mao et al. (2008)
Au	25–100 nm	1–60 μm	350 μm	Catalysis, biosensing	Cox and Dunand (2011)
TiO_2	6 nm		200 nm	Molecule detection, catalysis, dye-sensitized solar cells, nanobiotechnology, and nanomedicine	Henrist et al. (2013)
Al_2O_3–$Al_2(OH)_3$–Chitosan	2–50 nm	<500 μm	>500 μm	Biomaterials, filter, catalyst support	Salomão and Brandi (2013)
Alkali-bonded ceramics	0.1–0.3 μm	1–2 μm	200–1000 μm	Thermal insulators, catalysts, filters	Landi et al. (2013)
SiO_2 nanoparticles		50–200 nm	2–10 μm	Abortion, separations	Lind, Du Fresne Von Hohenesche, Smått, Lindén, and Unger (2003)
Carbon	4 nm		450 nm	Gas separation, catalyst supports, electrode materials, water purification, gas storage	Yang and Wang (2009)

References

Arora, P., & Zhang, Z. (2004). Battery separators. *Chemical Reviews, 104*(10), 4419–4462.

ASTM International. (2015). *ASTM52900-15 standard terminology for additive manufacturing—General principles—Terminology*. West Conshohocken, PA: ASTM International.

Aydogmus, T., & Bor, S. (2012). Superelasticity and compression behavior of porous TiNi alloys produced using Mg spacers. *Journal of the Mechanical Behavior of Biomedical Materials, 15*, 59–69.

Babcsán, N., Banhart, J., & Leitlmeier, D. (2003). Metal foams—manufacture and physics of foaming. Paper presented at the *Proceedings of the International Conference Advanced Metallic Materials*, 5–15.

Banhart, J. (2001). Manufacture, characterisation and application of cellular metals and metal foams. *Progress in Materials Science, 46*(6), 559–632.

Beachley, V., & Wen, X. (2010). Polymer nanofibrous structures: fabrication, biofunctionalization, and cell interactions. *Progress in Polymer Science, 35*(7), 868–892.

Bekoz, N., & Oktay, E. (2012). Effects of carbamide shape and content on processing and properties of steel foams. *Journal of Materials Processing Technology, 212*(10), 2109–2116.

Brothers, A. H., & Dunand, D. C. (2005). Plasticity and damage in cellular amorphous metals. *Acta Materialia, 53*(16), 4427–4440.

Brothers, A. H., & Dunand, D. C. (2006). Amorphous metal foams. *Scripta Materialia, 54*(4), 513–520.

Brothers, A. H., Scheunemann, R., Defouw, J. D., & Dunand, D. C. (2005). Processing and structure of open celled amorphous metal foams. *Scripta Materialia, 52*(4), 335–339.

Colombo, P. (2008). Engineering porosity in polymer-derived ceramics. *Journal of the European Ceramic Society, 28*(7), 1389–1395.

Cox, M. E., & Dunand, D. C. (2011). Bulk gold with hierarchical macro-, micro- and nano-porosity. *Materials Science and Engineering: A, 528*(6), 2401–2406.

Du, H., Song, G., Nakajima, H., Zhao, Y., Xiao, J., & Xiong, T. (2013). Study on lotus-type porous copper electroplated with a Ni coating on inner surface of pores. *Applied Surface Science, 264*, 772–778.

Du, Y. Q., Qiao, G. Y., Yin, Y. H., Li, T., Tong, H. Y., & Yu, X. G. (2020). Usefulness of 3D printed models in the management of complex craniovertebral junction anomalies: choice of treatment strategy, design of screw trajectory, and protection of vertebral artery. *World Neurosurgery, 133*, e722–e729.

Eliášová, P., Opanasenko, M., Wheatley, P. S., Shamzhy, M., Mazur, M., Nachtigall, P., ... Čejka, J. (2015). The ADOR mechanism for the synthesis of new zeolites. *Chemical Society Reviews, 44*(20), 7177–7206.

Elsayed, H., Colombo, P., & Bernardo, E. (2017). Direct ink writing of wollastonite-diopside glass-ceramic scaffolds from a silicone resin and engineered fillers. *Journal of the European Ceramic Society, 37*(13), 4187–4195.

Flanigen, E. M. (2001). Chapter 2 zeolites and molecular sieves: An historical perspective. In H. van Bekkum, E. M. Flanigen, P. A. Jacobs, & J. C. Jansen (Eds.), *Studies in Surface Science and Catalysis*. Elsevier. Available from: https://doi.org/10.1016/S0167-2991(01)80243-3.

Fleisher, A., Zolotaryov, D., Kovalevsky, A., Muller, G., Eshed, E., Kazakin, M., & Popov Jr, V. (2019). Reaction bonding of silicon carbides by binder jet 3D-printing, phenolic resin binder impregnation and capillary liquid silicon infiltration. *Ceramics International.*

Fricke, J., & Tillotson, T. (1997). Aerogels: Production, characterization, and applications. *Thin Solid Films, 297*(1–2), 212–223.

Gibson, L. J., & Ashby, M. F. (1999). *Cellular Solids: Structure and Properties.* Cambridge University Press.

Gladysz, G. M., & Chawla, K. K. (2002). Composite foams. *In Encyclopedia of Polymer Science and Technology.* New York: John Wiley & Sons, Inc.

Guo, Z., & Liu, W. (2007). Biomimic from the superhydrophobic plant leaves in nature: Binary structure and unitary structure. *Plant Science, 172*(6), 1103–1112.

Harvey, E., Bobyn, J., Tanzer, M., Stackpool, G., Krygier, J., & Hacking, S. (1999). Effect of flexibility of the femoral stem on bone-remodeling and fixation of the stem in a canine total hip arthroplasty model without cement. *Journal of Bone and Joint Surgery, 81*(1), 93–107.

Heinl, P., Müller, L., Körner, C., Singer, R. F., & Müller, F. A. (2008). Cellular Ti–6Al–4V structures with interconnected macro porosity for bone implants fabricated by selective electron beam melting. *Acta Materialia, 4*, 1536–1544. Available from: https://doi.org/10.1016/j.actbio.2008.03.013.

Henrist, C., Dewalque, J., Cloots, R., Vertruyen, B., Jonlet, J., & Colson, P. (2013). Hierarchical porous TiO_2 thin films by soft and dual templating: a quantitative approach of specific surface and porosity. *Thin Solid Films, 539*, 188–193.

Holzapfel, B. M., Reichert, J. C., Schantz, J., Gbureck, U., Rackwitz, L., Nöth, U., . . . Hutmacher, D. W. (2013). How smart do biomaterials need to be? A translational science and clinical point of view. *Advanced Drug Delivery Reviews, 65*(4), 581–603. Available from: https://doi.org/10.1016/j.addr.2012.07.009.

Hulbert, S., Young, F., Mathews, R., Klawitter, J., Talbert, C., & Stelling, F. (1970). Potential of ceramic materials as permanently implantable skeletal prostheses. *Journal of Biomedical Materials Research, 4*(3), 433–456.

Hyun, S., Ikeda, T., & Nakajima, H. (2004). Fabrication of lotus-type porous iron and its mechanical properties. *Science and Technology of Advanced Materials, 5*(1–2), 201–205.

Hyun, S., & Nakajima, H. (2002). Fabrication of lotus-structured porous iron by unidirectional solidification under nitrogen gas. *Advanced Engineering Materials, 4*(10), 741–744.

Irausquín, I., Pérez-Castellanos, J. L., Miranda, V., & Teixeira-Dias, F. (2013). Evaluation of the effect of the strain rate on the compressive response of a closed cell aluminium foam using the split Hopkinson pressure bar test. *Materials and Design, 47*, 698–705.

Jha, N., Mondal, D. P., Dutta Majumdar, J., Badkul, A., Jha, A. K., & Khare, A. K. (2013). Highly porous open cell Ti-foam using NaCl as temporary space holder through powder metallurgy route. *Materials and Design, 47*, 810–819.

Kashihara, M., Hyun, S. K., Yonetani, H., Kobi, T., & Nakajima, H. (2006). Fabrication of lotus-type porous carbon steel by unidirectional solidification in nitrogen atmosphere. *Scripta Materialia, 54*(4), 509–512.

Kennedy, A. R., & Lopez, V. H. (2003). The decomposition behavior of as-received and oxidized TiH_2 foaming-agent powder. *Materials Science and Engineering A, 357*(1), 258–263. Available from: https://doi.org/10.1016/S0921-5093(03)00211-9.

Koike, R., Matsumoto, T., Kakinuma, Y., Aoyama, T., Oda, Y., Kuriya, T., & Kondo, M. (2018). A basic study on metal foam fabrication with titanium hydride in direct energy deposition. *Procedia Manufacturing, 18*, 68–73. Available from: https://doi.org/10.1016/j.promfg.2018.11.009.

Kondo, S., & Miura, T. (2010). Reaction-diffusion model as a framework for understanding biological pattern formation. *Science (New York, N.Y.), 329*(5999), 1616–1620. Available from: https://doi.org/10.1126/science.1179047.

Kujime, T., Tane, M., Hyun, S. K., & Nakajima, H. (2007). Three-dimensional image-based modeling of lotus-type porous carbon steel and simulation of its mechanical behavior by finite element method. *Materials Science and Engineering: A, 460–461*, 220–226.

Landi, E., Medri, V., Papa, E., Dedecek, J., Klein, P., Benito, P., & Vaccari, A. (2013). Alkali-bonded ceramics with hierarchical tailored porosity. *Applied Clay Science, 73*, 56–64.

Lapidus, S. H., Halder, G. J., Chupas, P. J., & Chapman, K. W. (2013). Exploiting high pressures to generate porosity, polymorphism, and lattice expansion in the nonporous molecular framework zn (CN) 2. *Journal of the American Chemical Society, 135*(20), 7621–7628.

Lee, J., Hyun, S., Kim, M., Kim, M., Ide, T., & Nakajima, H. (2012). Elevated temperature compression behaviors of lotus-type porous NiAl. *Intermetallics, 29*, 27–34.

Lim, K. S., & Barigou, M. (2004). X-ray micro-computed tomography of cellular food products. *Food Research International, 37*(10), 1001–1012.

Lind, A., Du Fresne Von Hohenesche, C., Smått, J., Lindén, M., & Unger, K. K. (2003). Spherical silica agglomerates possessing hierarchical porosity prepared by spray drying of MCM-41 and MCM-48 nanospheres. *Microporous and Mesoporous Materials, 66*(2–3), 219–227.

Lindahl, J., Hershey, C., Gladysz, G., Mishra, V., Shah, K., & Kunc, V. (2019). Extrusion deposition additive manufacturing utilizing high glass transition temperature latent cured epoxy systems, United States. https://doi.org/10.33599/nasampe/s.19.1615.

Liu, X., Liu, X., & Xie, J. (2012). The investigation of fabrication processing for lotus-type porous magnesium by the in-situ reaction and unidirectional solidification method. *Procedia Engineering, 36*, 270–278.

Liu, X., Li, X., Jiang, Y., & Xie, J. (2012). Effect of casting temperature on porous structure of lotus-type porous copper. *Procedia Engineering, 27*, 490–501.

Maca, K., Dobsak, P., & Boccaccini, A. R. (2001). Fabrication of graded porous ceramics using alumina carbon powder mixtures. *Ceramics International, 27*(5), 577–584.

Manonukul, A., Muenya, N., Léaux, F., & Amaranan, S. (2010). Effects of replacing metal powder with powder space holder on metal foam produced by metal injection moulding. *Journal of Materials Processing Technology, 210*(3), 529–535.

Mao, X., Wang, S., & Shimai, S. (2008). Porous ceramics with tri-modal pores prepared by foaming and starch consolidation. *Ceramics International, 34*(1), 107–112.

Mills, N. (2007). *Polymer foams handbook: Engineering and biomechanics applications and design guide.* Butterworth-Heinemann.

Minami, H., Kobayashi, H., & Okubo, M. (2005). Preparation of hollow polymer particles with a single hole in the shell by SaPSeP. *Langmuir, 21*(13), 5655–5658.

Molitch-Hou, M. (2018). Overview of additive manufacturing process. *In Additive manufacturing* (pp. 1–38). Elsevier.

Morris, R. E., & Čejka, J. (2015). Exploiting chemically selective weakness in solids as a route to new porous materials. *Nature Chemistry, 7*(5), 381.

Myers, K., Cortes, P., Conner, B., Wagner, T., Hetzel, B., & Peters, K. (2015). Structure property relationship of metal matrix syntactic foams manufactured by a binder jet printing process. *Additive Manufacturing, 5*, 54–59.

Nakajima, H., Hyun, S. K., Ohashi, K., Ota, K., & Murakami, K. (2001). Fabrication of porous copper by unidirectional solidification under hydrogen and its properties. *Colloids and Surfaces A: Physicochemical and Engineering Aspects, 179*(2–3), 209–214.

Rasooli, A., Boutorabi, M., Divandari, M., & Azarniya, A. (2013). Effect of high heating rate on thermal decomposition behaviour of titanium hydride (TiH_2) powder in air. *Bulletin of Materials Science, 36*(2), 301–309.

Rietzel, D., Friedrich, M., & Osswald, T. A. (2017). Additive manufacturing. In T. A. Osswald (Ed.), *Understanding polymer processing (second edition)* (2nd ed., pp. 147–169). Munich: Hanser. Available from: https://doi.org/10.3139/9781569906484.007.

Salomão, R., & Brandi, J. (2013). Macrostructures with hierarchical porosity produced from alumina–aluminum hydroxide–chitosan wet-spun fibers. *Ceramics International, 39*(7), 8227–8235.

Santoliquido, O., Colombo, P., & Ortona, A. (2019). Additive manufacturing of ceramic components by digital light processing: A comparison between the "bottom-up" and the "top-down" approaches. *Journal of the European Ceramic Society, 39*(6), 2140–2148.

Schlögl, S., Reischl, M., Ribitsch, V., & Kern, W. (2012). UV induced microcellular foaming—A new approach towards the production of 3D structures in offset printing techniques. *Progress in Organic Coatings, 73*(1), 54–61.

Seo, J., & Shim, D. (2018). Effect of track spacing on porosity of metallic foam fabricated by laser melting deposition of Ti6Al4V/TiH_2 powder mixture. *Vacuum, 154*, 200–207.

Shapovalov, V. (1998). Formation of ordered gas-solid structures via solidification in metal-hydrogen systems. *In MRS Proceedings 1998* (p. 281) Cambridge: Cambridge University Press.

Shapovalov, V.I., & Serdyuk N. (1980) *Method of metal foam manufacturing*. USSR Patent No. 1725485.

Smith, B. H., Szyniszewski, S., Hajjar, J. F., Schafer, B. W., & Arwade, S. R. (2012). Steel foam for structures: A review of applications, manufacturing and material properties. *Journal of Constructional Steel Research, 71*, 1–10.

Soleimani Dorcheh, A., & Abbasi, M. H. (2008). Silica aerogel: synthesis, properties and characterization. *Journal of Materials Processing Technology, 199*(1–3), 10–26.

Sun, D. X., & Zhao, Y. Y. (2005). Simulation of thermal diffusivity of Al/NaCl powder compacts in producing Al foams by the sintering and dissolution process. *Journal of Materials Processing Technology, 169*(1), 83–88.

Tang, B., Tang, Y., Zhou, R., Lu, L., Liu, B., & Qu, X. (2011). Low temperature solid-phase sintering of sintered metal fibrous media with high specific surface area. *Transactions of Nonferrous Metals Society of China, 21*(8), 1755–1760.

Tappan, B. C., Huynh, M. H., Hiskey, M. A., Chavez, D. E., Luther, E. P., Mang, J. T., & Son, S. F. (2006). Ultralow-density nanostructured metal foams: combustion synthesis, morphology, and composition. *Journal of the American Chemical Society, 128*(20), 6589–6594.

Torres, Y., Pavón, J. J., & Rodríguez, J. A. (2012). Processing and characterization of porous titanium for implants by using NaCl as space holder. *Journal of Materials Processing Technology, 212*(5), 1061–1069.

Trousselet, F., Boutin, A., & Coudert, F. (2015). Novel porous polymorphs of zinc cyanide with rich thermal and mechanical behavior. *Chemistry of Materials, 27*(12), 4422–4430.

Turing, A. (1952). The chemical basis of morphogenesis. *Philosophical Transaction of the Royal Society of London B, 237*, 37.

Velasco, M. A., Lancheros, Y., & Garzón-Alvarado, D. A. (2016). Geometric and mechanical properties evaluation of scaffolds for bone tissue applications designing by a reaction-diffusion models and manufactured with a material jetting system. *Journal of Computational Design and Engineering, 3*(4), 385–397. Available from: https://doi.org/10.1016/j.jcde.2016.06.006.

Wadley, H. N. (2002). Cellular metals manufacturing. *Advanced Engineering Materials, 4*(10), 726–733.

Walker, E., & Guiang, C. (2007). Challenges in executing large parameter sweep studies across widely distributed computing environments. Paper presented at the *Proceedings of the 5th IEEE Workshop on Challenges of Large Applications in Distributed Environments* (pp. 11–18).

Wang, B., Matsumaru, K., Yang, J., Fu, Z., & Ishizaki, K. (2012). Mechanical behavior of cellular borosilicate glass with pressurized Ar-filled closed pores. *Acta Materialia, 60*(10), 4185–4193.

Wang, X., Ding, B., Yu, J., & Wang, M. (2011). Engineering biomimetic superhydrophobic surfaces of electrospun nanomaterials. *Nano Today, 6*(5), 510–530.

Xie, G., Fukuhara, M., Louzguine-Luzgin, D. V., & Inoue, A. (2010). Ultrasonic characteristics of porous Zr55Cu30Al10Ni5 bulk metallic glass fabricated by spark plasma sintering. *Intermetallics, 18*(10), 2014–2018.

Yan, Y., Gao, N., & Barthlott, W. (2011). Mimicking natural superhydrophobic surfaces and grasping the wetting process: A review on recent progress in preparing superhydrophobic surfaces. *Advances in Colloid and Interface Science, 169*(2), 80–105.

Yang, M., & Wang, G. (2009). Synthesis of hierarchical porous carbon particles by hollow polymer microsphere template. *Colloids and Surfaces A: Physicochemical and Engineering Aspects, 345* (1–3), 121–126.

Ye, B., & Dunand, D. C. (2010). Titanium foams produced by solid-state replication of NaCl powders. *Materials Science and Engineering: A, 528*(2), 691–697.

Zhang, Y., & Wang, J. (2017). Fabrication of functionally graded porous polymer structures using thermal bonding lamination techniques. *Procedia Manufacturing, 10*, 866–875. Available from: https://doi.org/10.1016/j.promfg.2017.07.073.

Zhao, Y., Han, F., & Fung, T. (2004). Optimisation of compaction and liquid-state sintering in sintering and dissolution process for manufacturing Al foams. *Materials Science and Engineering: A, 364* (1–2), 117–125.

Zou, R., Xia, Y., Liu, S., Hu, P., Hou, W., Hu, Q., & Shan, C. (2016). Isotropic and anisotropic elasticity and yielding of 3D printed material. *Composites B, 99*(2016), 506–513.

Further reading

Aguado, J., Serrano, D. P., & Rodríguez, J. M. (2008). Zeolite beta with hierarchical porosity prepared from organofunctionalized seeds. *Microporous and Mesoporous Materials, 115*(3), 504–513.

Bose, S., Roy, M., & Bandyopadhyay, A. (2012). Recent advances in bone tissue engineering scaffolds. *Trends in Biotechnology, 30*(10), 546–554.

Ortona, A., D'angelo, C., Gianella, S., & Gaia, D. (2012). Cellular ceramics produced by rapid prototyping and replication. *Materials Letters, 80*, 95–98.

Ryan, G. E., Pandit, A. S., & Apatsidis, D. P. (2008). Porous titanium scaffolds fabricated using a rapid prototyping and powder metallurgy technique. *Biomaterials, 29*(27), 3625–3635.

Wadley, H. N. G., Fleck, N. A., & Evans, A. G. (2003). Fabrication and structural performance of periodic cellular metal sandwich structures. *Composites Science and Technology, 63*(16), 2331–2343.

7 Techniques of introducing intentional voids into particles and fibers

7.1 Introduction

In this chapter we highlight two very important forms of materials, namely, hollow and porous particles and fibers and the techniques to make them. These two forms of materials find applications in biomaterials, drug release, thermal insulations, dielectrics, catalysis, molecular separations, lightweight structural reinforcements, etc. Many of the fabrication techniques are quite versatile in that we can use a single technique to make structures and voids over several length scales and geometries. We give examples of particles and fibers made from polymers, metals, and ceramics.

The field of nanomaterials has expanded rapidly since the last quarter of the 20th century and the variety of hollow nanostructures available is quite remarkable. However, there is a large amount of micrometer scale hollow particles and fibers available. In nanometer scale polymer particles, we commonly use the term hollow polymeric nanostructures. There is additional information about nanometer scale voids in Chapter 4, Nanometer scale porous structures and Chapter 5, Hollow and porous structures utilizing the Kirkendall effect. This refers to not only spherical and cylindrical structures but also other shapes, such as cuboids and ellipsoids. Furthermore, we include in this chapter:

1. length scales from 10^{-9} to 10^{-3} m,
2. material chemistries of polymers, metals, ceramics, and composites,
3. porous as well as hollow structures.

7.2 Hollow and porous particles

7.2.1 Introduction

We define a *hollow particle* as a particle containing a single, centralized void with a dense and continuous shell. A *porous particle*, on the other hand, has distributed voids throughout the entire volume of the particle. One can also have a *hollow, porous particle*, as well, which has a centralized void but the shell is not fully dense. In other words, the shell that surrounds the centralized void has a structure similar to a foam.

Voids in Materials. DOI: https://doi.org/10.1016/B978-0-12-819282-5.00007-7

These particles have interesting characteristics such as low density, tailored pore size/distribution, high surface area, low electrical and thermal conductivity, etc. They can be used both in a loose powder form and as a dispersed phase in a matrix material to form, for example, a composite foam. Examples of applications of loose powders are drug delivery and packed beds for thermal insulation, catalysis, distillation, etc. A common use of hollow particles is in syntactic foams. They are a type of composite material where hollow, more or less spherical, particles are dispersed in a matrix material.

7.2.2 Processing of porous particles

Porous particles can be manufactured from polymers, metals, and ceramics. Two common methods for making porous particles are *spray drying* and *heterogeneous polymerization*. Both methods can give control over the dimensions of pores as small as a nanometer.

Spray drying is a very common method of manufacturing particles in the food and pharmaceutical industries. The process involves atomizing a liquid stream containing the solute or suspended particles in a solvent, removing the solvent by evaporation, and collecting the remaining solid. The process of atomizing involves passage of the liquid through a *spray nozzle* or *atomizer*, which produces droplets. The residence time required to remove the solvent is a function of the solvent properties, droplet size, inlet gas temperature, and size of the spray dryer (Shabde et al., 2005). By adjusting airflow rates and manipulating airflow patterns within the spray dryer, one can vary the residence time of a droplet in the particle forming chamber. An important goal of spray drying is to ensure all solvent is removed from the droplet and the powder is dry when it exits. Fig. 7–1 shows a schematic of the spray drying process (Shabde et al., 2005). Spray drying typically can yield hollow/porous particles of size ranging from hundreds of nanometers to hundreds of micrometers. The particle size is controlled by the pressure, spray nozzle, and gas expansion (i.e., the volume increase) if a latent gas blowing agent is needed to create the hollow or porous particle. The choice of porogen material used to create the cell structure in the particle must be done with due consideration of the processing conditions in the spray dryer. It is common to use a latent gas material as a porogen because it decomposes at a specific temperature in the spraying process. Ammonium bicarbonate, $(NH)_4HCO_3$, is an example of a chemical blowing agent. It decomposes at 35°C (Speight, 2005), releasing CO_2 and NH_3 gases. Concomitant with the release of gas(es), it is important that the solid (continuous) phase in the droplet has the structural integrity to contain the gas thus creating the proper pore structure and particle density.

The porosity in the particle can result from addition of latent gas (Steckel & Brandes, 2004) or sacrificial pore forming material (Iskandar et al., 2009) into the liquid phase. A hierarchical pore structure can also be achieved by spray drying. A

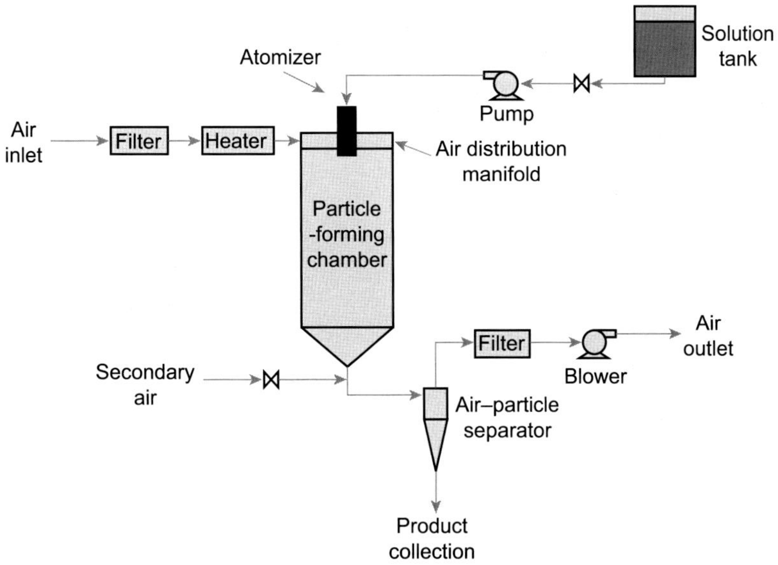

FIGURE 7–1 A schematic of the spray drying process. The essential parts include an atomizer, drying or particle forming chamber, and collection system. Porous and hollow particles are commonly made via spray drying. *Source: Adapted from Shabde V.S.; Emets S.V.; Mann U.; Hoo K.A.; Carlson N.N.; Gladysz G.M. (2005) Modeling a hollow micro-particle production process. Computers Chemical Engineering 29 (11–12) 2420–2428.*

spray dried, micrometer-scale hollow particle can have nanometer porosity/texturing and a central micrometer void (Okuyama, Abdullah, Lenggoro, & Iskandar, 2006).

Heterogeneous polymerization involves using two or more immiscible liquids stabilized into an emulsion and polymerizing dispersed droplets into a porous particle. When two immiscible liquids are combined, they form separate phases similar to oil in water. Phase separation of the liquids and subsequent stabilization of the dispersed droplets into an emulsion are the key steps as they determine the final particle size. Typically, the monomer is the nonaqueous phase and is subsequently polymerized. For more details on heterogeneous polymerization, one can consult Gokmen and Du Prez (2012).

The range of particle sizes produced by heterogeneous polymerization is very large, from several nanometers to several millimeters in diameter. Particle size distribution is generally very small when compared to a large industrial process like spray drying. The pore size can also vary over a wide range depending on the technique, from nanometers to hundreds of micrometers.

The pore formation can also result from the addition of porogens to the emulsion. Porogens can be solids, liquids, or latent gas materials. The mechanisms involved in forming voids in a dispersed phase are similar to the ones discussed in Chapter 6, Techniques for introducing intentional voids into materials, when discussing bulk materials. In other words, a solid *porogen* can also be thought of as a *placeholder*, to be removed once the polymerization is complete. Using a latent gas porogen, one of the reactants in the emulsion is activated, thus releasing a gaseous phase as a product, that is, a chemical blowing agent. Fig. 7–2 (Oh et al., 2011) shows three micrographs of particles produced with

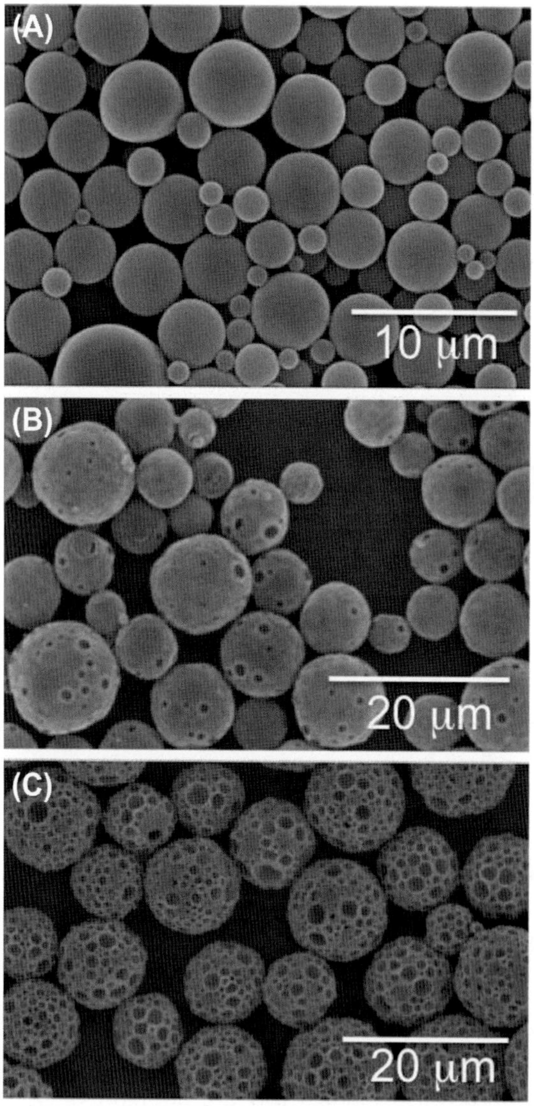

FIGURE 7–2 A spray-dried porous particle made from a water–oil–water emulsion. Ammonium bicarbonate loadings, by weight, of (A) 0%, (B) 1.5%, and (C) 3.0% were used as the chemical blowing agents SEM. *Source: Oh Y.J., Lee J., Seo J.Y., Rhim T., Kim S., Yoon H.J. and Lee K.Y., Preparation of budesonide-loaded porous PLGA microparticles and their therapeutic efficacy in a murine asthma model, Journal of Controlled Release 150 (1), 2011, 56–62.*

increasing amounts of a chemical blowing agent (0.0%, 1.5%, and 3.0% by weight) made with water–oil–water double emulsions method, described below. The blowing agent or porogen forming the voids in this case is ammonium bicarbonate, which, upon decomposition releases CO_2 and NH_3 gases.

We discussed in detail of solid (placeholder) and gas phase (gasar foams) porogens in Chapter 6, Techniques for introducing intentional voids into materials. Here we would like to point out more details of the liquid phase porogens as they relate to heterogeneous polymerization. They are classified into five types (Gokmen & Du Prez, 2012):

1. *υ-induced syneresis*—The "*υ*" refers to the crosslink density of the polymer network, which is an important characteristic that the resultant pore structure depends on. The method is defined by the use of excess solvent that serves as a diluent and swells the polymer phase as it polymerizes to a point where gelling occurs. Once the gel point is reached, phase separation follows, which in turn leads to the formation of porous particles.
2. *χ-induced syneresis*—The "χ" refers to the polymer–solvent interaction parameter and is defined by the use of a poor solvent, that is, one that will not cause swelling of the polymer. Voids form because of the *incompatibility* between the polymer network and the diluent molecules. In this case, phase separation occurs before the polymer gel point, forming discrete monomer droplets. These droplets then grow and polymerize in the discontinuous phase.
3. *Linear polymer*—The mechanism in this case is the same as the χ-induced syneresis except that a linear polymer is used instead of a solvent.
4. *Water*—This type has a double emulsion consisting of a water–oil–water. With the addition of an oil soluble surfactant, the water is taken into the monomer from the water phase of the emulsion. Stable droplets of water then form in the monomer. Once polymerized, the water is removed leaving voids.
5. High internal phase emulsion (HIPE)—This process is also called polyHIPE, when used in conjunction with polymerization. A HIPE forms when an aqueous phase is added to a continuous organic phase. The addition is done in a dropwise manner to at least 74% by volume or greater while mixing vigorously. With a polyHIPE, the external phase is polymerized and the dispersed, aqueous phase is removed forming a porous material, see Fig. 7–3. A surfactant is normally added to the continuous polymer phase to stabilize the bubble formation. This process can be used to make particles and bulk foams.

7.2.3 Hollow particles

Research on the use of hollow particles started in the 1950s and 1960s with the development of largescale processes to manufacture hollow particles. The most common hollow particle is a glass microballoon (GMB) or hollow glass microsphere (HGMS).

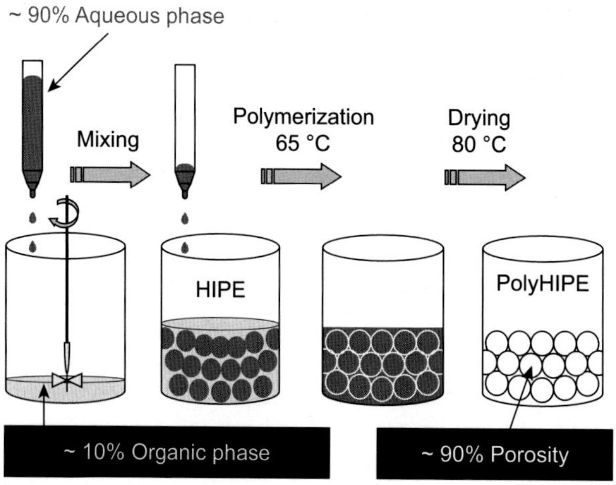

FIGURE 7–3 Schematic of the polyHIPE (high internal phase emulsion) process. *Source: Adapted from Silverstein M.S. (2014) PolyHIPEs: Recent advances in emulsion-templated porous polymers. Progress in Polymer Science 39 (1) 199–234.*

The process of making these particles consists of spray drying an aqueous solution of sodium silicate and a latent gas; the details of the composition are generally proprietary. The spray drying step consists of atomizing the aqueous solution into discrete droplets followed by evaporation of water, see Fig. 7–1. The atomized liquid droplets leave the outlet of the spray dryer as individual particles of water glass and latent gas. Water glass is the term used for a water soluble form of sodium silicate. A subsequent heating processes both expands the particle forming the hollow center and fuses the glass.

There are many types of hollow spheres (polymers, metals, and ceramics) that are available commercially for a wide variety of applications. Spray drying is the most economical way to manufacture hollow particles (Wang, Lu, & Sun, 2007), but other methods, such as emulsion, templating, and smelting have been developed for specialty applications needing specific material chemistries and particle size/distribution. A generalized model of the spray drying process for the manufacture of hollow particles has been developed (Shabde et al., 2005). Shabde et al. (2005) divided the process to form a hollow particle into two steps; starting with a liquid droplet and ending as a dry, hollow shell. The first step is evaporation of solvent from the droplet within the particle forming chamber. As the solvent evaporates, there is an accompanying reduction in the droplet diameter until all the solvent is removed. The second step is the increase in particle temperature until the blowing agent is activated (decomposed). This decomposition results in the release of gas that expands the particle diameter, thus forming the hollow shell.

Table 7–1 lists some of the micrometer sized particles, particle characteristics, and the processing methods used to make them.

Table 7–1 Hollow particle types: characteristics, processing methods, and manufacturers.

Particle type	Particle characteristics	Processing method	Reference/manufacturer
SiO_2	Average particle diameter ranging from 20 to 80 μm	Spray drying or milling, followed by furnace expansion	3M, Trelleborg AB, Potters
Carbon	1–2 μm in diameter, small particle size distributions	Ethanol decomposition over a Ni catalyst	Mi and Liu (2009)
Carbon	Preferential oxidation of carbon at the lower crystalline core leaving the higher crystalline shell	Heating a solid carbon sphere (made by chemical vapor deposition) in air	Yang, Liu, Luo, Jin, and Xu (2010)
Aluminosilicate, also known as cenospheres	Variable composition and mechanical properties; depends on coal chemistry and combustion conditions	By product of coal fired power plants, fly ash is floated and particles classified	Fillite—Omya AGUltraspheres-Tolsa Group
Phenolic	High fire resistance, low smoke and toxicity, and excellent ablative properties	Spray drying	Malayan Adhesives & Chemicals Sdn Bhd
Carbon	Nanocrystalline carbon	Carbonized hollow phenolic	Honeywell Gladysz, Perry, Mceachen, and Lula (2006), Carlisle, Lewis, Chawla, Koopman, and Gladysz (2007)
Hydroxyapatite	2–40 μm diameter, nanocrystalline	Spray drying	Luo and Nieh (1995), Jiao, Lu, Xiao, Xu, and Zhu (2012)
Poly(hydroxyethyl methacrylate–methyl methacrylate)	Monodispersed Diameter ranging from 17 to 42 μm	Emulsion templating	Zhang et al. (2009)
Bimetallic (Ni/Au, Ni/Ag, Ni/Pt, Ni/Pd) Noble Metal (Pt and Pd)	Average diameter 1.8 μm	Precipitation of metal hydroxides followed by decomposition and reduction	Yi et al. (2009)
MgS	Nanorod assemblies form 10 μm dandelion-like hollow spheres	Hydrothermal process	Ma et al. (2012)
Platinum, cobalt	1–2 μm	Replacement reaction route using metallic cobalt microspheres as sacrificial templates	Qin, Yang, Ma, and Lai (2011)

Spray drying of hollow particles as small as 300–400 nm is possible. Below this, however, other techniques are used. There are many techniques to make hollow particles from 1 to 500 nm, for example, plasma-enhanced processes (Hyodo, Murakami, Shimizu, & Egashira, 2005).

Plasma processing is a common technique for making hollow polymeric spheres, via plasma polymerization (Cao & Matsoukas, 2004). Hollow ceramic microspheres can be made by plasma enhanced chemical vapor deposition (CVD) (Hyodo et al., 2005) and by subjecting solid ceramic powders to thermal plasmas (Károly & Szépvölgyi, 2003). Plasma techniques typically have a microwave power unit to supply energy to make the plasma, a plasma reactor section with feedlines into the reactor, a particle collector, and an exhaust line. An example of a plasma system used to make hollow alumina nanospheres is shown in Fig. 7–4. One can feed a solid, gas, or liquid material into a plasma stream. Gaseous reactants are diluted with an inert gas and enter directly into the vacuum chamber and plasma zone. Liquid reactants are placed in a bubbler, where an inert gas, such

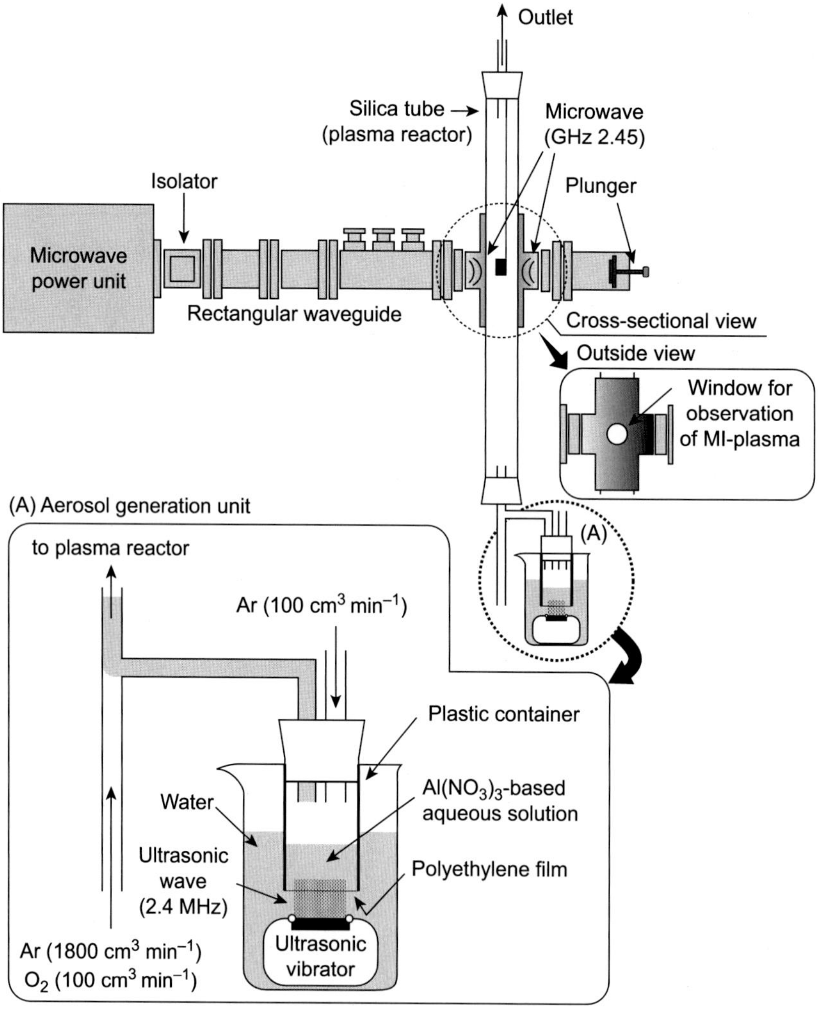

FIGURE 7–4 An example of a plasma system used to generate hollow alumina nanospheres. *Source: Adapted from Hyodo T.; Murakami M.; Shimizu Y.; Egashira M. (2005) Preparation of hollow alumina microspheres by microwave-induced plasma pyrolysis of atomized precursor solution. Journal of the European Ceramic Society 25 (16) 3563–3572.*

as argon or nitrogen, is bubbled into the reactant. Alternatively, liquids can be converted into an aerosol via an ultrasonic vibrator. As the liquid reactant enters the gaseous phase, it is carried into the chamber and plasma zone. The gas, which transports the liquid reactant, is called the *carrier gas.*

Other processes for making hollow nanospheres include (Wei et al., 2013):

1. Self-assembly processes
2. Hard templating: carbon, SiO_2, polystyrene
3. Soft templating: emulsions, droplets, surfactants, gas bubbles
4. Template-free
   a. Kirkendall effect
   b. Inside-out Ostwald ripening

In Chapter 6, Techniques for introducing intentional voids into materials, we discussed the use of templating or sacrificial pore formers. In the case of hollow particles, the template provides the surface for a shell to form and subsequent removal of the template will leave a hollow shell. Item numbers 2 and 3 in the above list are specific categories of the templating or sacrificial pore-forming process. Hard templates are solid materials used to make a sacrificial material. Soft templates can be liquids or gases.

Template free formation of a hollow particle using the Kirkendall effect is very interesting in that it does not require a sacrificial core or porogen to form a central void. We discussed the phenomenon of Kirkendall porosity in Chapter 5, Hollow and Porous Structures Utilizing the Kirkendall Effect. Kirkendall porosity form when two materials in contact have significantly different diffusivities. One can take advantage of this effect to form an intentional Kirkendall void in the form of a hollow nanoparticle (Zheng et al., 2007). Starting with a nanoparticle, one can react the particle with an appropriate gas, usually at high temperature, and form a hollow particle. The phenomenon is explained by the different diffusion rates of the gas versus the material making the particle, that is, Kirkendall effect. Yu, Zhu, and Ma (2013) demonstrated this process, making hollow cobalt compounds while Zheng et al. (2007) made hollow nanoaluminum nitride particles. Zheng's process involved taking nanometerscale (solid) aluminum particles and reacting them with NH_3. The proposed mechanism of forming hollow particles is shown in Fig. 7–5. After an initial surface layer of AlN is formed, the diffusion of Al to the surface is faster than the diffusion of nitrogen toward the center. This difference in diffusion rates allows the aluminum to diffuse and react at the surface leaving a hollow core.

7.2.4 Hollow, porous particles

In this section, we describe methods of making hollow particles with porous shells. Since porosities can exist at multiple length scales, this section is an example of

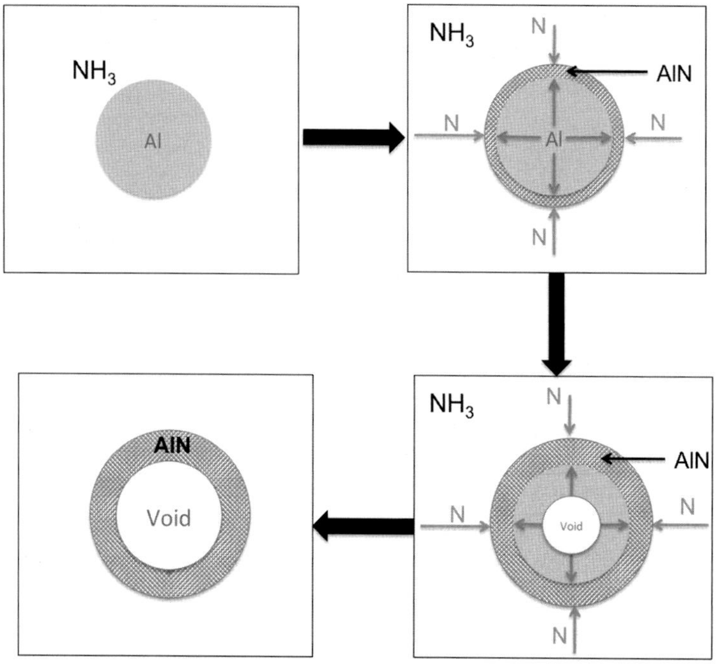

FIGURE 7–5 Schematic of the Kirkendall process for making hollow spheres. The process relies on the Al atoms diffusing out faster through the initial surface reaction layer (AlN) compared to nitrogen diffusing in through the surface layer. *Source: Adapted from Zheng J.; Song X.; Zhang Y.; Li Y.; Li X.; Pu Y. (2007) Nanosized aluminum nitride hollow spheres formed through a self-templating solid–gas interface reaction. Journal of Solid State Chemistry 180 (1) 276–283.*

hierarchical porosity. There are many applications for this type of structure in the area of drug delivery (Chen, Ding, Wang, & Shao, 2004) and gas sensors (Ma, Song, & Guan, 2013). One can use hierarchical porosity to design particles to maximize drug effectiveness. The design of the porosity ensures that the particle delivers the correct dose of the medicine, at the proper time and location in the body, and in a manner that optimizes effectiveness and minimizes side effects (Kim & Pack, 2006). The initial surface area to volume ratio as well as the evolving particle surface area (as the particle dissolves) are important features in how the drug will be released. For example, one drug may be more effective if there is an initial "burst" of release followed by an exponential decay (O'Donnell & Mcginity, 1997), whereas another may be more effective with a constant level of release with time (Kim & Pack, 2006). Fig. 7–6 shows such hierarchical porosity in a ZnO porous, hollow particle (X. Wang et al., 2013).

We describe an interesting example of a hollow porous particle made for drug delivery: PulmoSphere™. PulmoSpheres can be formulated with specific drugs and delivered into the lungs via a dry powder inhaler. For example, PulmoSpheres can

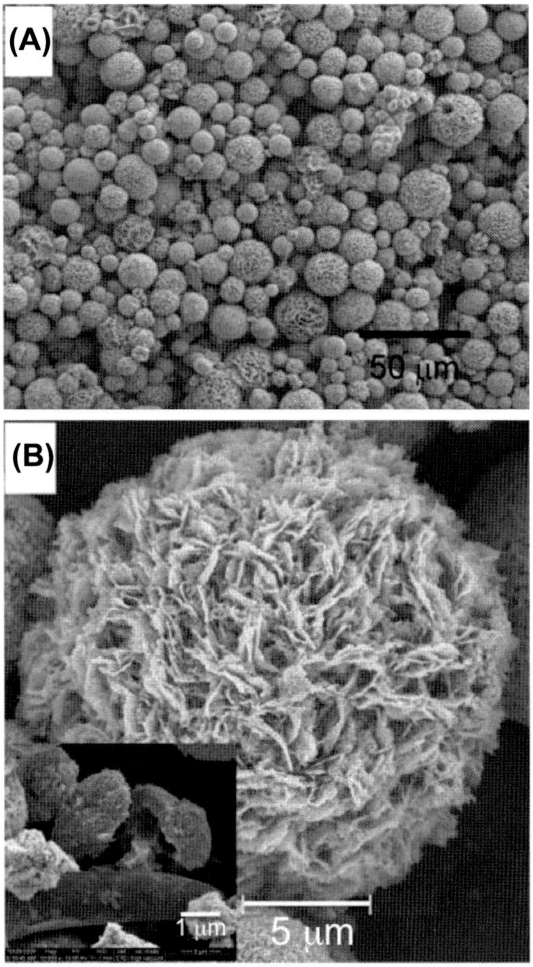

FIGURE 7–6 (A) General microstructure of a hollow porous ZnO particle. (B) High magnification view of a single sphere. With a large surface area, structures like this have potential for drug delivery. The inset shows the shell with the platelet structure on the surface. *Source: Adapted from Wang X.; Cai W.; Liu S.; Wang G.; Wu Z.; Zhao H. (2013) ZnO hollow microspheres with exposed porous nanosheets surface: structurally enhanced adsorption towards heavy metal ions. Colloids and Surfaces A: Physicochemical and Engineering Aspects 422 199–205.*

be formulated with a drug called Tobramycin and when delivered to the lungs, this drug would be effective in treating chronic pulmonary infections in people with cystic fibrosis (Geller, Weers, & Heuerding, 2011). The process of making PulmoSpheres is an emulsion-based, two-stage spray drying process. The atomized liquid is an oil-in-water emulsion. A schematic of the spray drying process is shown in Fig. 7–7. The dispersed phase is an oil that functions as the pore-forming material, that is, it is a liquid porogen, and is stabilized by a phospholipid. The phospholipid is a long-

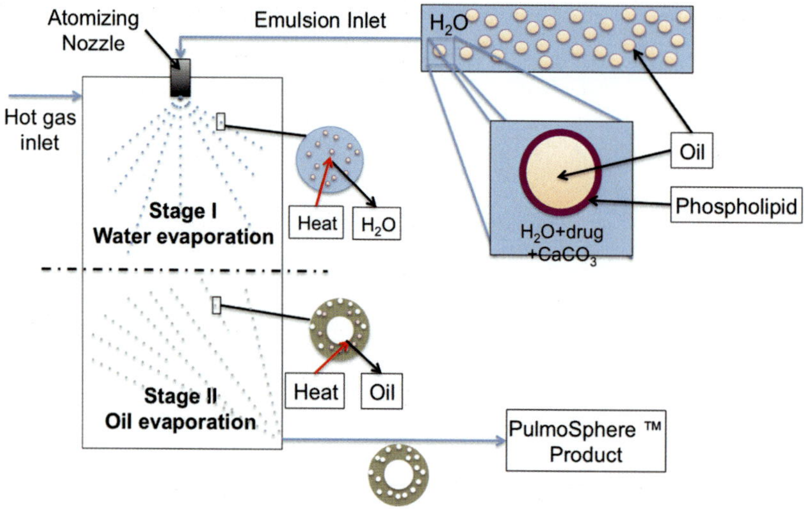

FIGURE 7–7 Schematic of the two-stage spray drying process to make PulmoSphere™. The continuous aqueous phase has calcium carbonate and drug (Tobramycin) in solution and the dispersed phase is oil stabilized by a phospholipid. Once atomized, the water first evaporates leaving behind the calcium carbonate, Tobramycin framework of the hollow shell with dispersed oil droplets. Further heating drives off the oil leaving the porosity within the shell. *Source: Adapted from Geller D.E.; Weers J.; Heuerding S. (2011) Development of an inhaled dry-powder formulation of tobramycin using PulmoSphere™ technology. Journal of Aerosol Medicine and Pulmonary Drug Delivery 24 (4) 175–182.*

chain molecule with one end being hydrophobic and the other hydrophilic. It forms a monolayer around the dispersed oil phase, thus stabilizing it. The continuous aqueous phase contains the drug Tobramycin and calcium carbonate both of which are initially in solution. The calcium carbonate is an excipient or an inactive filler or bulking agent. In Stage I of the process, the solid portion of the PulmoSphere structure is formed as water evaporates. This leaves a hollow shell containing calcium carbonate and Tobramycin surrounding the oil droplets. Further heating in Stage II vaporizes the oil, leaving behind the open cell foam shell, see Fig. 7–7. We discuss the details of design and functioning of PulmoSpheres in Chapter 10, Applications.

7.2.5 Porous and hollow macrometer scale particles

7.2.5.1 Hollow macrospheres

We define macrometer scale particles as having a diameter of 1000 μm (1 mm) or greater. Hollow structures having diameters of 1 mm and larger are very common in the food and sporting industries and even in nature. There are relatively few examples of highly engineered hollow macrosphere materials compared to micrometer and nanometer scale particles; most find important applications in buoyancy, ballistic, and impact and shock wave absorption.

In this section, we explore the processing of these large spherical particles. Such commercially available, engineered, hollow macrospheres can be made from polymers, metals, ceramics, and composites. These can be classified as either *seamless* or *bonded*. A seamless macrosphere is one that has been manufactured with a continuous shell and does not have a bond line. Bonded macrospheres are manufactured in sections and bonded together. Typically, seamless spheres have higher hydrostatic strength-to-weight ratio and are therefore superior to bonded spheres for high performance buoyancy applications. There are sometimes alignment difficulties when bonding the hemispheres that compromise strength. There can be defects in the adhesive layer that bonds the thin shells together, which can contribute to weakening of the spheres. In order to increase the integrity of the bond, each hemisphere can be made to have a thicker section (i.e., a band) at the equator; this increases the weight but makes for more surface area available for bonding.

Seamless hollow macrospheres are used in many buoyancy applications. An example of a high performing, seamless buoyancy sphere is a hollow alumina sphere (Weston, Stachiw, Merewether, Olsson, & Jemmott, 2005). These spheres, 100 mm in diameter, are made by rotational molding a ceramic slip *inside* a sacrificial hollow sphere followed by sintering of alumina. The mass of a fully sintered sphere is 140 g and such a macrosphere displaces over 400 g of seawater. These are used in deep sea exploration vehicles such as the remotely operated, unmanned *Nereus* hybrid vehicle, see Fig. 7–8. Nereus is designed

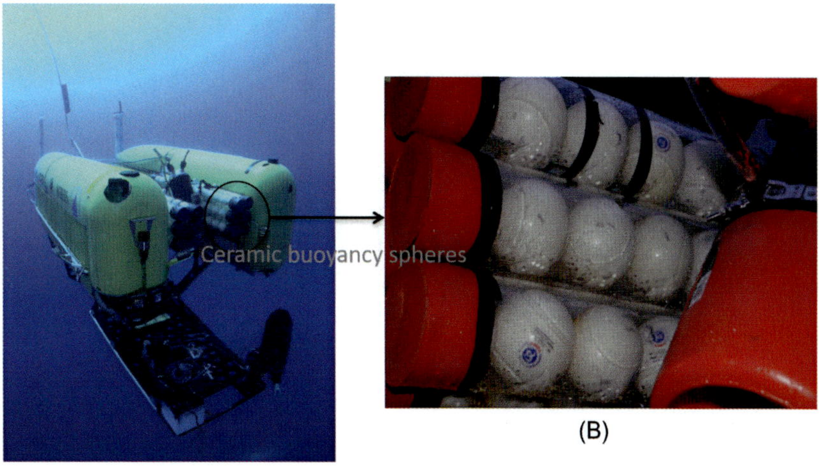

(A)

(B)

FIGURE 7–8 (A) Photograph of Nereus. The circled area highlights the placement of ceramic buoyancy spheres. (B) A close-up of the ceramic buoyancy spheres. These spheres are rotationally molded, hollow alumina spheres and provide the buoyancy for Nereus. Each sphere is 10 cm in diameter. *Source: (A) Photo by the Advanced Imaging & Visualization Lab © Woods Hole Oceanographic Institution. (B) Photo by Cathy Offinger © Woods Hole Oceanographic Institution.*

to operate at deep depths of the oceans, for example, at an operating depth of 11,000 m at the bottom of the Marianas Trench. The Marianas Trench is 2500 km long, crescent-shaped trench located in the western Pacific Ocean. The bottom of the trench is approximately 11,000 m deep, making it the deepest part of any ocean. The density of a hollow buoyancy sphere used for this application is 0.35 g/cm^3 (Weston et al., 2005), and at maximum operating depth it will withstand 107.9 MPa (1.079×10^8 Pa) of hydrostatic pressure.

Other less high-performing hollow spheres are used in composite foam formulations for providing buoyancy in support of offshore oil and gas exploration and production. Good examples are drill riser buoyancy modules, distributed buoyancy modules, and remotely operated subsea vehicles. All of these hollow macrospheres have a composite shell made from a thermosetting polymer, typically an epoxy reinforced with carbon or glass fibers. These hollow spheres can range in diameter from 3 mm to greater than 100 mm. These are rotationally molded, building up a fiber reinforced composite shell on a low density sacrificial core, layer-by-layer, until the appropriate density and/or strength are achieved.

Another method used to make hollow metal and ceramic seamless macrospheres involves a coaxial blowing nozzle, see Fig. 7–9. The film forming material is in the annulus and a gas in the center; the gas, at positive pressure, pushes the film forming material out of the nozzle and supports the formation of a hollow film. The film forming material

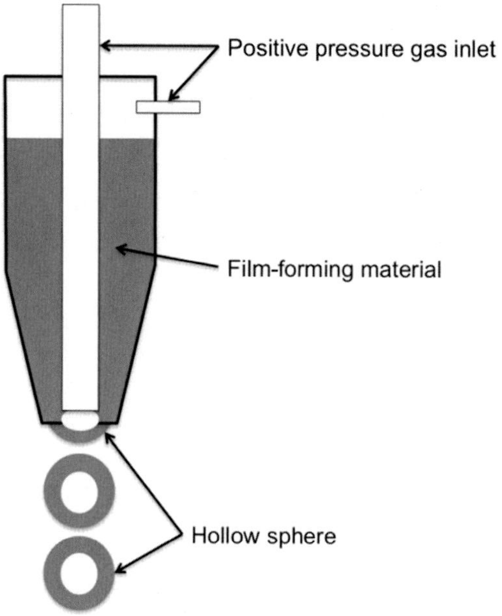

FIGURE 7–9 A schematic of a coaxial nozzle used to make hollow micrometer and macrometer scale spheres.

can be a preceramic polymer, ceramic slip, metal powder with a binder, etc., and post processing is required to convert the starting material into a metal or ceramic shell.

The general method to make bonded hollow macrospheres involves making hemispheres and in a subsequent processing step, bonding them together. The processing route of making the shells involves common processing methods for that type of shell material. For example, thermoplastic hemispheres are made by injection molding or compression molding. The joining of the hemispheres could involve an adhesive, friction welding, or localized melting process. Thermosetting polymers can be reaction injection molded, compression molded, followed by a bonding step. Metal hemispheres can be cast or powder processed followed by welding. To maximize the strength of the spheres, there must be a high degree of precision in the joining process.

7.2.5.2 Porous macrospheres

These can be fabricated using a pseudo-double-emulsion technique. It is called a double emulsion since there are two continuous phases and two dispersed phases, see Fig. 7–10. Lee and Park (2003) used this technique to produce porous mullite macrospheres with diameters in the range of 1–5 mm. One continuous phase within the particle contained a mullite suspension (dispersed phase) and a gelling agent. The dispersed phase contained bubbles whose size was controlled by the type of surfactant and impeller speed used for entraining air. The oil phase was liquid paraffin. The overall macrosphere size was controlled by the mixing speed when the mullite suspensions containing stabilized bubbles were introduced into the paraffin. Subsequent drying and sintering produced mullite powders containing 65–79 volume percent of porosity.

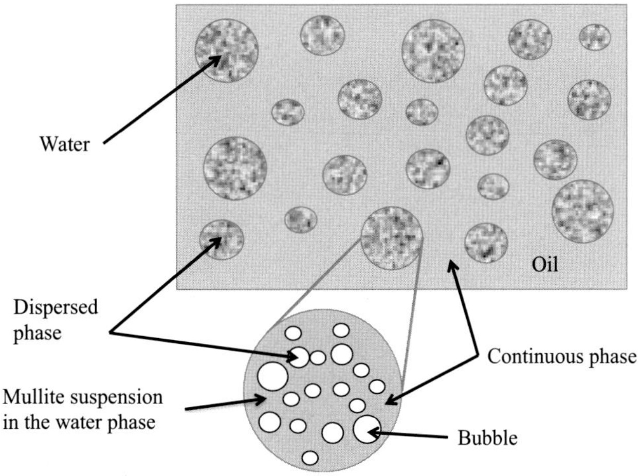

FIGURE 7–10 Schematic of the double water-in-oil emulsion that is used to make porous macrospheres. *Source: Adapted from Lee J.S.; Park J.K. (2003) Processing of porous ceramic spheres by pseudo-double-emulsion method. Ceramics International 29 (3) 271–278.*

7.3 Hollow and porous fibers

Hollow fibers are small and tubelike and most are polymeric with diameters of 200 μm or less. Ceramic (alumina) and metallic hollow fibers have also been made. These are useful in removal of contaminants from gaseous or aqueous streams. More common are polymeric hollow fibers. Examples of polymers commonly used to make hollow fibers are cellulose acetate, polyimide, polysulfone, polyethylene, and polytetrafluorethylene.

The polymeric hollow fiber membranes are produced by a technique called *diffusion* or *nonsolvent-induced phase separation*. One needs two key items for this technique, the first is a homogeneous polymer solution (also called dope). It is prepared by mixing a polymer, a suitable solvent, and some other additives. The second item is called the bore fluid or inner coagulant. The dope and the bore fluid are fed to a spinneret, see Fig. 7–11. A spinneret is a device commonly used to extrude fibers from a polymeric solution or melt (Chawla, 2016). It is a thimble-shaped, metal nozzle having fine holes through which the dope is forced to form a fiber. The key requirement of any fiber spinning process is that the jets emerging from the tiny holes must be stabilized against breaking into droplets. The emerging liquid jets are solidified by coagulation, evaporation, or cooling. Unlike conventional wet spinning of polymeric fibers, spinning of hollow fibers requires exposure to air for a certain period of time in what is called an "air gap" to evaporate solvent from dope, followed by immersion in the coagulant. This process is called dry–wet spinning. It is somewhat similar to the dry-jet wet spinning process used to make aramid fibers. The last step in any fiber spinning process is to collect the fibers on a bobbin after washing to remove any residual solvents.

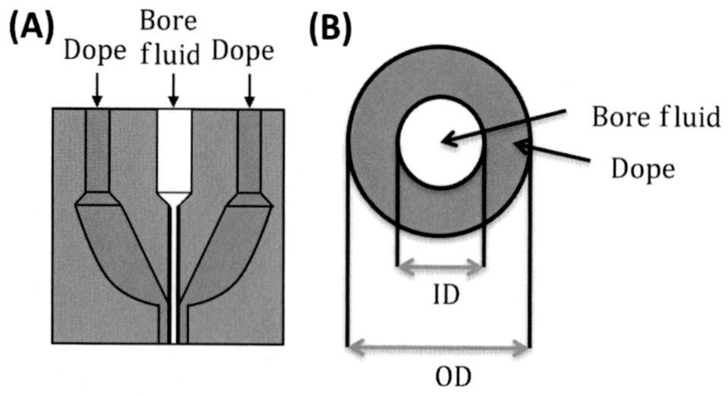

FIGURE 7–11 (A) Side view and (B) cross sectional view of a spinneret to manufacture hollow fibers. ID, inner diameter; OD, outer diameter. *Source: Adapted from Peng N.; Chung T.S. (2008) The effects of spinneret dimension and hollow fiber dimension on gas separation performance of ultra-thin defect-free Torlon® hollow fiber membranes. Journal of Membrane Science 310 (1–2) 455–465.*

Typically, the inner bore diameter is about 25 μm. Such fibers find applications in fields as diverse as biology, environmental cleanup, etc. These hollow fibers have a very large surface area. We generally characterize the surface area in units of square centimeter per milliliter (cm^2/mL). Note that 1 mL = 1 cm^3. The surface area/volume of hollow fibers is generally 100–200 cm^2/mL. Such large surface area allows culture of cells at in vivo like cell densities ($> 10^8$ cells per milliliter or even higher). In conventional practice, the cells are bound to a two-dimensional surface, such as in a petri dish. With hollow fibers, cells are bound to a three-dimensional porous structure. Microfiltration or ultrafiltration followed by reverse osmosis is an important technique for water treatment (desalination, treatment of effluents, wastewater treatment, etc.). Membranes made of hollow tubelike polymeric fibers are also used in dialysis. The cutoff molecular weight of the fiber can be controlled during manufacture. When we use cell culture with cartridges made of hollow fibers, cell splitting is not required. Hollow fiber membrane cartridges or modules can contain 3000–20,000 hollow fibers held together by a resin (epoxy or polyurethane). This has resulted in a new field called hollow fiber bioreactor technology (Fig. 7–12).

The permeation properties can be controlled via fiber morphology, which in turn can be controlled by changing spinning parameters, such as:

1. Geometry of spinneret (shear stress)
2. Temperature and composition of the dope
3. Temperature and composition of bore fluid
4. Flow rate of dope and bore fluid
5. Composition and temperature of coagulant
6. Spinning temperature and relative humidity of spinning room (dry–wet spinning)
7. Take-up speed

Techniques of making *membranes* from hollow porous fibers are based on traditional polymer spinning. For example, dry–wet spinning followed by a specific heat treatment (Luiten-Olieman et al., 2012) has been used to produce a variety of inorganic hollow, porous fibers with diameters less than 700 μm. Examples of inorganic membranes include stainless steel, nickel, silicon carbide, zirconia, alumina, and titania. See Fig. 7–12 for examples of the microstructure of porous hollow fibers made by dry–wet spinning.

Another very versatile technique with an ability to produce nanometer- and micrometer-scale hollow, porous, organic and inorganic fibers is electrospinning (Zhang, Li, An, & He, 2010). A wide variety of raw materials can be electrospun. Examples include polymers by solvent and melt processing (Huang, Zhang, Kotaki, & Ramakrishna, 2003), sol–gel (J. Wang et al., 2013), preceramic polymers, and

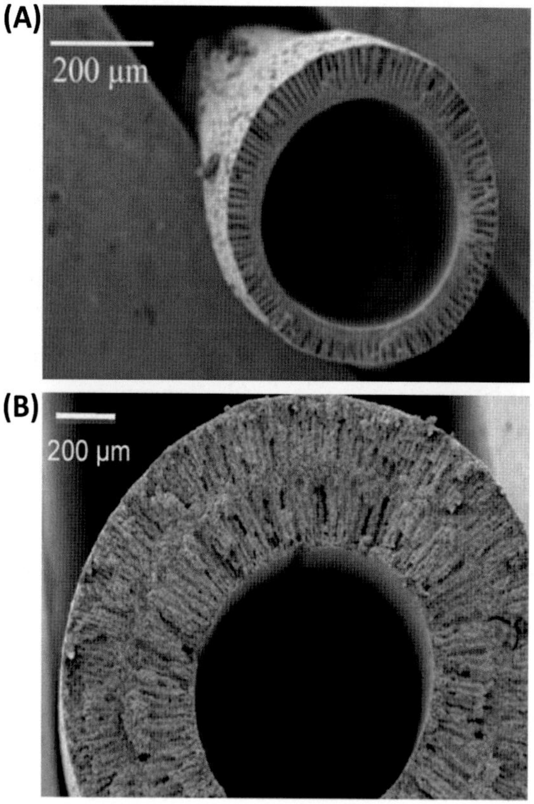

FIGURE 7–12 (A) Micrometer scale yttria stabilized zirconia and (B) Stainless steel porous, hollow fibers (membranes) made by dry-wet spinning. *Source: (A) Luiten-Olieman M.W.J.; Raaijmakers M.J.T.; Winnubst L., Bor T.C.; Wessling M.; Nijmeijer A. and Benes N.E. (2012) Towards a generic method for inorganic porous hollow fibers preparation with shrinkage-controlled small radial dimensions, applied to Al2O3, Ni, SiC, stainless steel, and YSZ, Journal of Membrane Science 407–408, 155–163. (B) Luiten-Olieman M.W.J.; Winnubst L., Nijmeijer A.; Wessling M. and Benes N.E. (2011) Porous stainless steel hollow fiber membranes via dry-wet spinning, Journal of Membrane Science 370 (1-2), 124–130*

organometallic mixtures with binders. Fig. 7–13 shows a schematic of electrospinning process. The majority of the fiber diameter reduction is done once the fiber leaves the needle. In this region, there is rapid acceleration and stretching of the fiber resulting in a reduction of fiber diameter (Fu et al., 2013).

There are four essential components in electrospinning fibers:

1. High voltage DC power supply
2. Spinneret
3. Syringe pump
4. Collection substrate

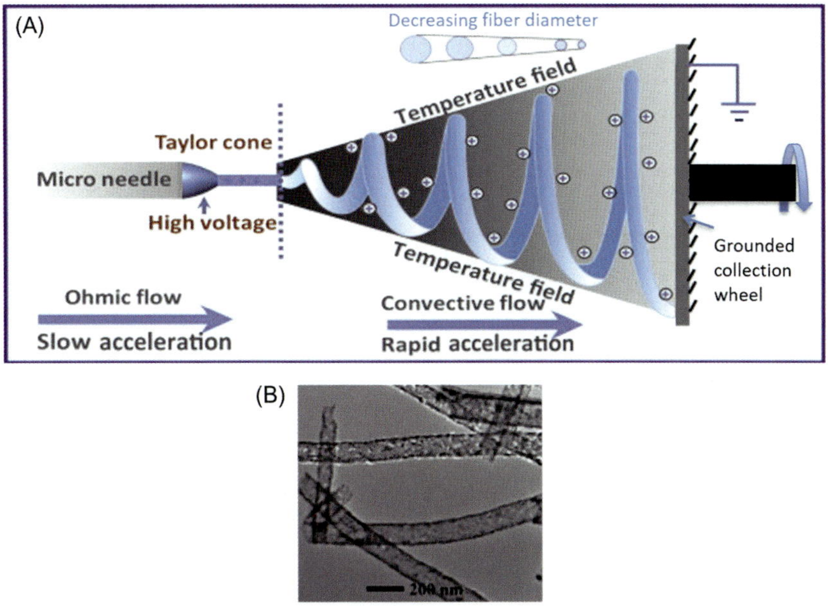

FIGURE 7–13 (A) An electrospinning process capable of making hollow nanometer and micrometer scale fibers. In the conical section, there is a significant reduction in the fiber diameter between the microneedle and the grounded collection wheel. (B) $MgFe_2O_4$ nanotube made by electrospinning. *Source: Adapted from Fu J.; Zhang J.; Zhao C.; Peng Y.; Li X.; He Y.; . . . Xie E. (2013) Solvent effect on electrospinning of nanotubes: The case of magnesium ferrite. Journal of Alloys and Compounds 577 (0) 97–102.*

The collection substrate can be a stationary plate, which will yield webs of randomly oriented fibrous structures, or a rotating wheel that allows for aligned fibers. Structures with complex porosity such as tube-in-tube, rod-in-tube, compartmentalized tube can be made via electrospinning, some of which are shown in Fig. 7–14 (X. Wang, Ding, Yu, & Wang, 2011; Z. Wang, Luan, Boey, & Lou, 2011). Examples of hollow fibrous materials made by electrospinning include SnO_2, TiO_2, ZnO, CeO, and poly(glycolic acid)/poly(L-lactic acid).

We described several additive manufacturing process in Chapter 6, Techniques for introducing intentional voids into materials, as a versatile method for making stochastic and nonstochastic foams. Some of these technologies can also be used in making hollow, porous fibers. *Robocasting* is a materials extrusion additive manufacturing process in which a paste is extruded through a syringe and deposited into a predefined geometry. The paste is prepared by mixing a polymer, which serves as a binder, a solvent, and any appropriate powder. To be suitable for extrusion through the syringe opening, the paste must have shear-thinning properties. Michielsen et al. (2013) used this technique to make hollow, porous stainless steel fibers, see Fig. 7–15.

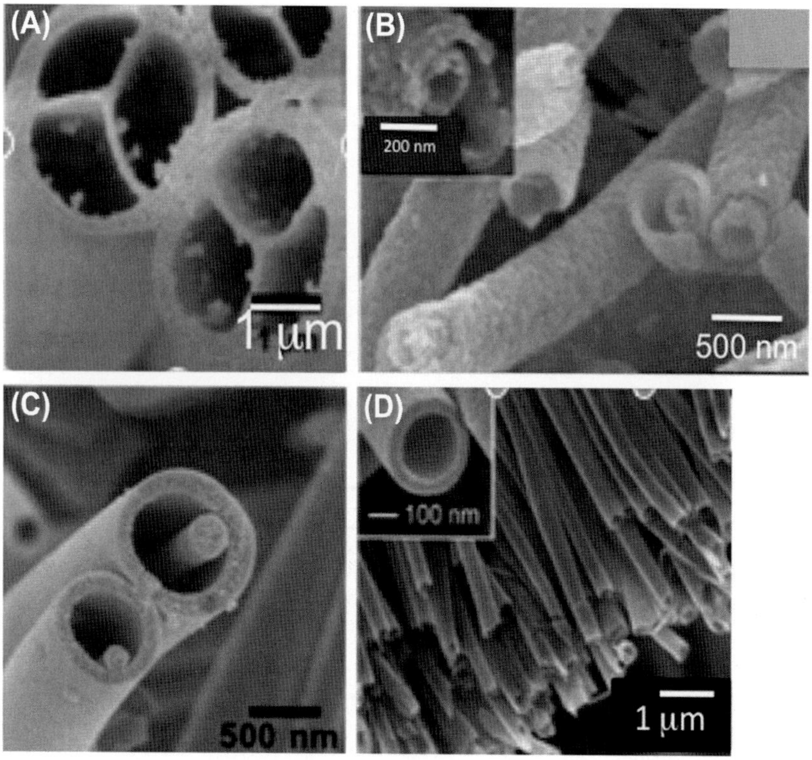

FIGURE 7–14 Micrographs of complex hollow fibrous structures made by electrospinning. (A) Compartmentalized, (B) tube-in-tube, (C) fiber-in-tube, and (D) hollow tube. *Source: (A) Zhao Y., Cao X. and Jiang L., Bio-mimic multichannel microtubes by a facile method, Journal of the American Chemical Society 129 (4), 2007, 764–765. (B) Mou F., Guan J., Shi W., Sun Z. and Wang S., Oriented contraction: a facile nonequilibrium heat-treatment approach for fabrication of maghemite fiber-in-tube and tube-in-tube nanostructures, Langmuir 26 (19), 2010, 15580–15585, (C) Chen H., Wang N., Di J., Zhao Y., Song Y. and Jiang L., Nanowire-in-microtube structured core/shell fibers via multifluidic coaxial electrospinning, Langmuir 26 (13), 2010, 11291–11296. (D) Li D. and Xia Y., Direct fabrication of composite and ceramic hollow nanofibers by electrospinning, Nano Letters 4 (5), 2004, 933–938.*

7.3.1 Carbon nanotubes

Carbon has many forms: amorphous, graphite, diamond, fullerene, nanotubes, and graphene. Fullerene is a spherical molecule of carbon, for example, C_{60} is a carbon molecule containing 60 carbon atoms. If we stretch the fullerene sphere, we will get an elongated, cylindrical cage made of carbon atoms. These are called carbon nanotubes (CNTs). One can have a *single walled nanotube* (SWNT) or *multiwalled nanotube* (MWNT). The first MWNTs, were synthesized and identified in early 1990s (Iijima, 1991; Iijima & Ichihashi, 1993). An SWNT is a hollow fiber with the wall being one atom thick. One can visualize a SWCNT as forming a seamless cylinder from a single layer of graphite as shown in

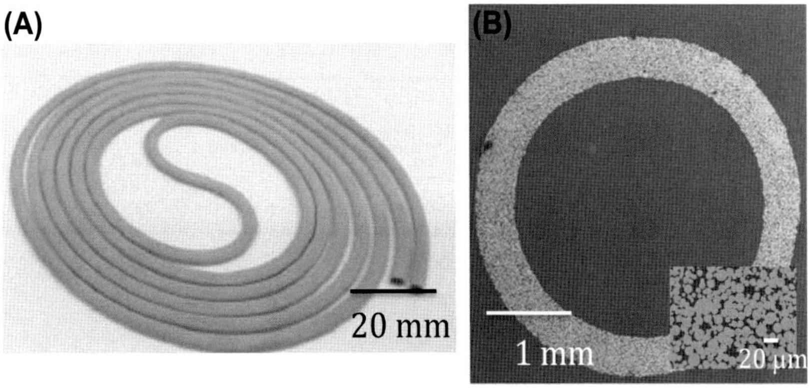

FIGURE 7–15 (A) Low magnification and (B) medium magnification of porous, hollow fibers, also called membranes. The inset in (B) is a high magnification image of the microstructure of porous, hollow stainless steel fibers. *Source: Adapted from Michielsen B.; Chen H.; Jacobs M.; Middelkoop V.; Mullens S.; Thijs I.; ... Snijkers F. (2013) Preparation of porous stainless steel hollow fibers by robotic fiber deposition. Journal of Membrane Science 437 17–24.*

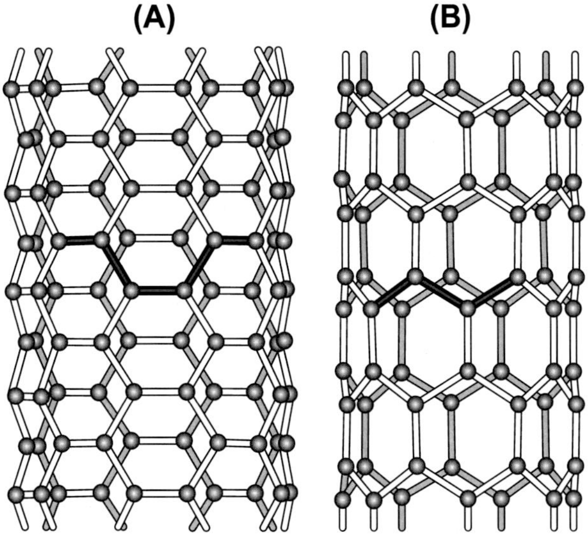

FIGURE 7–16 Two different types of single walled carbon nanotubes (SWCNT). (A) Armchair and (B) zigzag nanotube. *Source: Adapted from Thostenson E.T.; Ren Z.; Chou T. (2001) Advances in the science and technology of carbon nanotubes and their composites: A review. Composites Science and Technology 61 (13) 1899–1912.*

Fig. 7–16. A single atom layer of carbon is also known as graphene. By using graphene and the addition of another layer of a concentric seamless tube on top of an SWCNT, one can form a *double walled carbon nanotube*. Further successive additions of graphene layers result in *multiwalled carbon nanotubes*. The terminal ending of the SWCNT either can be open or capped. A capped ending is a hemisphere of a fullerene molecule having 30 carbon atoms. There can be different orientations of SWCNT, armchair and zigzag, depending on how the six-member ring is oriented with respect to the longitudinal axis of the nanotube, see Fig. 7–16.

There are several methods to make CNTs (Merchan-Merchan, Saveliev, Kennedy, & Jimenez, 2010):

1. Chemical vapor deposition (CVD)
2. Arc discharge
3. Pulsed laser vaporization
4. Flame synthesis

The vast majority of CNTs, whether single or multiwalled, is made by CVD on a metal catalyst nanoparticle. Typically the carbon source is a gas phase hydrocarbon, such as methane or ethylene diluted in a carrier, such as hydrogen. The catalysts commonly used to make CNTs in a CVD reactor are Ni-based alloys and ferrocene (an organometallic compound).

Because of the small size, unique geometry, and perfect covalent bonding of CNTs, they have high modulus and strength as well as excellent electrical and thermal conductivities. These properties are dependent on processing and structure and details of the morphology. Small changes in the atomic structure, specifically the orientation in which the graphene sheet is rolled up (i.e., zigzag or armchair, see Fig. 7–16), can lead to a SWCNT having different behavior. For example, depending on the orientation, a SWCNT can behave as a metal or a semiconductor. Coupled with the difficulties of controlling the structural defects, this can prevent CNTs from realizing their theoretical properties. Carbon nanotube market has grown by some estimates at 33% a year since 2014. As world-wide CNT production has increased, so have applications and products. Initial applications were related to enhancing mechanical and other properties by adding small amounts of CNT to thermosetting and thermoplastic polymers. Application have expanded into markets such as integrated circuits, field emission displays, hydrogen storage, lithium batteries, solar photovoltaic cells, fuel cells, and drug delivery.

7.4 Nonspherical hollow particles

A quick investigation of manufactured particles (hollow or solid) will show that they are overwhelmingly approximately spherical. Here we are not including the powders that have been reduced in size by, for example, milling or some other mechanical means. Of all types of particles, a sphere has the lowest surface area to volume ratio; therefore, a sphere is the favored geometry for the particles formed from a liquid state, for example, when using techniques such as spray drying or emulsions. A spherical shape is preferred, thermodynamically, in order to minimize surface area per unit volume, that is, to minimize the surface energy.

When we discuss nonspherical particles, there is a functionality that the shape brings in when compared to a spherical particle. An example of an additional functionality that shape can bring in is that of an ellipsoid compared to a sphere. *Random close packing* of mono modal spheres is 62.5% (McGeary, 1961), but ellipsoids can randomly assemble into packing arrangements greater than 74% (Delaney, Weaire, Hutzler, & Murphy, 2005; Donev et al., 2004). The random close packed structure of a mono modal ellipsoid-shaped particles can be as high as the *theoretical close packing* of single-sized spheres, namely, 74%. Hollow ellipsoids have been made both in nanometer and millimeter sizes via sacrificial core (Carlisle, Brito, Gladysz, Ricci, & Koopman, 2009; Liu, Yang, & Wang, 2008).

An important feature that nonspherical particles have is a higher surface area to volume ratio compared to spherical particles. When combined with techniques that introduce hierarchical porosity, functionalities such as photocatalysis and drug delivery are enhanced. Fig. 7–17 shows a cage or rubik cube structure made of TiO_2. Titania is known to have important photocatalytic characteristics. Table 7–2 provides information on other nonspherical particles including materials, functionality, and fabrication methods. The majority of the nonspherical structures is made by a

FIGURE 7–17 SEM micrographs and a schematic showing the hollow cage structure of a TiO_2 rubik made by templating on a Cu_2O core. *Source: Adapted from Jiang G.; Wang X.; Zhou Y.; Wang R.; Hu R.; Xi X.; Chen W. (2012) Hollow TiO2 nanocages with rubik-like structure for high-performance photocatalysts. Materials Letters 89 59–62.*

Table 7–2 Some nonspherical hollow particles and their characteristics.

Shape	Size	Functionality	Processing method	Material	Reference
Ellipsoid	<100 nm	Photonic crystals	Self-templating—wet etch	SiO_2	Zhang, Zhou, Li, Bandosz, and Akins (2012)
Ellipsoid	3 mm	Increased packing efficiency	Sacrificial core/rotational molding	Glass fiber/ epoxy	Carlisle et al. (2009)
Cube	300 nm	Ferromagnetic behavior	Self-assembly/Ostwald ripening	Cobalt-iron cyanide	Zhai, Du, Zhang, and Yang (2011)
Cube	200 nm to 1 μm	Improved electrochemical performance (anodes in lithium ion batteries)	Sacrificial Cu_2O core	SnO_2	X. Wang, Ding, Yu, and Wang (2011); Z. Wang, Luan, Boey, and Lou (2011)
Cage/rubik	300 nm	Improved photocatalytic degradation of phenols	Sacrificial Cu_2O core	TiO_2	Jiang et al. (2012)
Urchin	0.8 to 1.2 μm	Improved electrochemical performance (lithium-ion batteries)	Ostwald ripening	α-MnO_2	Li, Rong, Xie, Huang, and Feng (2006)
Calabash	>0.8 μm	Gas/energy storage	Mixed valence oxide catalytic carbonization	Carbon	Wang and Yin (1998)

solid or hard templating method. One of the methods in Table 7–2 that we have not discussed in detail is Ostwald ripening (Ostwald, 1896). This phenomenon was first described by W. Ostwald in 1896. Basically, it says that when a phase precipitates out of a solid solution, energy considerations will cause large precipitates to grow at the expense of smaller precipitates. This thermodynamically driven spontaneous Ostwald ripening occurs when we have agglomerates of small particles. The smaller particles combine to form larger particles in order to minimize surface energy. As the particles grow larger, a void can form at the center of the particle cluster forming a hollow particle.

References

Cao, J., & Matsoukas, T. (2004). Synthesis of hollow nanoparticles by plasma polymerization. *Journal of Nanoparticle Research, 6*(5), 447–455.

Carlisle, K. B., Brito, V., Gladysz, G. M., Ricci, W., & Koopman, M. (2009). Fabrication and finite element modeling of ellipsoidal macro-shells. *Journal of Materials Science, 44*(6), 1449–1455.

Carlisle, K. B., Lewis, M., Chawla, K. K., Koopman, M., & Gladysz, G. M. (2007). Finite element modeling of the uniaxial compression behavior of carbon microballoons. *Acta Materialia, 55* (7), 2301–2318.

Chawla, K. K. (2016). *Fibrous materials* (2nd ed.). Cambridge, UK: Cambridge University Press.

Chen, H., Wang, N., Di, J., Zhao, Y., Song, Y., & Jiang, L. (2010). Nanowire-in-microtube structured core/shell fibers via multifluidic coaxial electrospinning. *Langmuir, 26*(13), 11291–11296.

Chen, J., Ding, H., Wang, J., & Shao, L. (2004). Preparation and characterization of porous hollow silica nanoparticles for drug delivery application. *Biomaterials, 25*(4), 723–727.

Delaney, G., Weaire, D., Hutzler, S., & Murphy, S. (2005). Random packing of elliptical disks. *Philosophical Magazine Letters, 85*(2), 89–96.

Donev, A., Cisse, I., Sachs, D., Variano, E. A., Stillinger, F. H., Connelly, R., . . . Chaikin, P. M. (2004). Improving the density of jammed disordered packings using ellipsoids. *Science, 303* (5660), 990–993.

Fu, J., Zhang, J., Zhao, C., Peng, Y., Li, X., He, Y., . . . Xie, E. (2013). Solvent effect on electrospinning of nanotubes: The case of magnesium ferrite. *Journal of Alloys and Compounds, 577*(0), 97–102.

Geller, D. E., Weers, J., & Heuerding, S. (2011). Development of an inhaled dry-powder formulation of tobramycin using PulmoSphere™ technology. *Journal of Aerosol Medicine and Pulmonary Drug Delivery, 24*(4), 175–182.

Gladysz, G., Perry, B., Mceachen, G., & Lula, J. (2006). Three-phase syntactic foams: Structure-property relationships. *Journal of Materials Science, 41*(13), 4085–4092.

Gokmen, M. T., & Du Prez, F. E. (2012). Porous polymer particles—A comprehensive guide to synthesis, characterization, functionalization and applications. *Progress in Polymer Science, 37*(3), 365–405.

Huang, Z., Zhang, Y., Kotaki, M., & Ramakrishna, S. (2003). A review on polymer nanofibers by electrospinning and their applications in nanocomposites. *Composites Science and Technology, 63*(15), 2223–2253.

Hyodo, T., Murakami, M., Shimizu, Y., & Egashira, M. (2005). Preparation of hollow alumina microspheres by microwave-induced plasma pyrolysis of atomized precursor solution. *Journal of the European Ceramic Society, 25*(16), 3563–3572.

Iijima, S. (1991). Helical microtubules of graphitic carbon. *Nature, 354*(6348), 56–58.

Iijima, S., & Ichihashi, T. (1993). Single-shell carbon nanotubes of 1-nm Diameter. *Nature, 363*(430), 603–605.

Iskandar, F., Nandiyanto, A. B. D., Widiyastuti, W., Young, L. S., Okuyama, K., & Gradon, L. (2009). Production of morphology-controllable porous hyaluronic acid particles using a spray-drying method. *Acta Biomaterialia, 5*(4), 1027–1034.

Jiang, G., Wang, X., Zhou, Y., Wang, R., Hu, R., Xi, X., & Chen, W. (2012). Hollow TiO_2 nanocages with rubik-like structure for high-performance photocatalysts. *Materials Letters, 89*, 59–62.

Jiao, Y., Lu, Y., Xiao, G., Xu, W., & Zhu, R. (2012). Preparation and characterization of hollow hydroxy-apatite microspheres by the centrifugal spray drying method. *Powder Technology, 217*, 581–584.

Károly, Z., & Szépvölgyi, J. (2003). Hollow alumina microspheres prepared by RF thermal plasma. *Powder Technology, 132*(2–3), 211–215.

Kim K.K. and Pack D.W., Microspheres for drug delivery, in: Ferrari M., Lee A.P., Lee L.J. (Eds.), *BioMEMS and Biomedical Nanotechnology*. 2006, Springer, Boston, MA,19-50.

Lee, J. S., & Park, J. K. (2003). Processing of porous ceramic spheres by pseudo-double-emulsion method. *Ceramics International, 29*(3), 271–278.

Li, B., Rong, G., Xie, Y., Huang, L., & Feng, C. (2006). Low-temperature synthesis of α-MnO2 hollow urchins and their application in rechargeable Li + batteries. *Inorganic Chemistry, 45*(16), 6404–6410.

Li, D., & Xia, Y. (2004). Direct fabrication of composite and ceramic hollow nanofibers by electrospinning. *Nano Letters, 4*(5), 933–938.

Liu, G., Yang, X., & Wang, Y. (2008). Synthesis of ellipsoidal hematite/silica/polymer hybrid materials and the corresponding hollow polymer ellipsoids. *Langmuir, 24*(10), 5485–5491.

Luiten-Olieman, M. W. J., Raaijmakers, M. J. T., Winnubst, L., Bor, T. C., Wessling, M., Nijmeijer, A., & Benes, N. E. (2012). Towards a generic method for inorganic porous hollow fibers preparation with shrinkage-controlled small radial dimensions, applied to Al_2O3, Ni, SiC, stainless steel, and YSZ. *Journal of Membrane Science, 407–408*, 155–163.

Luiten-Olieman, M. W. J., Winnubst, L., Nijmeijer, A., Wessling, M., & Benes, N. E. (2011). Porous stainless steel hollow fiber membranes via dry–wet spinning. *Journal of Membrane Science, 370*(1–2), 124–130.

Luo, P., & Nieh, T. G. (1995). Synthesis of ultrafine hydroxyapatite particles by a spray dry method. *Materials Science and Engineering: C, 3*(2), 75–78.

Ma, W., Chen, G., Zhang, D., Zhu, J., Qiu, G., & Liu, X. (2012). Shape-controlled synthesis and properties of dandelion-like manganese sulfide hollow spheres. *Materials Research Bulletin, 47* (9), 2182–2187.

Ma, X., Song, H., & Guan, C. (2013). Interfacial oxidation–dehydration induced formation of porous SnO_2 hollow nanospheres and their gas sensing properties. *Sensors and Actuators B: Chemical, 177*, 196–204.

McGeary, R. K. (1961). Mechanical packing of spherical particles. *Journal of the American Ceramic Society, 44*(10), 513–522.

Merchan-Merchan, W., Saveliev, A. V., Kennedy, L., & Jimenez, W. C. (2010). Combustion synthesis of carbon nanotubes and related nanostructures. *Progress in Energy and Combustion Science, 36*(6), 696–727.

Mi, Y., & Liu, Y. (2009). A simple solvothermal route to synthesize carbon microspheres. *New Carbon Materials, 24*(4), 375–378.

Michielsen, B., Chen, H., Jacobs, M., Middelkoop, V., Mullens, S., Thijs, I., ... Snijkers, F. (2013). Preparation of porous stainless steel hollow fibers by robotic fiber deposition. *Journal of Membrane Science, 437*, 17–24.

Mou, F., Guan, J., Shi, W., Sun, Z., & Wang, S. (2010). Oriented contraction: a facile nonequilibrium heat-treatment approach for fabrication of maghemite fiber-in-tube and tube-in-tube nanostructures. *Langmuir, 26*(19), 15580–15585.

O'Donnell, P. B., & Mcginity, J. W. (1997). Preparation of microspheres by the solvent evaporation technique. *Advanced Drug Delivery Reviews, 28*(1), 25–42.

Oh, Y. J., Lee, J., Seo, J. Y., Rhim, T., Kim, S., Yoon, H. J., & Lee, K. Y. (2011). Preparation of budesonide-loaded porous PLGA microparticles and their therapeutic efficacy in a murine asthma model. *Journal of Controlled Release, 150*(1), 56–62.

Okuyama, K., Abdullah, M., Lenggoro, I. W., & Iskandar, F. (2006). Preparation of functional nanostructured particles by spray drying. *Advanced Powder Technology, 17*(6), 587–611.

Ostwald, W. (1896). *Lehrbruck der Allgemeinen Chemie,* (vol. 2), part 1 Leipzig, Germany: Engelmann.

Peng, N., & Chung, T. S. (2008). The effects of spinneret dimension and hollow fiber dimension on gas separation performance of ultra-thin defect-free Torlon® hollow fiber membranes. *Journal of Membrane Science, 310*(1–2), 455–465.

Qin, W., Yang, C., Ma, X., & Lai, S. (2011). Selective synthesis and characterization of metallic cobalt, cobalt/platinum, and platinum microspheres. *Journal of Alloys and Compounds, 509* (2), 338–342.

Shabde, V. S., Emets, S. V., Mann, U., Hoo, K. A., Carlson, N. N., & Gladysz, G. M. (2005). Modeling a hollow micro-particle production process. *Computers Chemical Engineering, 29* (11–12), 2420–2428.

Speight, J. G. (2005). *Lange's handbook of chemistry* (vol. 1). New York: McGraw-Hill.

Silverstein, M. S. (2014). PolyHIPEs: Recent advances in emulsion-templated porous polymers. *Progress in Polymer Science, 39*(1), 199–234.

Steckel, H., & Brandes, H. G. (2004). A novel spray-drying technique to produce low density particles for pulmonary delivery. *International Journal of Pharmaceutics, 278*(1), 187–195.

Thostenson, E. T., Ren, Z., & Chou, T. (2001). Advances in the science and technology of carbon nanotubes and their composites: A review. *Composites Science and Technology, 61*(13), 1899–1912.

Wang, A., Lu, Y., & Sun, R. (2007). Recent progress on the fabrication of hollow microspheres. *Materials Science and Engineering: A, 460–461*, 1–6.

Wang, J., Gao, Q., He, H., Li, X., Ren, Z., Liu, Y., . . . Han, G. (2013). Fabrication and characterization of size-controlled single-crystal-like PZT nanofibers by sol–gel based electrospinning. *Journal of Alloys and Compounds, 579*, 617–621.

Wang, X., Cai, W., Liu, S., Wang, G., Wu, Z., & Zhao, H. (2013). ZnO hollow microspheres with exposed porous nanosheets surface: Structurally enhanced adsorption towards heavy metal ions. *Colloids and Surfaces A: Physicochemical and Engineering Aspects, 422*, 199–205.

Wang, X., Ding, B., Yu, J., & Wang, M. (2011). Engineering biomimetic superhydrophobic surfaces of electrospun nanomaterials. *Nano Today, 6*(5), 510–530.

Wang, Z. L., & Yin, J. S. (1998). Graphitic hollow carbon calabashes. *Chemical Physics Letters, 289* (1–2), 189–192.

Wang, Z., Luan, D., Boey, F. Y. C., & Lou, X. W. (2011). Fast formation of SnO_2 nanoboxes with enhanced lithium storage capability. *Journal of the American Chemical Society, 133*(13), 4738–4741.

Wei, W., Wang, Z., Liu, Z., Liu, Y., He, L., Chen, D., . . . Li, J. (2013). Metal oxide hollow nanostructures: fabrication and Li storage performance. *Journal of Power Sources, 238*, 376–387.

Weston, S., Stachiw, J., Merewether, R., Olsson, M., Jemmott, G. (2005). Alumina ceramic 3.6 in flotation spheres for 11 km ROV/AUV systems, OCEANS, 2005. *Proceedings of MTS/IEEE 2005*, 1 172–177.

Yang, Y., Liu, X., Luo, Q., Jin, L., & Xu, B. (2010). Structure evolution of carbon microspheres from solid to hollow. *New Carbon Materials, 25*(6), 431–437.

Yi, R., Shi, R., Gao, G., Zhang, N., Cui, X., He, Y., & Liu, X. (2009). Hollow metallic microspheres: fabrication and characterization. *Journal of Physical Chemistry C, 113*(4), 1222–1226.

Yu, M., Zhu, P., & Ma, Y. (2013). Effects of particle clustering on the tensile properties and failure mechanisms of hollow spheres filled syntactic foams: a numerical investigation by microstructure based modeling. *Materials and Design, 47*, 80–89.

Zhai, C., Du, N., Zhang, H., & Yang, D. (2011). Cobalt–iron cyanide hollow cubes: Three-dimensional self-assembly and magnetic properties. *Journal of Alloys and Compounds, 509* (33), 8382–8386.

Zhang, H., Ju, X., Xie, R., Cheng, C., Ren, P., & Chu, L. (2009). A microfluidic approach to fabricate monodisperse hollow or porous poly(HEMA–MMA) microspheres using single emulsions as templates. *Journal of Colloid and Interface Science, 336*(1), 235–243.

Zhang, H., Zhou, Y., Li, Y., Bandosz, T. J., & Akins, D. L. (2012). Synthesis of hollow ellipsoidal silica nanostructures using a wet-chemical etching approach. *Journal of Colloid and Interface Science, 375*(1), 106–111.

Zhang, Y., Li, J., An, G., & He, X. (2010). Highly porous SnO_2 fibers by electrospinning and oxygen plasma etching and its ethanol-sensing properties. *Sensors and Actuators B: Chemical, 144*(1), 43–48.

Zhao, Y., Cao, X., & Jiang, L. (2007). Bio-mimic multichannel microtubes by a facile method. *Journal of the American Chemical Society, 129*(4), 764–765.

Zheng, J., Song, X., Zhang, Y., Li, Y., Li, X., & Pu, Y. (2007). Nanosized aluminum nitride hollow spheres formed through a self-templating solid–gas interface reaction. *Journal of Solid State Chemistry, 180*(1), 276–283.

8 Void characterization techniques

8.1 Introduction

In this chapter we focus on the techniques that are used to directly characterize and quantify voids. Frequently, the amount of voids is measured indirectly. Take for example density; it is a measure of the amount of the solid present in a material. Based on the total volume and the theoretical density of the solid, one can indirectly calculate the volume of void that is present.

In this chapter we focus on those techniques that directly give quantitative information about void characteristics. This includes intentional and intrinsic voids. We have discussed these void characteristics throughout this book, for example, volume fraction, structure (open or closed cell), size/distribution, and shape. We will outline these techniques, discuss the type of data they generate, and explain the application of those data to characterize the voids.

8.2 Microscopy

The use of microscopy is invaluable in imaging voids. Whether intrinsic or intentional, just seeing the image of voids is vital to understanding material behavior. Optical and electron microscopy equipment is commonplace. As such, these are frequently the first tools used when initiating a material-related project or investigation. In addition to qualitatively seeing voids, there are techniques that can be used to quantify characteristics of the voids.

8.2.1 Optical microscopy

Although the inventor of the optical microscope is debatable, the basic instrument has been in use since the 1600s. Optical microscopy uses light and lenses to magnify an object. The main components of an optical microscope are the light source, ocular lens, objective lens, focus knobs (coarse and fine focus), mechanical sample stage, diaphragm, and condenser. Most optical microscopes have a computer interface and a digital camera that allow for electronic image capture. The limitations are the resolution limit of $\sim 0.2\ \mu m$, limited depth of focus causing reduced image clarity, and difficulty in imaging light, weakly refracting objects, such as glass.

Voids in Materials. DOI: https://doi.org/10.1016/B978-0-12-819282-5.00008-9

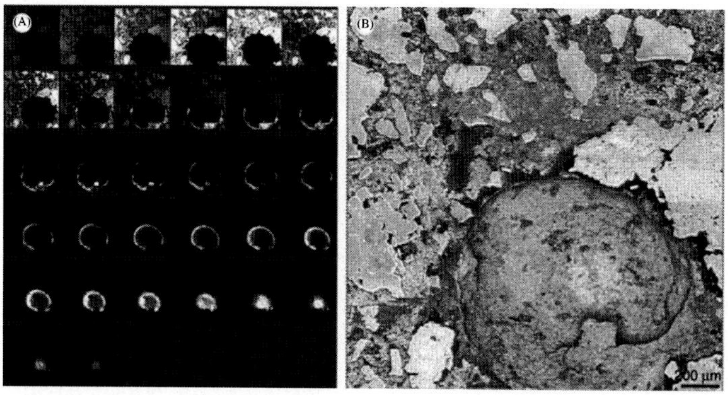

FIGURE 8–1 (A) A series of two-dimensional confocal images of a single pore in concrete at consecutively deeper focal planes. (B) The reconstruction of the images in (A) can provide the volume of the void. *Source: Adapted from Kurtis K.E.; EL-Ashkar N.H.; Collins C.L.; Naik N.N. (2003) Examining cement-based materials by laser scanning confocal microscopy. Cement and Concrete Composites 25 (7) 695–701.*

In 1958 Minsky patented the process of confocal imaging using a point source of light in a microscope that scans or rasters over a surface for increased resolution and decreased blurring (Minsky, 1957). Improvements were made and in 1980s the use of a laser as a light source led to laser scanning confocal microscopy (LSCM).

Confocal microscopy finds application in imaging and characterizing materials, because of better depth of field and elimination or reduction of background information away from the focal plane compared to conventional optical microscopy. LSCM is quite common and used widely in imaging emulsions, biological samples, medicine, and food samples, and to some extent in composites materials, geomaterials, and cement. Fig. 8–1A shows a series of images of a void in cement collected using LSCM. The series was made by consecutively changing the focal planes deeper into the void (Kurtis, EL-Ashkar, Collins, & Naik, 2003). The image in Fig. 8–1B is the reconstruction of the series of images in Fig. 8–1A. This technique of changing the focal planes can be used to determine the volume of individual surface voids and quantify surface roughness and wear in concrete, ceramics, and textiles (Becker, Grousson, & Jourlin, 2001).

8.2.2 Electron microscopy

8.2.2.1 Scanning electron microscopy (SEM)

An SEM uses electrons, rather than light, to image samples. The magnification in an SEM can be 250 times greater than that in an optical microscope. The SEM reaches greater magnification because the wavelength of an electron beam is smaller than the wavelength of light. The wavelength of visible light is ~550 nm. The wavelength (λ)

in nanometers of an electron beam can be calculated based on the accelerating voltage (V) of electron gun by the equation:

$$\lambda = \frac{1.23}{V^{1/2}}. \tag{8.1}$$

Typically one can vary the accelerating voltage from $\sim$2 to 100 kV. The electron beam in an SEM rasters or scans across a specific area, building an image pixel by pixel. This is in contrast to optical microscopy and transmission electron microscopy (TEM) where the entire field of view is captured instantaneously. To increase the magnification, the scan area is simply decreased. The depth of field in an SEM is much better than in an optical microscope, meaning different levels of a sample can be in focus at the same time.

The basic components of an SEM are an electron gun, condenser lens, scanning coils, and objective lens. The two principal modes of imaging in an SEM are: secondary electrons (SEs) and backscattered electrons (BSEs). As the high-energy electron (primary electrons) beam rasters across the sample, it ejects electrons from atoms on or near the sample surface. The ejected SEs have lower energy than the primary electrons. The SE detector has a positive charge bias, which attracts the SEs and converts them into pixels, which are then assembled into images. SE images generally contain better topographical information than other modes. BSEs are also ejected from the sample, but they come from a larger material volume below the sample surface. BSEs have higher energy than SEs so they can overcome the positive charge bias of the SE detector. BSEs can give more information on chemical composition than other modes. BSEs from heavier atoms will give a stronger signal than BSEs from lighter atoms. For example, molybdenum, atomic number 42, will appear brighter than aluminum, atomic number 13. This is called Z (atomic number) contrast. Since SEs are more effective for imaging topographic information and surface features, they are generally used to image voids and porosity in a sample. Fig. 8–2 is a secondary electron SEM of a silica aerogel. As illustrated in Fig. 8–2, aerogels are highly porous and the silica aerogel density can be as low as 0.003 g/cm^3. The SEM image shows both the nanometer-scale interstitial voids between the silica particles and the larger micrometer-scale porosity.

8.2.2.2 Transmission electron microscopy (TEM)

TEM is similar to SEM in that it uses electrons rather than photons to image materials. Unlike SEM, TEM does not raster across a sample, creating SEs and BSEs. Instead, the electron beam passes through the sample and creates a snapshot of it. The main components of a TEM are the electron gun, condenser lens, diffraction lens, projector lens, and viewing screen.

A high accelerating voltage is used to generate the electrons with the electron gun. General purpose transmission electron microscopes have an accelerating voltage of

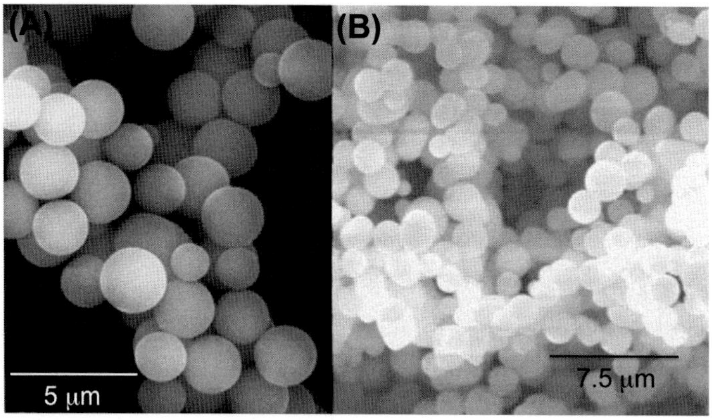

FIGURE 8–2 Silica network of particles make the silica aerogel at (A) high magnification and (B) low magnification. Note the presence of nanometer and micrometer pore structure. SEM. *Source: Adapted from Soleimani Dorcheh A.; Abbasi M.H. (2008) Silica aerogel; synthesis, properties and characterization. Journal of Materials Processing Technology 199 (1–3) 10–26.*

approximately 100–300 kV. The highenergy beam of electrons travels in vacuum through a column having multiple electromagnetic lenses and the beam interacts with the solid sample. In order for the beam to be transmitted through the sample, the sample is usually very thin. In practice it is about 5–10 nm thick. The electron beam strikes the sample and the transmitted electrons are converted to light and form an image.

TEM can be used to image intentional as well as an intrinsic porosity. In general, the lighter areas in a TEM image represent the areas where more electrons can pass through, whereas the darker regions represent denser areas of the sample where fewer electrons could pass. This contrast can provide information about topography, shape, and size of the features.

Fig. 8–3 shows intentional voids in an alumina aerogel (Poco, Satcher, & Hrubesh, 2001). The alumina aerogel was made by sol–gel process and the microstructure is described as acicular or needlelike. It is straightforward to distinguish between the alumina and void in the microstructure; however, quantitative assessment of pore characteristics is difficult in an aerogel sample. However, pore characteristics of traditional foams and cellular materials are more easily quantifiable. Additionally, it should be noted that in a typical TEM image one will be looking at a very small area of the sample and care must be taken to ensure that it is representative of the bulk material.

High resolution transmission electron microscopy (HRTEM) is capable of resolving details down to 0.1 nm. With this high resolution, one can obtain quantitative information at the atomic level; atomic structure, atomic-scale intrinsic defects, defect dimensions, and lattice parameters (Meyer, 2014). Fig. 8–4A is an HRTEM image of graphene, a single atomic layer of connected carbon atoms arranged in a

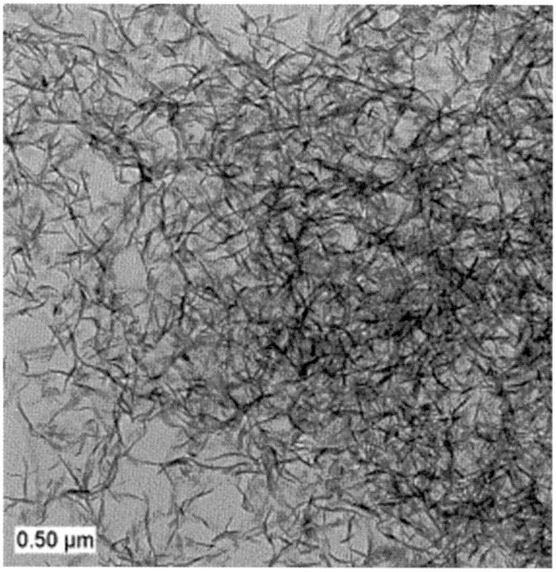

FIGURE 8–3 Image of alumina aerogel. Since a TEM detects transmitted electrons, the darker areas are alumina while the grey background is void within the aerogel structure. TEM. *Source: Adapted from Poco J.F.; Satcher J.H. Jr.; Hrubesh L.W. (2001) Synthesis of high porosity, monolithic alumina aerogels. Journal of Non-Crystalline Solids 285 (1–3) 57–63.*

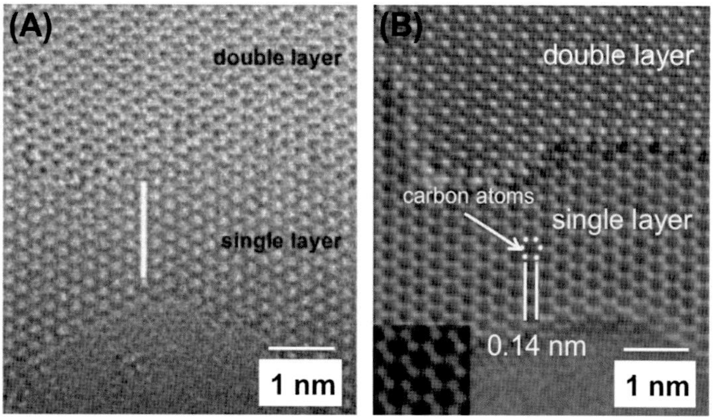

FIGURE 8–4 Image of (A) a single and double layer of graphene. (B) A three dimensional reconstruction of a series of through-focal HRTEM images showing clearly the single and double layer, the 0.14 nm distance between the carbon atoms and the central intrinsic void. The inset in (B) shows the hexagonal atomic arrangement of the carbons in the single layer graphene structure. *Source: Adapted from Jinschek J.R.; Yucelen E.; Calderon H.A.; Freitag B. (2011) Quantitative atomic 3-D imaging of single/double sheet graphene structure. Carbon 49 (2) 556–562.*

honeycomb pattern. Taking a series of HRTEM through-focal images and reconstructing them into a three-dimensional image, one can get an improved image of the atomic structure. Fig. 8–4B is a three-dimensional image of graphene constructed from a series of 19 through-focal images. This technique of HRTEM atomic-scale construction of three-dimensional images is known as *exit wave reconstruction.* The exit wave reconstruction in Fig. 8–4B allows one to see the atomic structure in three dimensions, the 0.14 nm distance between carbon atoms in the honeycomb, and the intrinsic void in the center of the honeycomb.

8.3 Positron annihilation lifetime spectroscopy (PALS)

A positron (e^+) is the subatomic antiparticle of an electron. Predicted by Dirac (1930); it was first observed 3 years later by Anderson (1933). Since the 1960s, PALS has been used to investigate porous materials (Gidley, Peng, & Vallery, 2006), defects, and free volumes in materials. PALS is not the only positron technique to characterize materials, there are several other techniques also. Examples of other techniques that use positrons are transmission positron microscopy and positron reemission microscopy (Oka, Jinno, & Fujinami, 2009).

Positrons have a charge of +1 and are typically generated using a Na source. In the PALS technique the Na decays to Ne and emits a positron beam. When positrons enter a material, they quickly loose energy and slow down. They can combine with an electron in the material to form a positronium. There are two types of positroniums: *ortho-positronium* (*o*-Ps) and *para-positronium* (*p*-Ps). A detailed discussion of positroniums, their interactions, and possible paths of annihilation is beyond the scope of this book. We will focus on the lifetime of *o*-Ps because by monitoring the number of *o*-Ps created and their annihilation time, one can measure void size, void connectivity, and void volume fraction in a solid (Gidley et al., 2006).

The most important parameter for characterizing voids via PALS is the positron lifetime; that is, the time from positron creation to annihilation. The characteristic lifetime of *o*-Ps is 242 ns in vacuum. This annihilation results in three γ-photons being emitted. When *o*-Ps are created in a solid, they attracted to areas of nuclear low charge density, such as vacancies, voids, surfaces, and free volumes, see Fig. 8–5. In this low charge density area, *o*-Ps will move through the void, deflect off the wall, and interact with the bound electrons, annihilate and emit two photons. It has been estimated (Gidley et al., 2006) that *o*-Ps can undergo 10^6 collisions before annihilation and emit two γ-photons.

A positron that enters a solid will scatter and quickly lose energy. A positron could then combine with an electron and form o-Ps. The o-Ps is attracted to and will get trapped in the low charge density regions, such as a void. Within this region, o-Ps will interact with bound electrons and shorten its lifetime. The o-Ps will be annihilated and emit two γ-photons.

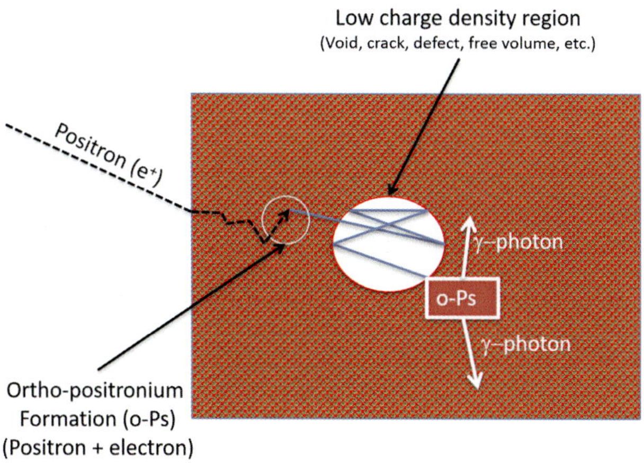

FIGURE 8–5 Schematic of the positron annihilation lifetime spectroscopy method.

Annihilation time of *o*-Ps within a void is on the order of 1–2 ns compared to its characteristic lifetime in vacuum of 242 ns. This activity results in the positron undergoing annihilation with an electron and emitting two γ-photons. The larger the volume of void, the longer the time it takes to reach the $\sim 10^6$ collisions. Assuming the void is spherical, one can calculate the radius of the void, R (expressed in Å), by the following semiempirical equation (Eldrup, Lightbody, & Sherwood, 1981; Nakanishi, Wang, Jean, & Sharma, 1988):

$$\tau_3 = \frac{1}{2}\left[1 - \frac{R}{R_0} + \frac{1}{2\pi}\sin\left(\frac{2\pi R}{R_0}\right)\right]^{-1} \tag{8.2}$$

and $R_0 = R + \Delta R$, where τ_3 is the *o*-Ps lifetime in nanoseconds and ΔR is electron layer thickness ($=1.66$ Å).

The volume of the void, $V\upsilon$, is then calculated using R from Eq. (8.2) as:

$$V_\upsilon = \frac{4}{3}\pi R^3. \tag{8.3}$$

The volume percent of void, $f\upsilon$, can be calculated using the following semiempirical equation (Wang, Nakanishi, Jean, & Sandreczki, 1990):

$$f_v = CV_v I_3, \tag{8.4}$$

where C has been empirically determined to be 0.0018, $V\upsilon$ is the volume of the void calculated from Eq. (8.3), and I_3 is the intensity (%) of the τ_3 peak taken from the lifetime spectra data.

Winberg, Eldrup, and Maurer (2004) investigated the change in void volume (free volume) with temperature of polydimethylsiloxane (PDMS) using PALS. Fig. 8–6

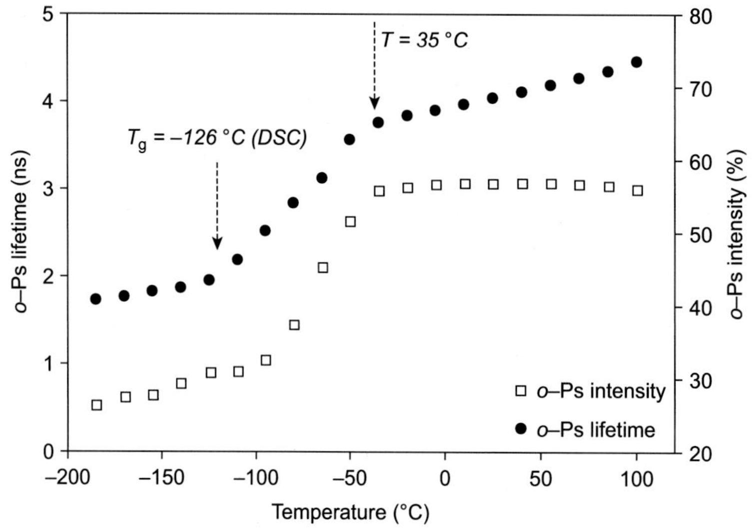

FIGURE 8–6 *o*-Ps lifetime and intensity of polydimethylsiloxane as a function of temperature. Note the rapid increase in intensity and lifetime immediately above the *Tg* of − 126°C. PALS. *Source: Adapted from Winberg P.; Eldrup M.; Maurer F.H.J. (2004) Nanoscopic properties of silica filled polydimethylsiloxane by means of positron annihilation lifetime spectroscopy. Polymer 45 (24) 8253–8264.*

shows a graph of the *o*-Ps lifetime (τ_3) and intensity (I_3) as a function of the temperature of PDMS. The temperature range of the PDMS includes the glass transition temperature (T_g) of − 126°C. There is a discontinuity at − 126°C, above which there is an increased slope in both sets of data. The increase in *o*-Ps intensity indicates an increase in the number of voids while the increase in *o*-Ps lifetime indicates that the volume of the individual voids increases with temperature. At − 35°C there is another discontinuity in the data. Above − 35°C the *o*-Ps remains approximately constant indicating the number of voids remains constant. In this same range, *o*-Ps lifetime increases slightly, indicating a slower increase in void volume with temperature.

Taking the *o*-Ps lifetime (τ_3) values from Fig. 8–6, Winberg et al. calculated the void volume in the PDMS as a function of temperature using Eqs. (8.2) and (8.3). Fig. 8–7 is a plot of data (squares) from Winberg et al. (2004). The trend in void volume with temperature is similar to *o*-Ps lifetime with temperature. The PDMS void volume increases by a factor of approximately three between − 126 and − 35°C.

8.4 Three-dimensional imaging

There have been many advances in sensing and visualization of data. These have allowed researchers to make advances examining the internal structure of matter and obtain three-dimensional images, which provide much more useful information

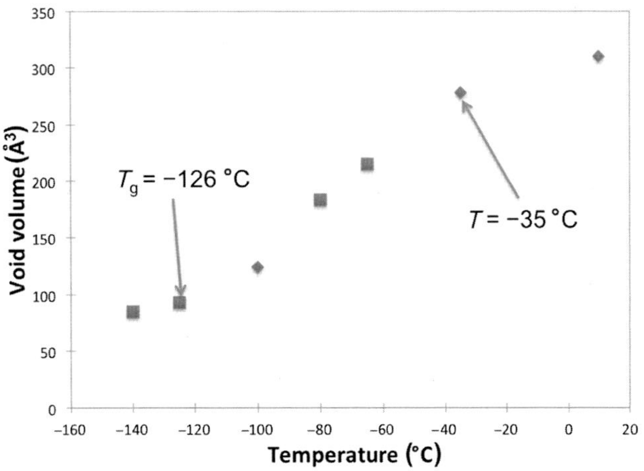

FIGURE 8–7 Void volume in polydimethylsiloxane as a function of temperature. The (■) represent calculation by Winberg et al. while the (♦) represent the void volumes that were calculated by using their data. *Source: Adapted from Winberg P.; Eldrup M.; Maurer F.H.J. (2004) Nanoscopic properties of silica filled polydimethylsiloxane by means of positron annihilation lifetime spectroscopy. Polymer 45 (24) 8253–8264.*

than the conventional two-dimensional imaging techniques. We describe below the basic principles involved in such techniques.

There are many different types of three-dimensional imaging techniques used to investigate material characteristics. Some of these techniques are destructive while others are nondestructive. Destructive techniques involve serial sectioning or slicing of the sample, two-dimensional imaging each slice and reconstructing a three-dimensional image. Sectioning involves removal of a sample, layer-by-layer. This can involve grinding, polishing, ion milling, and microtomy. The choice of the techniques used to image each slice is determined by the information ones needs to collect during the reconstruction. Optical microscopy, SEM, TEM, and atomic force microscopy can be used if topographical and porosity information are needed. If chemical composition information is required, energy dispersive X-ray analysis or other analytical techniques can be used. Reconstruction involves assembling a series of two-dimensional slices and transforming them, with the aid of a computer, into a three-dimensional image.

Nondestructive imaging requires probing of the material at specific depths without serial sectioning. Fig. 8–8 plots spatial resolution against imaging depth for various mechanical and electromagnetic energies for different techniques that have been used to probe and characterize materials (Appel, Anastasio, Larson, & Brey, 2013). The techniques include confocal, nuclear, X-ray, ultrasound, photoacoustic microscopy, magnetic resonance imaging, multiphoton fluorescence microscopy, and optical

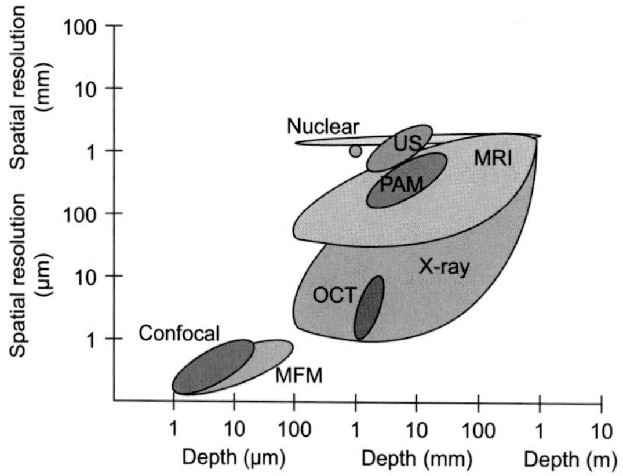

FIGURE 8–8 Approximate ranges of spatial resolution and imaging depth achievable by imaging techniques. Ultrasound (US), photoacoustic microscopy (PAM), magnetic resonance imaging (MRI), multiphoton fluorescence microscopy (MFM), and optical coherence tomography (OCT). *Source: Adapted from Appel A.A.; Anastasio M.A.; Larson J.C.; Brey E.M. (2013) Imaging challenges in biomaterials and tissue engineering. Biomaterials 34 (28) 6615–6630.*

coherence tomography. As Fig. 8–8 indicates, X-rays have a large range of imaging depth (100 μm to 1 m) and resolution (1 μm to several millimeters). As such, X-rays represent a very valuable tool for nondestructive, three-dimensional imaging of a wide range of materials including foams, biomaterials, and composites. X-ray computer tomography or CT-scan is frequently used to nondestructively characterize porosity size, size distribution, shape, and location within a sample.

There are two major types of X-ray sources in CT scans, conventional X-ray tube and a synchrotron source. Laboratory scale CT systems use X-ray tubes. In this system an electron beam is accelerated by a large potential difference onto a small target. When the electrons strike the target, X-rays of a characteristic energy are emitted. In a synchrotron, X-rays are produced in a fundamentally different way. Synchrotrons are designed to increase the intensity of the X-ray beam. Electrons travel near the speed of light in a storage ring of an electromagnetic field. When the electrons accelerate by changing direction within this field, a continuous, high intensity X-ray beam is emitted. The synchrotron-based scanning provides a much higher image contrast than laboratory-based scanners. The scanning times are comparable. The major drawbacks of a synchrotron are the cost and the difficulty in imaging high Z materials.

In addition to foams, biomaterials, and composites, X-ray tomography can be used to image processing (intrinsic) voids in solders. The presence of porosity in solders is very important to characterize. The volume percent, size, shape, clustering, and the position of voids can lead to drastically different behavior of a solder joint (Padilla, Jakkali, Jiang, & Chawla, 2012). Padilla et al. (2012) used X-ray tomography to characterize, in three

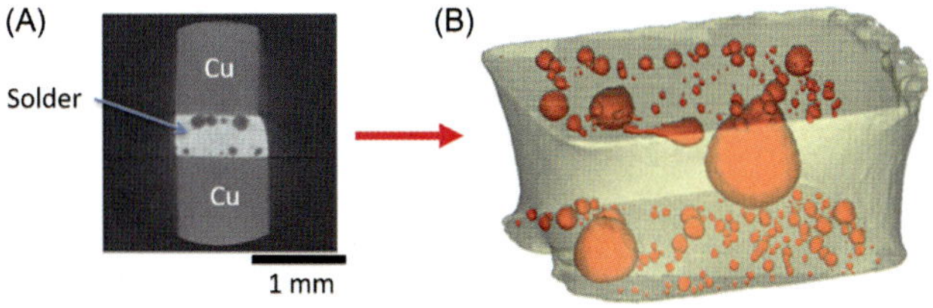

FIGURE 8–9 (A) CT-scan and (B) three-dimensional rendering of a single-lap shear joint of Sn-3.9Ag-0.7Cu alloy between two copper bars. *Source: Adapted from Padilla E.; Jakkali V.; Jiang L.; Chawla N. (2012) Quantifying the effect of porosity on the evolution of deformation and damage in Sn-based solder joints by X-ray microtomography and microstructure-based finite element modeling. Acta Materialia 60 (9) 4017–4026.*

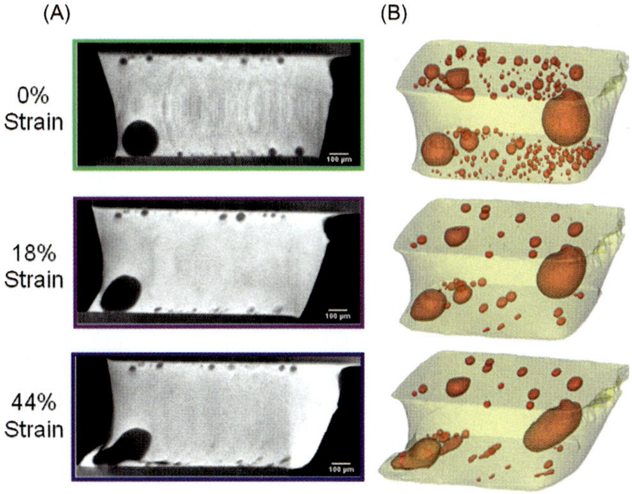

FIGURE 8–10 Tin-based solder joint (A) two-dimensional cross-sectional view and (B) three-dimensional reconstruction of the X-ray tomography images. The series of images at 0%, 18%, and 44% shear strain makes a four-dimensional microstructural study on the deformation of voids.

dimensions, the initial location and void characteristics of a lead-free solder, see Fig. 8–9. Including time as another variable, we can call it a four-dimensional study; it is only possible because of the nondestructive nature of X-ray tomography.

In addition to the initial condition of the sample, X-ray tomography was used to track the evolution of the solder microstructure during shear deformation. The joint was sheared at a constant strain rate of 10^{-3}/s and once a given strain was applied, the sample was removed from the test fixture and imaged again using X-ray tomography. A total of 745 slices of the solder joint was used at each of the five shear strain increments (0, 0.05, 0.12, 0.18, and 0.44). Fig. 8–10 shows X-ray tomography images of the solder joint at shear strains of 0%, 18%, and 44%. The images give insight into

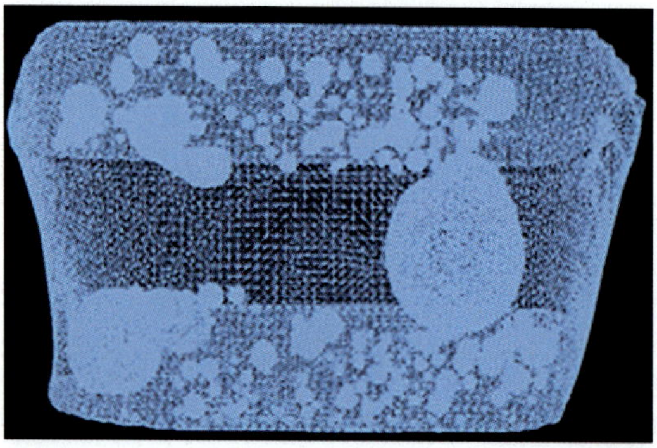

FIGURE 8–11 Meshed three-dimensional image created from an X-ray tomography image of a Sn-based solder joint (0% shear strain). *Source: Adapted from Padilla E.; Jakkali V.; Jiang L.; Chawla N. (2012) Quantifying the effect of porosity on the evolution of deformation and damage in Sn-based solder joints by X-ray microtomography and microstructure-based finite element modeling. Acta Materialia 60 (9) 4017–4026.*

localized deformation as well as damage evolution (void formation) and failure mechanisms.

Another benefit of using X-ray tomography is that the three-dimensional image can be imported into finite element modeling (FEM) software for simulations (Maire & Withers, 2014). Fig. 8–11 is the solder joint at 0% shear strain generated by X-ray tomography and meshed by FEM software (Padilla et al., 2012). The experimental shear strain can then be simulated by the software. A detailed comparison of phenomena such as the evolution of localized strain, damage, voiding, failure, etc., can be done between the simulation and X-ray tomography images.

8.5 Gas adsorption

The method of gas adsorption is credited to Brunauer et al. in 1938 based on the work of Stephen Brunauer, P.H. Emmet, and Edward Teller (Brunauer, Emmett, & Teller, 1938). The method is based on the theory of gas adsorption onto surfaces and commonly used to characterize porosity in materials in a wide variety of applications. For example, gas adsorption is used to characterize heterogeneous catalysis, pharmaceuticals, filters, adsorbents, and molecular sieves such as zeolites used in separation processes. By monitoring gas adsorption and desorption characteristics from a solid surface, one can quantify characteristics of porosity at the nanometer scale. This phenomenon is based on adsorption of gas molecules on the surface of a porous solid to form a monolayer or multilayers. How a gas adsorbs/desorbs from a sample will be a function of the open cell porosity, pore size, capillary forces, etc. (Brunauer et al., 1938). A standard was developed by the

International Union of Pure and Applied Chemists for naming porosity based on three pore size regimes (Pierotti & Rouquerol, 1985):

1. Macropores—greater than 50 nm
2. Mesopores—between 2 and 50 nm
3. Micropores—less than 2 nm.

The pore size categories are somewhat arbitrary in that they were developed by using nitrogen as the adsorbing gas. They are based on transport mechanisms of the adsorbed nitrogen gas at its normal boiling point through the porosity. For example, macroporosity is indicative of bulk diffusion and viscous flow. Flow through mesopores is described by Knudsen diffusion, which is based upon the collisions between molecules and the pore walls at low pressure and the mean free path restricted by the geometry of the void space. In microporosity, the movement of nitrogen molecules is characterized by activated gas transport. When another gas having a different molecular size is used, the transitions between the porosity categories will change.

There are many types of analyses based on adsorption/desorption experiments, a common one is known as Brunauer–Emmett–Teller (BET) analysis. In BET analysis, a sample is placed in vacuum and a known amount of gas, generally nitrogen, is introduced at a constant temperature, typically 77 K. BET analysis is based on the characteristics of the nitrogen gas adsorbed as a function of pressure (i.e., *adsorption isotherm*) profile as the sample approaches equilibrium pressure. Then depending on the shape of the gas adsorption and desorption isotherm, see Fig. 8–12, one can interpret pore size distribution and geometry (Pierotti & Rouquerol, 1985; Sing, 1984). For example, Type II isotherm indicates that adsorption is reversible (i.e., desorption of the gas retraces the adsorption curve) and indicative of a nonporous or macroporous surface. The point labeled “B” is interpreted as the point at which a monolayer coverage is complete and multilayer formation begins. Although Type IV has a similar shape and point “B” when compared to Type II, the Type IV isotherm is not reversible. This isotherm has a hysteresis loop. The hysteresis loop is caused by capillary condensation in mesoporosity. There are other isotherms that we have omitted from Fig. 8–9.

One can calculate the radius of the mesospore in a material exhibiting a Type IV isotherm by using the Kelvin equation to first calculate the Kelvin radius (Thomson, 1871):

$$r_K = \frac{2\gamma V_m}{RT \ln \frac{p^0}{p}}, \tag{8.5}$$

where γ is the surface tension of the liquid condensate, V_m is the molar volume, R is the universal gas constant, T is the temperature (K), and p^0/p is the relative pressure at which condensation occurs. The shape of pore is typically considered to be either

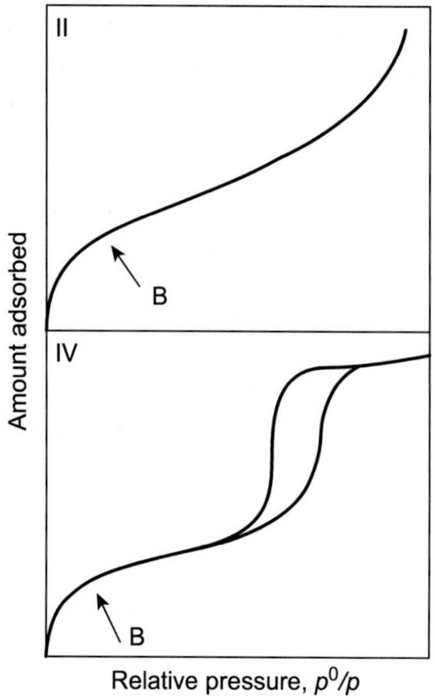

FIGURE 8–12 Characteristic curves, Type II and IV, of adsorption isotherm of gas onto solids with nanometer-scale surface porosity. The point labeled "B" in the Type II and IV isotherms indicates that a monolayer coverage is complete and multilayer adsorption begins.

slit-shaped or cylindrical. One can use r_K to calculate either the pore radius (r_p) or slit width (d_s). Assuming a spherical pore, r_p is calculated by:

$$r_p = r_K + t, \tag{8.6}$$

or if it is a slit-shaped pore:

$$d_s = r_K + 2t, \tag{8.7}$$

where t in both of these equations is the thickness of the adsorbed layer. For detailed interpretation of isotherms and calculations the reader is referred to Pierotti and Rouquerol, 1985

8.6 Chromatographic porosimetry

8.6.1 Introduction

In the most general terms, chromatography is a separation technique composed of a stationary phase, a mobile phase, and an interaction phenomenon between these phases that is the basis for the separation process. The phenomenon separating unlike molecules can be, for example, polarity, charge, binding affinity, or size differential. The final component of this technique is an appropriate detector placed at the exit of the stationary phase to identifying the mobile phase molecules entering the detector (with time) creating a chromatogram.

There are many different types of chromatography. The specific type of chromatography depends on the components present in stationary and mobile phases, whether it is a solid, liquid, or gas, and on the basis of separation. For example, one of the techniques discussed below is a solid stationary phase with a liquid mobile phase and the porosity present in the stationary separates the molecules in the mobile phase based on size. This technique in called size exclusion chromatography (SEC). The second type we will discuss is an absorption technique called gas chromatography (GC) where there is a gaseous mobile phase, solid stationary phase separating based on solid surface interaction with the gas phase (i.e., polarity, acidity, molecular area, and electron donor/acceptor number). In traditional SEC and GC the solid stationary phase is well-characterized and a separation is carried out on molecules in the mobile phase. To make chromatography a useful tool in characterizing the voids in solids, the focus of this book, these chromatographic techniques no longer are used to identify components in the mobile phase. Instead, they use molecules in the mobile phase to probe the porosity present in the stationary phase. When this occurs and the molecules in the mobile phase are well-characterized and solid stationary phase is then uncharacterized, it is considered an "inverse" technique. Therefore, *inverse* size exclusion chromatography (ISEC) uses a well-characterized probe polymer molecule(s) in the mobile phase to probe and characterize the pore structure of a solid stationary phase. In *inverse* gas chromatography (IGC), a well-characterized gaseous probe molecule is used to investigate the pore structure in a solid stationary phase, see Fig. 8–13.

8.6.2 Inverse gas chromatography (IGC)

IGC is useful in analyzing porosity of synthetic polymers, biological polymers, copolymers, polymer blends, glasses, carbon fibers, solid foods, and pharmaceutical powders and can be done in a wide range of temperatures (Giridhar, Manepalli, & Apparao, 2017). There are two type of IGC, and the naming of them depends of the stationary phase. If the stationary phase is a liquid the technique is inverse gas-liquid chromatography, IGLC. If the stationary phase is a solid, it is referred to as inverse gas-solid chromatography, IGSC. This section will focus on IGSC. The following basic conditions should be satisfied: (1) no adsorption interactions between probe molecule and porous material: (2) chromatography to be carried out in a quasi-equilibrium mode; and (3) no intermolecular interactions in the solution or in the stationary phase (Gorbunov, Solovyova, & Pasechnik, 1988).

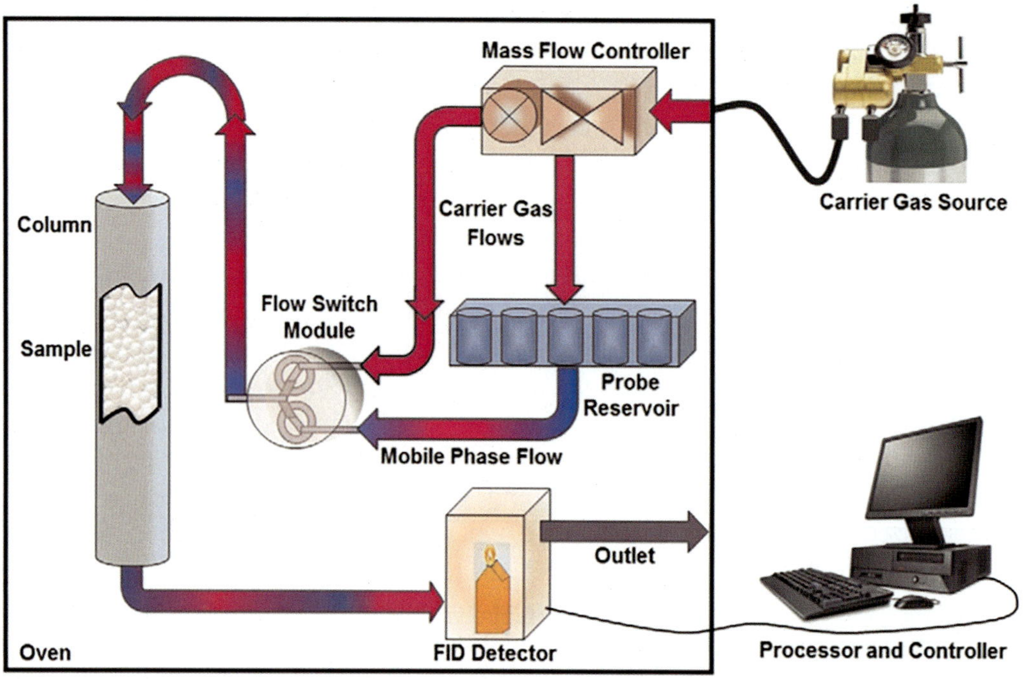

FIGURE 8–13 Schematic of an inverse gas chromatography analysis for characterizing the porosity in the stationary phase contained in the sample column. *Source: Adapted from Mohammadi-Jam, S., & Waters, K.E. (2014). Inverse gas chromatography applications: A review advances in colloid and interface, Science* ***212****, 21–44. https://doi.org/10.1016/j.cis.2014.07.002.*

The mobile phase in an IGSC system is typically a probe gas that is injected into a carrier gas. The probe is carried through the sample column by a constant flow of carrier gas. The most common carrier gases are high purity inert gases such as helium, argon, and nitrogen. Since there is expected to be absorption between the probe gas and the solid within the sample column, it is common that the mobile phase contain a tracer gas that does not interact to determine the *dead time* in the system. The dead time is defined as the time a probe molecule would travel through the column without interacting with the solid. A common tracer gas is methane, argon, nitrogen, and hydrogen (Thielmann, 2004).

The IGSC technique involves filling a column with a stationary phase of the solid material one would like to characterize. The solid material may be in the form of a thin coating on an inert substrate, a finely divided solid, strands of fiber, or a thin polymeric coating on the column wall.

As depicted in Fig. 8–13, a stream of carrier gas is saturated with a probe molecule by passing the carrier gas through a liquid-phase reservoir of the probe molecule. The concentration of the probe molecule can be changed by controlling the temperature of the system. This gaseous stream is injected into the sample column.

The carrier gas should be inert and not interact with the solid within the column while the probe molecule should be absorbed onto and then desorbed from the solid surface. The higher the surface area and energy, the higher the gross retention time. The detection device measures this gross retention time of the probe molecule. Some common detection instruments are flame ionization detector, thermal conductivity detector, and mass spectrometric detectors (Thielmann, 2004).

8.6.3 Inverse size exclusion chromatography

Using IGC for porosity characterization has been successfully demonstrated on synthetic polymers, biological polymers, copolymers, polymer blends, glasses, carbon fibers, solid foods, and pharmaceutical powders (Giridhar et al., 2017). The carrier liquid can either be aqueous or solvent-based. The process of size exclusion is shown in Fig. 8–14. Larger components of the liquid probe mixture will enter either a smaller number of pores than smaller probe components. Because of this process, the larger molecules will exit the column first and followed later in time by the smaller molecules. Elution in

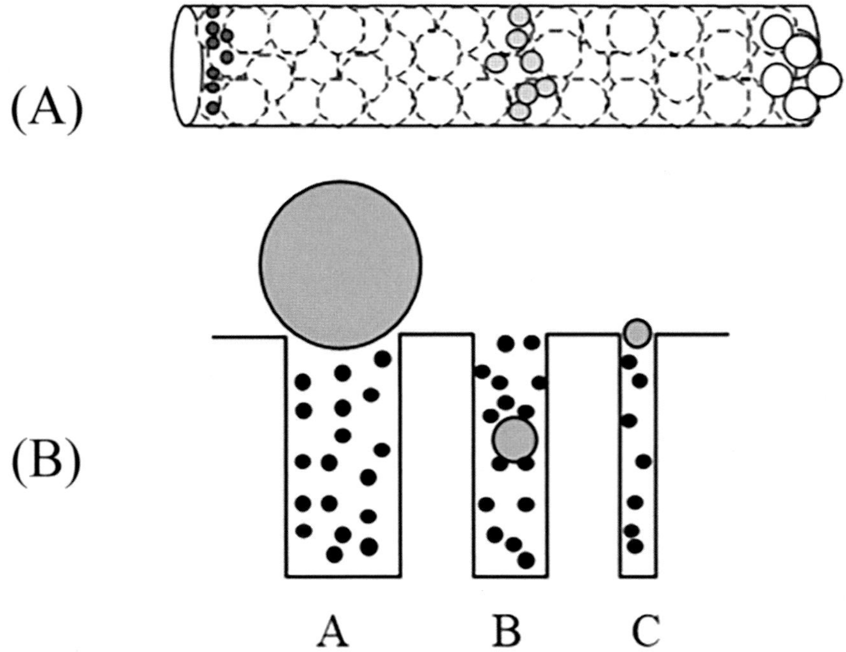

FIGURE 8–14 Schematic representation of the size separation in an ISEC sample column. Pores A, B, and C are decreasing in size. As such, they allow only those probe molecules of appropriate molecular diameters to enter. Black circles represent carrier liquid and the gray circles represent probe molecules of various molecular diameters. *Source: Adapted from Kongdee, A., Bechtold, T., Burtscher, E., & Scheinecker, M. (2003). Inverse size exclusion chromatography—A technique of pore characterisation of textile materials. Lenzinger Berichte, 82(96-101), 193.*

ISEC is in the reverse order of molecular size, and the method may be thought of as an "inverse-sieving" technique (Striegel, 2017).

The stationary phase in ISEC is the material being characterizing. The solid material may be in the form of a thin coating on an inert substrate, a finely divided solid, strands of fiber, or a thin polymeric coating on the column wall (Giridhar et al., 2017). In ISEC, the size differential between the probe molecule(s) will be what drives the separation process and the basis for the characterization of the pore size and size distribution. Fig. 8–14 shows ability of a probe molecule to enter that pore based on the interaction of the pore size and probe molecule diameter. As a general rule (Nelson & Oliver, 1971), the pore diameter should be three times larger than the largest molecular probe that can enter that pore.

The raw data generated by ISEC systems are elution times and concentrations. Specifically, the times and concentrations of probe molecules reaching the detector at the exit of the sample column. This data can be used to determine mean void size and void volume (mL/g). One of the advantages of ISEC is the fact that it is a wet method. Many of the applications where pore dimensions are important, such and the treatments and uptake of dyes and pigments (Sun, Zhou, & Xing, 2016), into yarns, fabrics, and textiles (Kongdee, Bechtold, Burtscher, & Scheinecker, 2003), happen in aqueous or solvent environments. Eq. (8.1) is used for calculating accessible pore volume, V_i (mL/g)

$$V_i = \frac{(T_e - T_o) \times F}{W} \tag{8.8}$$

Where T_e is the elution time of each probe molecule (min), T_o is the elution time of the largest probe molecule (min), F is the flowrate of the mobile phase and W is the weight of the material in the column (g).

Kongdee et al. (2003) used ISEC to investigate the pore characteristics of cellulose-based yarns with various treatments, see Table 8–1. Five different yarns, all of them

Table 8–1 Mean pore sizes for various yarns and treatments.

	Mean pore size (Å)		
		Treatment	
Cellulose yarn type	**As received**	**No tension**	**Tension**
Non-crosslinked	38	33	30
Crosslinked (Type 1)	32	32	30
Crosslinked (Type 2)	34	34	33
Micromodal	37	40	30
Viscose	30	31	32

Source: Adapted from Kongdee, A., Bechtold, T., Burtscher, E., & Scheinecker, M. (2003). Inverse size exclusion chromatography—A technique of pore characterisation of textile materials. *Lenzinger Berichte*, 82(96-101), 193.

cellulose-derived and similar to rayon, were tested using ISEC. Each of the five yarns were treated in a water bath and dried at 105°C. Additionally, in the water bath, each yarn type was treated with and without being under a tensile load. The yarn in this case is the static phase and was packed into a stainless steel column. The probe molecules had molecular weights ranging from 62 to 4.5×10^6 daltons. Representative curves for the elution time for six probe molecules in the mobile phase are shown in Fig. 8–15. There were three different probe molecular weights of (poly)ethylene glycol (PEG) molecules, PEG 400, PEG 282, and PEG 194; one triethylene glycol, TEG 152; diethylene glycol, DEC 108; and ethylene glycol, EC 62; the number value after each of the chemical names indicates the molecular weight (MW). Knowing the molecular weight, the molecular diameter ($d_{molecular}$) of various forms of ethylene glycols can determined by the following equation:

$$d_{molecular} = 1.74(MW)^{0.4}. \tag{8.9}$$

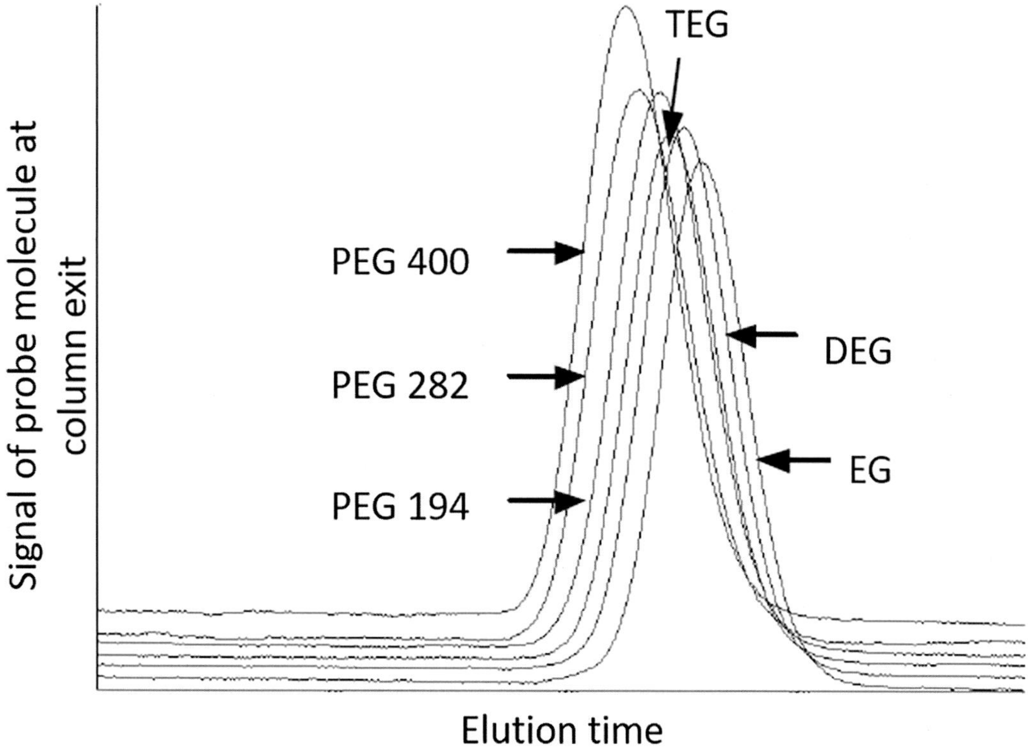

FIGURE 8–15 Representative chromatograph for ISEC. The elution times for various probe molecules, having specific molecular diameters, are an important parameter in determining the mean pore size of the stationary phase.

Knowing the elution times, molecular weights, and accessible pore volumes (using Eq. 8.8), one can determine the pore characteristics (Kongdee et al., 2003). A summary of pore size results for each of the yarns and treatments are show in Table 8–1.

References

Anderson, C. D. (1933). The positive electron. *Physical Review, 43*(6), 491.

Appel, A. A., Anastasio, M. A., Larson, J. C., & Brey, E. M. (2013). Imaging challenges in biomaterials and tissue engineering. *Biomaterials, 34*(28), 6615–6630.

Becker, J., Grousson, S., & Jourlin, M. (2001). Surface state analysis by means of confocal microscopy. *Cement and Concrete Composites, 23*(2–3), 255–259.

Brunauer, S., Emmett, P. H., & Teller, E. (1938). Adsorption of gases in multimolecular layers. *Journal of the American Chemical Society, 60*(2), 309–319.

Dirac, P. A. M. (1930). Note on exchange phenomena in the Thomas atom. *Mathematical Proceedings of the Cambridge Philosophical Society, 26*(03), 376–385.

Eldrup, M., Lightbody, D., & Sherwood, J. (1981). The temperature dependence of positron lifetimes in solid pivalic acid. *Chemical Physics, 63*(1), 51–58.

Gidley, D. W., Peng, H., & Vallery, R. S. (2006). Positron annihilation as a method to characterize porous materials. *Annual Review of Materials Research, 36*, 49–79.

Giridhar, G., Manepalli, R. K. N. R., & Apparao, G. (2017). In S. Thomas, R. Thomas, A. K. Zachariah, & R. K. Mishra (Eds.), *Chapter 3—Size-exclusion chromatography* (pp. 51–65). Cambridge, USA: Elsevier. Available from https://doi.org/10.1016/B978-0-323-46139-9.00003-7.

Gorbunov, A. A., Solovyova, L. Y., & Pasechnik, V. A. (1988). Fundamentals of the theory and practice of polymer gel-permeation chromatography as a method of chromatographic porosimetry. *Journal of Chromatography A, 448*, 307–332. Available from https://doi.org/10.1016/S0021-9673(01)84594-5.

Jinschek, J. R., Yucelen, E., Calderon, H. A., & Freitag, B. (2011). Quantitative atomic 3-D imaging of single/double sheet graphene structure. *Carbon, 49*(2), 556–562.

Kongdee, A., Bechtold, T., Burtscher, E., & Scheinecker, M. (2003). Inverse size exclusion chromatography—A technique of pore characterisation of textile materials. *Lenzinger Berichte, 82*(96-101), 193.

Kurtis, K. E., EL-Ashkar, N. H., Collins, C. L., & Naik, N. N. (2003). Examining cement-based materials by laser scanning confocal microscopy. *Cement and Concrete Composites, 25*(7), 695–701.

Maire, E., & Withers, P. J. (2014). Quantitative X-ray tomography. *International Materials Reviews, 59*(1), 1–43.

Meyer, J. C. (2014). 5-Transmission electron microscopy (TEM) of graphene. In V. Skákalová, & A. B. Kaiser (Eds.), *Graphene* (pp. 101–123). Massachusetts: Woodhead.

Minsky, M. (1957). US Patent# 3013467, Microscopy Apparatus.

Mohammadi-Jam, S., & Waters, K. E. (2014). Inverse gas chromatography applications: A review advances in colloid and interface. *Science, 212*, 21–44. Available from https://doi.org/10.1016/j.cis.2014.07.002.

Nakanishi, H., Wang, S., Jean, Y., & Sharma, S. (1988). *Positron annihilation studies of fluids* (p. 292) Singapore: World Science.

Nelson, R., & Oliver, D. W. (1971). Study of cellulose structure and its relation to reactivity. Paper presented at the *Journal of Polymer Science Part C: Polymer Symposia, 36*(1) 305–320.

Oka, T., Jinno, S., & Fujinami, M. (2009). Analytical methods using a positron microprobe. *Analytical Sciences, 25*(837), 837–844.

Padilla, E., Jakkali, V., Jiang, L., & Chawla, N. (2012). Quantifying the effect of porosity on the evolution of deformation and damage in Sn-based solder joints by X-ray microtomography and microstructure-based finite element modeling. *Acta Materialia, 60*(9), 4017–4026.

Pierotti, R., & Rouquerol, J. (1985). Reporting physisorption data for gas/solid systems with special reference to the determination of surface area and porosity. *Pure and Applied Chemistry, 57*(4), 603–619.

Poco, J. F., Satcher, J. H., Jr, & Hrubesh, L. W. (2001). Synthesis of high porosity, monolithic alumina aerogels. *Journal of Non-Crystalline Solids, 285*(1–3), 57–63.

Sing K.S. (1984). Reporting physisorption data for gas/solid systems. *Fundamentals of Adsorption: Proceedings of the Engineering Foundation Conference Held at Schloss Elmau*, Bavaria, West Germany, May 6–11, 1983. American Institute of Chemical Engineers, p. 567.

Soleimani Dorcheh, A., & Abbasi, M. H. (2008). Silica aerogel; synthesis, properties and characterization. *Journal of Materials Processing Technology, 199*(1–3), 10–26.

Striegel, A. M. (2017). In S. Fanali, P. R. Haddad, C. F. Poole, & M. Riekkola (Eds.), *Chapter 10—Size-exclusion chromatography* (pp. 245–273). Cambridge, USA: Elsevier. Available from https://doi.org/10.1016/B978-0-12-805393-5.00010-5.

Sun, Z., Zhou, X., & Xing, Z. (2016). Effect of liquid ammonia treatment on the pore structure of mercerized cotton and its uptake of reactive dyes. *Textile Research Journal, 86*(15), 1625–1636.

Thielmann, F. (2004). Introduction into the characterisation of porous materials by inverse gas chromatography. *Journal of Chromatography A, 1037*(1), 115–123. Available from https://doi.org/10.1016/j.chroma.2004.03.060.

Thomson, W. (1871). On the equilibrium of vapour at a curved surface of liquid. *The London, Edinburgh, and Dublin Philosophical Magazine and Journal of Science, 42*(282), 448–452.

Wang, Y., Nakanishi, H., Jean, Y., & Sandreczki, T. (1990). Positron annihilation in amine-cured epoxy polymers—Pressure dependence. *Journal of Polymer Science Part B: Polymer Physics, 28*(9), 1431–1441.

Winberg, P., Eldrup, M., & Maurer, F. H. J. (2004). Nanoscopic properties of silica filled polydimethylsiloxane by means of positron annihilation lifetime spectroscopy. *Polymer, 45*(24), 8253–8264.

9 Characteristics and properties of porous materials

9.1 Introduction

Porous materials are composed of a continuous solid phase and either a continuous or discontinuous, generally, gaseous phase. These are very important materials in the worldwide economy. They are widely used in market sectors, such as the construction, automotive and aeronautical industries, renewable energy, sporting goods and leisure, cushioning and packaging, and biotechnology. Polymeric foams hold the largest global market share of any other foam material and is approximately 10% of annual consumption of polymer materials (Notario, Pinto, & Rodriguez-Perez, 2016). The global polymer foam market size was valued at $109.71 billion (US) in 2018 and is projected to expand at an annual growth rate of 4.0% from 2019 to 2025. Polymer foams that are most widely manufactured globally are polyurethane, polystyrene, polyolefin, melamine, phenolic, and polyvinyl chloride (Grand View Research, 2019). The metal foam market is projected to grow from $84 million (US) in 2019 to $103 million (US) by 2024, an annual growth rate of 4.2%. The metals making up the largest portion of that market are aluminum, nickel, and copper (Markets & Markets, 2019). The ceramic foams market was valued at $327.6 million (US) in 2017 and is projected to reach $441.7 million (US) by 2023, an annual growth rate of 5.2%. Silicon carbide, alumina, and zirconia are the most common ceramic foams manufactured globally (Markets & Markets, 2018).

Foams are the cellular form of materials wherein one uses voids by design to obtain the desired properties. The properties of such foams depend on the topology (shape and connectivity of cells) as well as the specific material that comprises the foam. In the most general sense, foams can be gaseous, liquid, or solid. The gaseous foams are part of astrophysics (e.g., galaxies), whereas liquid foams are common in the food and beverage industry (e.g., think of head of the beer). We will skip these two and concentrate on solid foams. The solid foams can be made from polymers, metals, and ceramics. Composite foams can be any mixture of these materials. In addition, the porosity can be a blown, unreinforced phase, and/or hollow particle phase.

This chapter discusses foams, which by definition, have intentional voids engineered into the structure. The mere addition of intentional voids is typically required

Voids in Materials. DOI: https://doi.org/10.1016/B978-0-12-819282-5.00009-0

to gain a functionality or property. One would not go through the process, time, cost, and effort of introducing voids into a material unless it served a function. As such, there are certain concepts and characteristics that are important when describing foams so they can be compared and distinguished. We first discuss some of these characteristics that are used to describe foams and then discuss the properties that result and relevant expressions.

9.2 General characterization

9.2.1 Cell size

The cell size is an important microstructural parameter. Many properties of foam depend on the cell size. Throughout this book, the reader will encounter intentional voids that range in size from one nanometer to tens of millimeters and these differences lead to performance differences and even different target applications. In Chapter 2, Intrinsic voids in crystalline materials: Ideal materials and real materials, we discussed the size effects crystalline domains can have on the performance of materials. The same can be said about the void domains as well. Miller and Kumar (2011) investigated the size effect of voids on impact energy per unit thickness of polyetherimide (PEI) foam with micrometer and nanometer scale porosity. The PEI with nanometer scale porosity had a cell size ranging from 50-100 nm while the foam with micrometer scale porosity had a cell size ranging from 2-5 μm. Miller and Kumar measured impact energy as a function of relative density and compared the nanocellular and microcellular materials by normalizing impact energy absorbed by sample thickness. The test was performed in accordance with ASTM D5420 (ASTM International, 2016) and the results for foamed PEI as well as fully dense PEI are presented in Fig. 9−1. These results indicate that at the same relative density, PEI with nanometer scale porosity absorbed more energy compared to PEI with micrometer scale porosity. Furthermore, these trendlines are opposite; nanocellular material absorbed more energy with increasing relative density while microcellular material absorbed less energy with increasing relative density. For all relative densities tested, the nanocellular foams outperformed the fully dense PEI, while all microcellular foams absorbed less energy than fully dense PEI.

In Chapter 10, Applications, we will discuss ceramic macrospheres used in deep-sea applications with a diameter as high as 100 mm. When characterizing foams with micrometer- and nanometer-sized porosity, one typically examines the cell structure from two-dimensional micrographs. The average diameter of a circular segment, *d*, can be obtained by taking the average of nominal cell diameters in a given micrograph. It is important to recognize that the average spherical diameter, *D*, is greater than the average circular segment diameter, *d*, obtained from a two-dimensional micrograph.

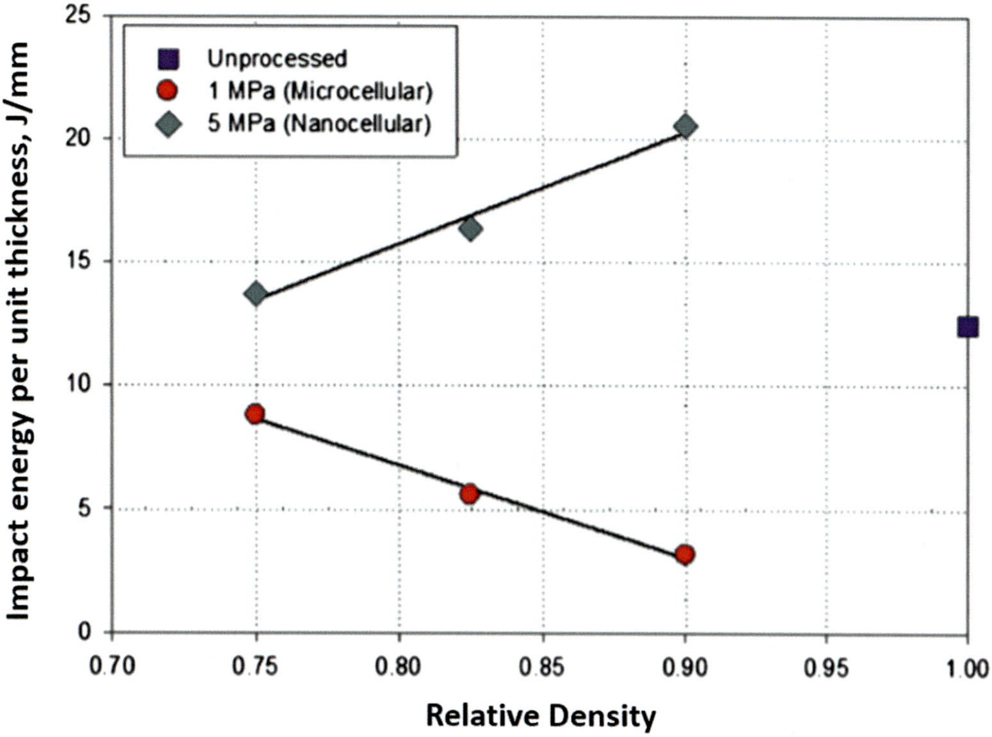

FIGURE 9–1 Porosity size effects on impact energy for nanocellular and microcellular PEI foams.

This is because the cells are randomly truncated along the depth. The ASTM standard D3576 gives the following expression relating d to D:

$$D = \frac{4d}{\pi}. \tag{9.1}$$

Computer tomography is a technique used to obtain three-dimensional images of cell structures in foams. The technique assembles individual two-dimensional slices of the foam and forms a three-dimensional image. Unlike the optical and electron microscopy two-dimensional imaging, there is very little sample preparation to obtain cell size and size distribution in three dimensions. Direct characterization of voids using both two- and three-dimensional techniques is discussed in Chapter 8, Void characterization techniques.

9.2.2 Open versus closed cell

The cells in a foam can be interconnecting (open cell), nonconnecting (closed cell), or a mixture of the two. An understanding of this structure is essential, as it will determine the performance in many applications. Shutov (2004) outlined the criteria to be met for a predominantly open cell structure:

1. Cells must have at least two discontinuous or broken faces.
2. An overwhelming majority of the cell struts must be shared by at least three cells.

When these conditions are not met then we have either a 100% closed cell structure or some mixture or ratio of open to closed cells. One can design a foam along this continuum of open cell/closed cell ratio resulting in properties that can vary widely. To quantify the percentage of closed cell and open cell porosity, we can follow ASTM standard D2662. A three-phase syntactic foam can also be assessed by ASTM D2662. There are two types of voids in a three-phase syntactic foam. A 100% closed cell, reinforced voids, defined by the wall of the hollow microballoon and interstitial or unreinforced voids, located between the microballoons. The interstitial void can be open or a closed cell depending on the volume percent of binder phase that is present.

The procedure to determine open/closed cell porosity in the ASTM D2662 standard involves the use of a gas pycnometer, which is used to determine the density of the sample. More specifically, the volume, V, of the sample can be determined to a high degree of precision. If the mass (m) of the sample is known, one can determine the density (m/V). To determine the open and closed cell content, the volume of two samples is first determined in a pycnometer. The geometries of the two samples are typically cubes but could also be cylinders. The volume percent of open cell porosity is calculated by using the following expression:

$$\text{Volume \% open cell porosity} = \frac{V_{\text{geo1}} - V_{\text{pyc1}}}{V_{\text{geo1}}} \times 100 \tag{9.2}$$

where V_{geo1} is the total geometric volume of both cubes and V_{pyc1} is the volume measured by the pycnometer.

During specimen preparation, porosity exposed on the surface will lead to a lower V_{pyc1}. This volume loss would be reflected in higher volume percent of open cell porosity. In principle, all foams will encounter this effect. There is a correction procedure if one is concerned that the V_{pyc1} reflects a significantly less volume because of this exposed surface porosity. To account for this loss in volume, each sample is sectioned into 8 smaller samples and the volume of all 16 smaller samples is again determined geometrically (V_{geo2}) and volumetrically (V_{pyc2}) in a pycnometer. The corrected open cell porosity is then calculated by the following expression:

$$\text{Volume \% open cell porosity} = \frac{V_{\text{geo2}} - 2V_{\text{pyc1}} + V_{\text{pyc2}}}{V_{\text{geo2}}} \times 100. \tag{9.3}$$

9.2.3 Reinforced versus unreinforced voids

Reinforced voids are present in composite foams, such as syntactic foams. A microballoon is a reinforced void and is a constituent in all syntactic foams. In a three-phase syntactic foam, by definition, there are also present unreinforced voids; these are the interstitial voids, between microballoons. One can use a three-phase syntactic foam to demonstrate how unreinforced and reinforced voids (Gladysz, Perry, Mceachen, & Lula, 2006) change mechanical properties. The system we will discuss is shown in Fig. 9–2; a three-phase syntactic foam system with carbon microballoons (CMBs) with a bismaleimide (BMI) binder.

To change the volume fraction of unreinforced/reinforced voids, we:

1. keep the syntactic foam density constant (keep the mass fractions of the CMB and binder constant),
2. keep the volume fraction of the binder phase constant, and
3. change the true particle density of the microballoon in the syntactic foam.

This is shown schematically in Fig. 9–3. The reinforced and unreinforced void volume fractions are altered by using either a high density CMB (0.45 g/cm^3) or a low density CMB (0.34 g/cm^3) to make the syntactic foams. While the CMB density changes, one needs to keep the binder volume percent constant (8.5%) and the syntactic foam density constant (0.30 g/cm^3). Note that even though the fraction of reinforced and unreinforced voids changes with the density of the CMBs, total void volume fraction is constant. Furthermore, the syntactic foam density remains 0.30 g/cm^3; therefore, the mass ratio of the microballoons and BMI is constant. Essentially, we are changing the packing arrangement of the CMBs. The syntactic foam with the 0.34 g/cm^3 CMB will have greater packing factor (volume fraction) than the syntactic foams made with the 0.45 g/cm^3 CMBs.

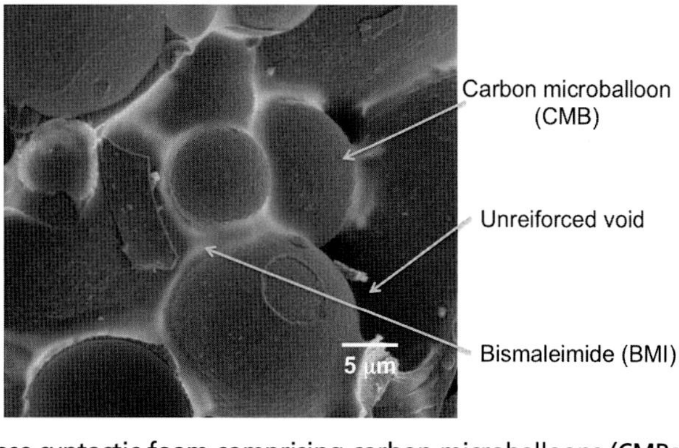

FIGURE 9–2 Three-phase syntactic foam comprising carbon microballoons (CMBs), bismaleimide (BMI), and unreinforced or interstitial void.

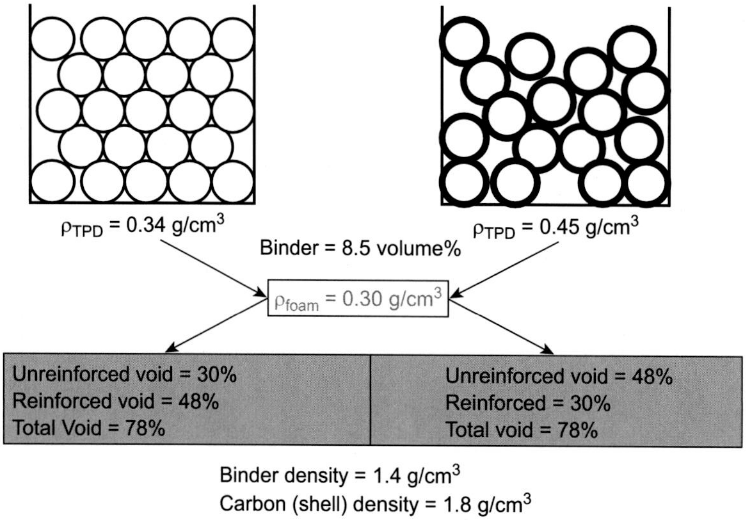

FIGURE 9–3 Two designs of a 0.30 g/cm^3 three-phase syntactic foam. The material on the left hand side has lower true particle density (TPD) microballoons resulting in a more tightly packed arrangement compared to the material on the right. The structure on the left has a lower unreinforced void content and greater reinforced void content resulting in enhanced mechanical properties.

We now discuss the compressive strength of the two syntactic foams discussed above; one made with 0.34 g/cm^3 CMBs and the other with 0.45 g/cm^3 CMBs. The effect of the reinforced/unreinforced void content on compressive strength is shown in Fig. 9–4. Let us follow the vertical dashed lineup from the x-axis at a total void fraction of ~0.791 until we reach the hollow circle (○). This hollow circle represents the total reinforced void content of the syntactic foam made with 0.45 g/cm^3 CMBs. Following the horizontal dotted line to the secondary y-axis (on the right), this sample has 0.37 reinforced void volume fraction. If we again follow the vertical dashed lineup to the solid circle (●), this sample with 0.37 reinforced void fraction has a corresponding compressive strength of 5.5 MPa. Following the vertical dashed lineup, we next intersect the solid square (■) indicating the syntactic foam made with 0.34 g/cm^3 microballoons. This syntactic foam has a compressive strength of 7.7 MPa. Following the vertical dashed lineup to the hollow square (□), we find that the sample made with the 0.34 g/cm^3 CMBs has 0.52 reinforced void fraction. Comparing these two syntactic foams, there is a 40% increase in strength when one increases the reinforced void content (using a lower-density microballoon) by 15%.

A similar comparison is made with regard to the compressive modulus of these two syntactic foams when both have the same density, see Fig. 9–5. At 0.791 total void volume fraction, the sample with the microballoon density of 0.45 g/cm^3 has

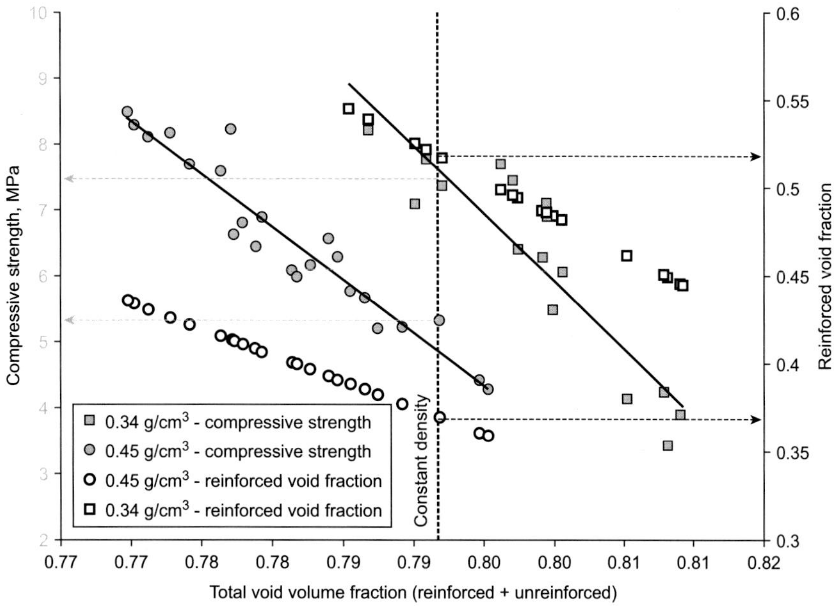

FIGURE 9–4 Three-phase syntactic foams using two different microballoon true particle densities, 0.34 and 0.45 g/cm^3. The hollow circles and squares correspond to the reinforced void volume fraction and the solid circles and squares correspond to the corresponding compressive strength. Results indicate, at a constant density, a greater than 40% increase in compressive strength by increasing the amount of reinforced void by 15%.

reinforced void fraction of 0.37 and a corresponding compressive modulus of ~570 MPa. The sample made with the 0.34 g/cm^3 microballoon density has a reinforced void volume fraction of 0.53 corresponding to compressive modulus of ~880 MPa. At this constant density-vertical line, this represents a 56% increase in compressive modulus with a 15% increase in reinforced void content. Another way to view this result is that with fewer unreinforced voids there are more microballoon pathways to carry the compressive forces, resulting in the higher strength and modulus.

9.2.4 Density

The density of a single-phase or composite foam material, ρ_f, can be related to the density of the constituents and the void volume fraction by a simple rule-of-mixtures type of expression:

$$\rho_f = \frac{m_1 + m_2 + m_3 \ldots + m_i}{V_f}, \tag{9.4}$$

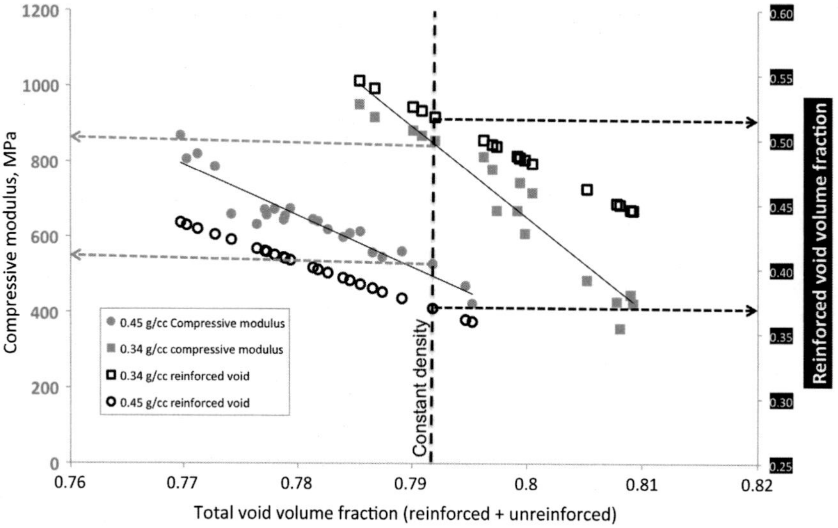

FIGURE 9–5 Three-phase syntactic foams using two different microballoon densities, 0.34 and 0.45 g/cm^3. The hollow circles and squares correspond to the reinforced void volume fraction and the solid circles and squares correspond to the corresponding compressive modulus. Results indicate, at a constant density, a 56% increase in compressive modulus by increasing the volume of reinforced void by 15%. In other words, at a constant density, by shifting unreinforced void to reinforced void one gets a higher performing foam.

where m is mass of constituent 1, 2, 3... i, and V_f is the bulk volume of foam that can usually be measured directly or calculated by the following equation:

$$V_f = V_v + V_1 + V_2 + V_3 \ldots V_i, \tag{9.5}$$

where the subscript V indicates void and the numbers 1, 2, 3... indicate the constituent. In a single-phase foam one only needs to consider a single solid phase constituent and the voids when applying the rule-of-mixtures equation. In a composite foam there can be several solid phases such as microballoons (sometimes of different compositions), fibers, nanoreinforcements, binder phase, etc., in addition to interstitial voids.

9.2.5 Relative density

One very important parameter concerning any kind of foam or material containing voids is its relative density, ρ_r, which is defined as the ratio of the foam density, ρ_f, to the theoretical density of the solid material, ρ_s:

$$\rho_r = \frac{\rho_f}{\rho_s}. \tag{9.6}$$

As the cell walls of the foam become thicker, its relative density, ρ_r, increases. In the case of a syntactic foam, a higher relative density implies a higher aspect ratio (wall thickness/sphere radius) of the microsphere. Gupta, Woldesenbet, and Mensah (2004) have referred to the aspect ratio as the radius ratio of a microballoon. Typically, when discussing foams or microballoons, there is a broad distribution of cell wall thickness or radius ratios, and we use a single value as the average across a broad distribution of cells or microballoons.

In a conventional foam (Gibson & Ashby, 1997), above a relative density of about 0.3, there occurs a transition from an open cell foam to a closed cell foam. In engineering practice, the parameter relative density, ρ_r, becomes very useful one because it is easy to determine in single-phase foams and gives a measure of the aspect ratio (radius ratio) of the hollow microsphere:

$$\rho_r = \frac{\rho_f}{\rho_s} = C\frac{t}{D}, \tag{9.7}$$

where ρ_f is the density of the foam, ρ_s is the density of cell wall material, C is a constant, t is the cell wall thickness, and D is the cell diameter. Almost all properties of a foam are affected by and are a function of its relative density. Typically, the stiffness and strength increase with increasing relative density.

9.2.6 Energy absorption

Foams of all sorts (conventional, syntactic, honeycomb) can absorb impact energy either through collapse and plastic deformation in compression or brittle failure in foam layers as in a crumple or crush zone. The cellular structure of these materials allows them to undergo large strains at constant stress. In very simple terms, one understands this exceptional behavior by examining Fig. 9–6. After the initial elastic loading, there is a extended plateau region followed by a densification region. The area under the plateau region of this plot represents most of the mechanical energy per unit volume approximately equal to $\sigma_c\varepsilon_{crush}$, where σ_c is the compressive strength at the onset of the crushing regime or the plateau region while ε_{crush} is the strain at the end of the plateau. It is easy to see that one can increase the energy absorbed per unit volume by increasing σ_c or ε_{crush} or both.

The value of stress in the plateau region is a very important design criterion. One would always want this value to be as high as possible to maximize the energy absorbed. Fig. 9–6 shows an idealized situation; however, it must be emphasized that some form of constant compressive stress over large strains is a feature of all foams. The plateau may not be as horizontal or well developed as shown in the figure. The force corresponding to the crushing strength, σ_c, is transmitted through the foam to the structure being protected.

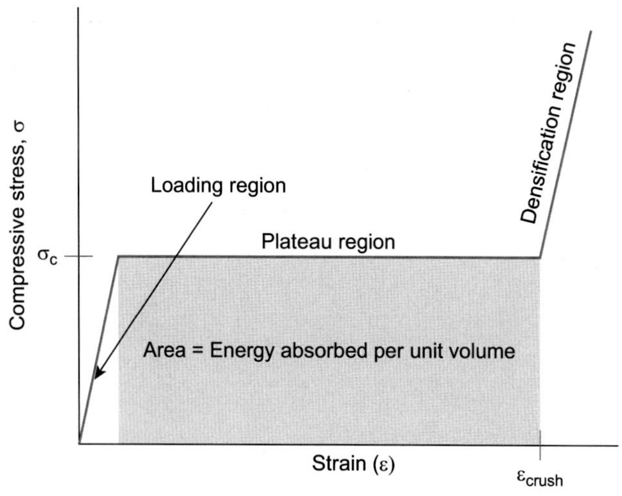

FIGURE 9–6 Idealized compressive behavior of a cellular energy absorbing material. The shaded area under the plateau region represents most of the energy absorbed per unit volume.

One must keep in mind what exactly is being protected, that is, a piece of infrastructure or a human body, and forces it can withstand. This design criterion usually requires:

1. Setting limits to the maximum plateau stress.
2. Ensuring all the required energy will be absorbed before the onset of densification.

We provide an example of how to apply the above criterion in using foams to capture high velocity projectiles. These projectiles can be automobiles, components such as valves, handles, and lids in high-pressure lines/chambers or even munitions. The kinetic energy of such projectile is given by $\frac{1}{2}mv^2$, where m is the mass and v is the velocity. The area under the plateau region of the curve shown in Fig. 9–6 must be equal to this kinetic energy to bring the projectile to a stop. One must remember that this is typically a high strain-rate loading. If the material is strain-rate sensitive (as most polymers are), high strain-rate testing must be done to quantify this behavior. Typically, a polymer based foam, whether single-phase or composite, will have a greater modulus and compressive strength at high strain rates.

The example we consider is a lid of a high pressure vessel, such as that used in a compressed gas cylinder or test chamber. For safety reasons, one would like to capture this lid if it were to fail catastrophically, as shown in Fig. 9–7. This lid can have angular features that will lead to stress concentration, so typically there is a load spreader. A load spreader is located adjacent to foam surface that is initially impacted by a projectile. It is a simple plate designed to engage the entire cross-sectional area of the energy absorbing foam.

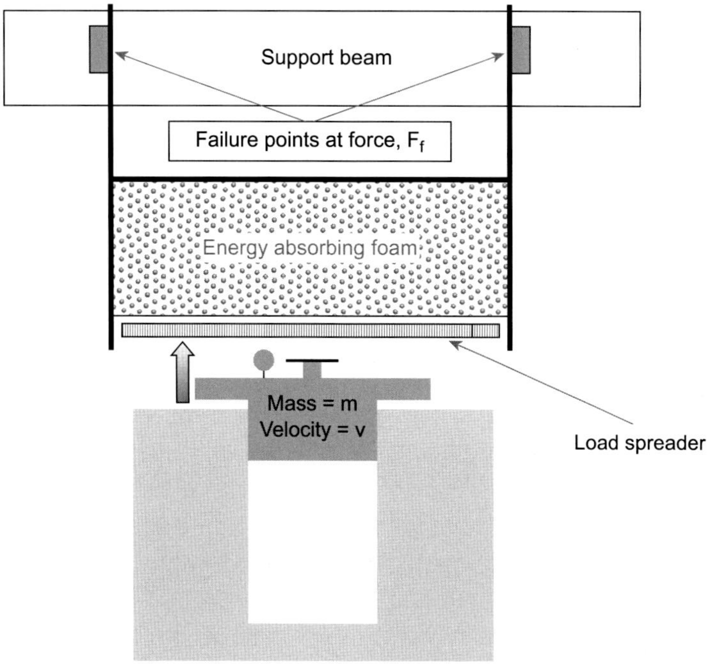

FIGURE 9–7 Example of an application of an energy absorbing foam.

In order to design the crush zone for such a projectile, one could choose a material with a crush strength such that when it is multiplied by the surface area, the transmitted force is less than of some critical value, F_f, which will lead to a structural failure; housing attachment in this case. Once that is made sure of, we can calculate the length of the crush zone, keeping in mind not to approach strains where densification starts. Typically, there are safety factors used with the crush strength of the foam and the crush zone length. For a catastrophic incident as described, the foam would be used only one time.

Compressive strength will depend on the strength of the strut material or face material, and in the case of a composite foam, the binder strength. The crush strain will depend on the void volume fraction in the cellular solid. The higher the volume fraction of voids, the higher the ε_{crush} will be before densification starts. For more detailed treatments on this subject involving the yield surface, the reader is referred to Lemaitre and Lippmann (1996) and Deshpande and Fleck (2000).

9.2.7 Cell size distribution and regularity

When characterizing a foam using the imaging techniques discussed in Chapter 8, Void characterization techniques, one can get cell size information and produce a statistical distribution of the cell sizes present. The consequence of using a combination of blowing agents on the cell size distributions of polystyrene(PS) is shown Fig. 9-8 (Nistor et al, 2017). The PS was first impregnated with concentrations of n-pentane (nC5), ranging from 0 - 7.0 g nC5/100g PS. Once the PS contained the appropriate amount of nC5, the PS was then saturated with CO_2. This was performed by placing the impregnated PS in a pressure vessel at 28 MPa (280 bar) of CO_2, at 70 °C until equilibrium was reached. The cell formation occurred in the PS when the chamber was depressurized to atmospheric pressure in 5s producing stochastic foams. The stochastic monomodal cell structure in Fig 9-8A was created by using only supercritical CO_2 as the blowing agent. The PS exhibits a monomodal cell size distribution with an average cell size of ~1.1 mm. Two blowing agents, supercritical CO_2 and n-pentane, nC5, were needed to create the distributions in Fig 9-8 B and C. In this case, the distribution was bimodal meaning that there were two distributions or modes in cell sizes. The average cell size of the two modes were 1.5 μm and 25 μm. Note that as concentration of nC5 increased from 2.0 to 3.6 g nC5/100 g PS there was a shift to a greater frequency of the larger diameter cells. Finally, at the highest concentration of nC5, 4.6 g nC5/100g PS, there was again a single mode but of the larger cell size, averaging ~20 μm in diameter. The cell size distribution for each of these foams are quantified graphically by number frequency in Fig. 9-8E.

We refer to the cell structure shown in Fig. 9–8A and C as stochastic because the cell structure is not regular or repeatable. Despite the stochastic nature of the cell structure and the variability in performance that it causes, supercritical CO_2 and the

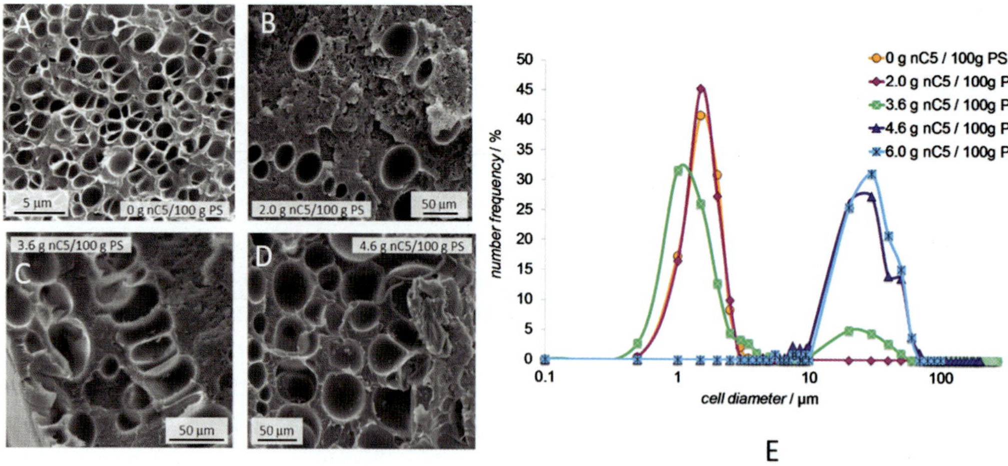

FIGURE 9–8 (A) SEM image of the monomodal size distribution of a polystyrene foam using only supercritical CO_2 as a blowing agent. SEM images of polystyrene co-blowing agents; supercritical CO_2 and (B) 2.0 nC5/100 g PS and C) 3.6 g nC5/100 g PS. The graph (E) quantifies the distribution frequency of the cell sizes present in A-D. *Source: Nistor, A., Topiar, M., Sovova, H., & Kosek, J. (2017). Effect of organic co-blowing agents on the morphology of CO2 blown microcellular polystyrene foams. The Journal of Supercritical Fluids, 130, 30-39.*

other pore forming processes make up the bulk of manufactured foamed products. In practice, knowing that the pore structure will not be identical from part-to-part, one must design a process with sufficient controls so the variability in the pore structure will be small enough so the vast majority of parts manufactured will satisfy performance requirements.

The concept of regularity, *R*, is important in describing cell size distribution for stochastic and nonstochastic foams. The value of *R* can range from 0, completely random/stochastic, to 1, ordered/nonstochastic. These two limiting cases of $R = 0$ and $R = 1$ correspond to there being no minimum imposed distance between any two nuclei, and one where there is a maximum imposed distance, respectively. An example of cellular material with $R = 1$ is a regular hexagonal honeycomb (Christodoulou & Tan, 2013), see Fig. 9–9D. The regularity value can be illustrated as the distance between points (compared to the maximum possible distance) along a line in one dimension or in a two-dimensional plane, as the distance between two cell nuclei, see Fig. 9–9. When considering materials and structures, Fig. 9–9 depicts Voronoi cells with regularities of (A) 0.0, (B) 0.5, (C) 0.7 and, (D) 1.0. One can extend the concept of regularity into three-dimensional

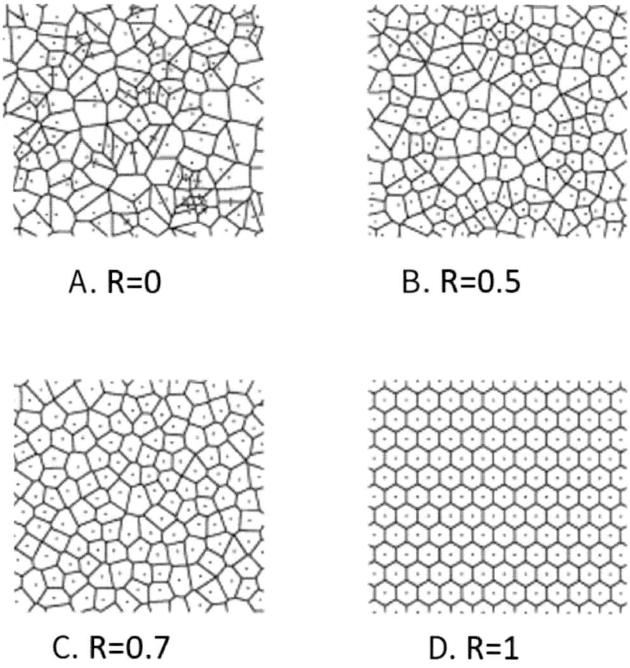

FIGURE 9–9 Voronoi cells with a range of regularity, *R*, (A) R = 0, (B) R = 0.5, (C) R = 0.7 and (D) R = 1. Each lattice comprises of approximately 150 cells. The nucleus of each cell is indicated by a dot. *Source: Christodoulou, I., & Tan, P. J. (2013). Crack initiation and fracture toughness of random Voronoi honeycombs. Engineering Fracture Mechanics, 104, 140–161.*

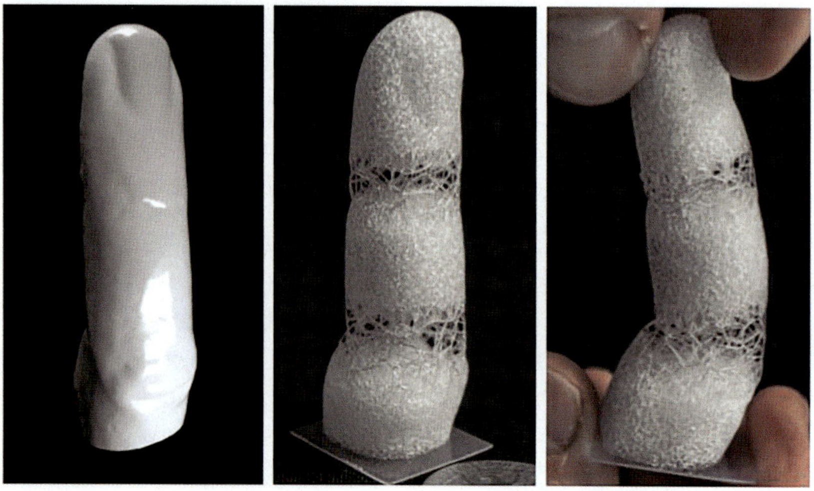

FIGURE 9–10 A Voronoi foam articulating finger designed by regionally tailoring the foam modulus. *Source: Martínez, J., Dumas, J., & Lefebvre, S. (2016). Procedural voronoi foams for additive manufacturing. ACM Transactions on Graphics (TOG), 35(4), 44.*

foam as done by Duan et al. (2019) in Figure 1–10. Voronoi foams were discussed in Chapter 1, Introduction, and describe a foam material designed with a specific regularity value. Martínez, Dumas, and Lefebvre (2016) used regularity and cell structure to tailor the modulus within a single structure. The functionality of articulation can be designed into a Voronoi foam structure by selectively creating regions of high and low modulus foam, see Fig. 9–10.

9.3 Conventional foams

9.3.1 Stress–strain behavior in compression

A representative schematic of the compressive stress–strain behavior of an elastic–plastic foam under uniaxial compression is shown in Fig. 9–6. The curve shows a series of regimes. The first regime occurs at small strains and is characterized by the elastic bending of struts. The long plateau region at medium strains represents plastic yielding or crushing in the case of a brittle material. The final regime is densification where the slope of the stress–strain curve increases rapidly. In this region, cells are collapsing, that is, the top of a cell is touching its bottom.

Fig. 9–11 shows an aluminum foam where the cell structure was made by the removal of a sacrificial template or porogen. In this case the sacrificial material was a NaCl, salt, preform. The NaCl preform was infiltrated with molten aluminum followed by leaching of the salt. The foam produced was an open cell porous

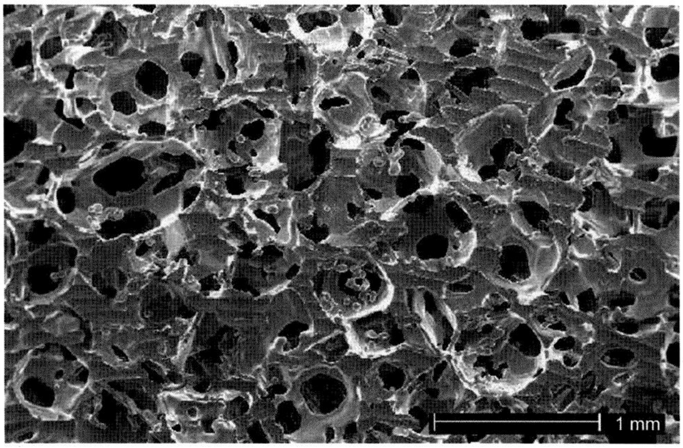

FIGURE 9–11 Aluminum foam where the cell structure was made by removal of a sacrificial NaCl template. Molten aluminum was infiltrated into the porous NaCl preform followed by the removal of an NaCl preform by dissolution with water. *Courtesy: A. Mortensen, EPFL.*

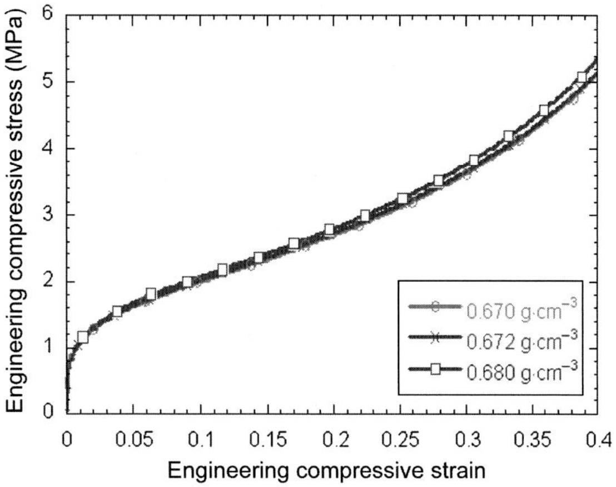

FIGURE 9–12 Stress–strain curves in the compression of aluminum foam.

aluminum featuring very fine pore size (down to a few micrometer) and a small pore size distribution. These microcellular metals show a highly reproducible and regular mechanical behavior. The stress–strain behavior of this foam in compression is shown in Fig. 9–12.

9.3.2 Elastic constants

An open cell foam made up of struts joined at nodes is shown in Fig. 9–13. The length of cell strut, *l*, is much greater than the strut thickness, *t*. We will relate this structure of a cell to the regions identified in Fig. 9–6. The characteristic feature of

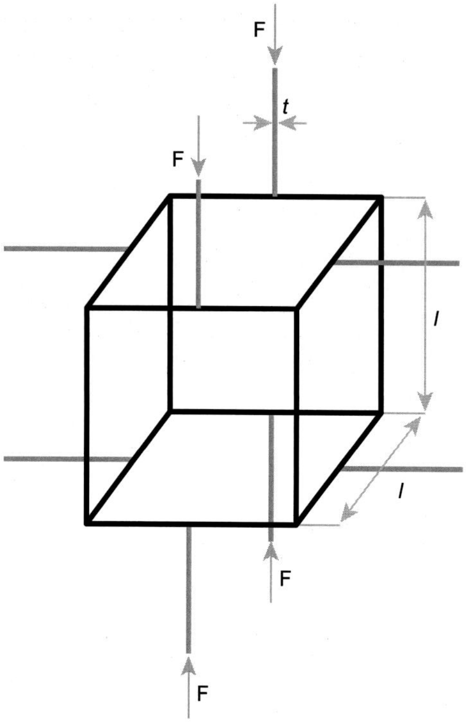

FIGURE 9–13 Idealized cell structure of an open cell foam made of only struts under compressive load.

such an open cell foam is that a very high portion of cells have connectivity with neighboring cells. For an open cell foam of this type the relative density is typically ~0.1 and is proportional to the square of the ratio of t to l (Gibson & Ashby, 1997):

$$\frac{\rho_f}{\rho_s} \propto \left(\frac{t}{l}\right)^2, \tag{9.8}$$

where t is the strut thickness, which is much smaller than the side of the unit cell (l). Inserting a proportionally constant, C_1, in the above expression, allows us to rewrite the Eq. (9.8) as follows:

$$\frac{\rho_f}{\rho_s} = C_1 \left(\frac{t}{l}\right)^2. \tag{9.9}$$

The elastic properties at small deformation in such an open cell foam can be calculated from the simple theory of linear elastic bending of a beam or strut. Following Gibson and Ashby (1997), we treat the open cell foam in the form of a cubic array of cells. This will allow us to relate the relative density to the Young's modulus. When a compressive load is initially applied to the cell structure as shown in an idealized manner in Fig. 9–13, there occurs a deflection, δ. This initial deflection corresponds to the elastic loading portion of the curve in Fig. 9–6. Elastic beam theory gives this deflection as:

$$\delta = C_2 \frac{Fl^3}{E_s I}, \tag{9.10}$$

where C_2 is a proportionally constant, E_S is the Young's modulus of the solid, F is the applied force, and I is the second moment of area of the beam. For a strut with a square cross section, the second moment of area is given by

$$I = \frac{t^4}{12}. \tag{9.11}$$

When we apply a uniaxial compressive stress to the foam, the force, F, is related to the compressive stress (Gibson & Ashby, 1997) by

$$\sigma = \frac{F}{l^2}. \tag{9.12}$$

Rearranging, we get the strain on the cell as:

$$\varepsilon = \frac{\delta}{l} = C_2 \frac{12Fl^2}{E_S t^4}. \tag{9.13}$$

The Young's modulus (stress/strain) of the foam, E_f is obtained by using Eqs. (9.12) and (9.13):

$$E_f = \frac{\sigma}{\varepsilon} = C_2 \frac{\left(\frac{F}{l^2}\right)}{\left(\frac{12Fl^2}{E_S t^4}\right)} = \frac{E_S t^4}{12C_2 l^4}. \tag{9.14}$$

One can derive the relationship between relative density and relative modulus from Eq. (9.9):

$$\frac{E_f}{E_S} = \frac{C_1}{12C_2}\left(\frac{\rho_f}{\rho_s}\right)^2. \tag{9.15}$$

Experimentally it has been shown that the quantity $C_1/12C_2$ is ~1 (Gibson & Ashby, 1997). Eq. (9.15) can then be simplified to

$$\frac{E_f}{E_S} \propto \left(\frac{\rho_f}{\rho_S}\right)^2, \tag{9.16}$$

where the left-hand side is the relative stiffness and the right-hand side is the relative density, E and ρ represent the modulus and density, respectively, while the subscripts f and S designate the foam and solid form of the material, respectively.

Eq. (9.16) can be further modified based on the deformation characteristics of the cell's struts. The elastic modulus or stiffness of a cellular solid or foam scales with the density of the cellular solid. The manner in which the struts deform has an important bearing on the scaling relationship. The following relationship between the tensile (or compressive) modulus and the relative density of the foam summarizes the scaling law:

$$\frac{E_f}{E_S} \propto \left(\frac{\rho_f}{\rho_S}\right)^m, \tag{9.17}$$

where E_f/E_s is the relative stiffness and ρ_f/ρ_s is the relative density, E and ρ represent the modulus and density, respectively, while the subscripts f and s designate the foam and solid form of the material, respectively. The exponent m can vary with the deformation mode. We have shown above that for the bending mode, $m = 2$. For the stretching mode, $m = 1$ (not shown here).

In the plateau region the cells collapse, layer-by-layer. Once the plateau region is reached the struts will begin to deform plastically if the material of the strut can undergo plastic deformation; the stress will then correspond to the plastic yield stress, σ_p.

9.3.3 Dielectric constant

Among the physical properties of a foam, dielectric properties are very important. Although the term *insulator* implies low electrical conduction, "dielectric" is typically used to describe materials with a high *polarizability*. When a dielectric is placed in an electric field, electric charges do not flow through the material as they would in a metallic conductor. In insulators, these charges only slightly shift from their average equilibrium positions causing dielectric polarization. During dielectric polarization, positive charges are displaced toward the electric field and negative charges shift in the opposite direction. This creates an internal electric field that partly compensates for the external field within the dielectric.

The dielectric constant, κ, is a ratio defined as,

$$\kappa = \frac{C_x}{C_0}, \tag{9.18}$$

where C_x is the capacitance of the dielectric material. We can measure the dielectric constant as follows: first the capacitance of a test capacitor, C_0, is measured with vacuum between its plates. Then, using the same capacitor and distance between its plates, the capacitance C_x with a dielectric between the plates is measured. The ratio of the two measurements gives us the dielectric constant. Ashby (2006) found the dielectric constant for foams is proportional to the relative density, ρ_f/ρ_s, by the following equation:

$$\kappa_f = 1 + (\kappa_s - 1)\left(\frac{\rho_f}{\rho_s}\right), \tag{9.19}$$

where k_f is the dielectric constant of the foams and k_s is the dielectric constant of the solid material of which the cell walls are made from.

When a foam is used as a dielectric, we find an approximately linear decrease in dielectric constant with decreasing relative density, see Fig. 9–14. Foams are commonly used to reduce signal loss in communication cables (Strååt, Chmutin, & Boldizar, 2008), coaxial cables, communication buoys, radomes, and radio frequency (RF) and microwave devices.

Dissipation factor, μ, is a measure of the dielectric loss in the dielectric material when used in an alternating electric field and of the energy dissipated as heat. Fig. 9–14 shows the dissipation factor as a function of relative density for a polymer-based foam. A low dissipation factor indicates low AC dielectric losses. Dissipation factor is useful as a means of quality control and monitoring changes in performance due to contamination and physical aging.

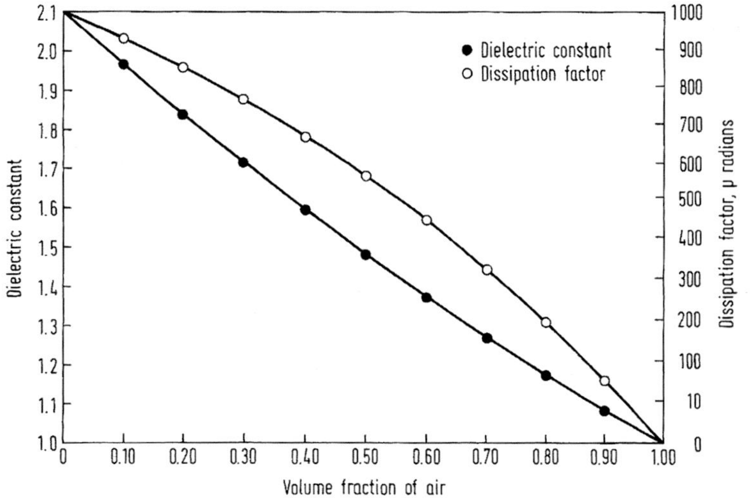

FIGURE 9–14 Dielectric constant and dissipation factor for perfluoroethylene–polypropylene copolymer as a function of void content. *Source: Adapted from Klempner, D. and Frisch, K.C. (1991) Handbook of polymeric foams and foam technology. Hanser: Munich.*

9.4 Syntactic foams

9.4.1 Growth and performance

The field of syntactic foams has grown rapidly since the 1960s. It should be pointed out that these composite foams are multifunctional, as such; the goal is not always to maximize only mechanical properties. In addition to mechanical properties, optimization may involve several variables such as electrical and thermal properties or even flammability and smoke emission characteristics.

For information on mechanical properties such as flexure, see Maharsia, Gupta, and Jerro (2006), Tagliavia, Porfiri, and Gupta (2012), Gupta, Gupta, and Mueller (2008), Gupta and Woldesenbet (2005), Gladysz et al. (2006), Benderly, Rezek, Zafran, and Gorni (2004), Koopman, Chawla, Carlisle, and Gladysz (2006) and for tensile properties, see Gupta, Ye, and Porfiri (2010), Gupta and Nagorny (2006), Maharsia and Jerro (2007), Huang, Vaikhanski, and Nutt (2006), Wouterson, Boey, Hu, and Wong (2007), Zhang and Ma (2010). Most of these data are about compressive properties, which we explain later.

One way of assessing and comparing the performance of materials is to consider *specific properties. Specific*, in this case, means the property of interest divided by density, ρ. Specific strength, $\frac{\sigma}{\rho}$, and specific modulus, $\frac{E}{\rho}$, and these are commonly used to rate the mechanical performance of materials and guide design engineers in material selection. Ashby and Medalist (1983) studied specific properties at great length and constructed graphs that are commonly known as *Ashby plots*. These plots are of specific properties and exist for many different classes of material. The data in these plots come from various sources in literature.

Experimental data on syntactic foams can be traced back to 1970s. In this section, we provide a compilation of these data. The majority of the work involves glass microballoons in epoxy, but the data reflect all microballoons and binder phases. Examples of other microballoons included are cenospheres, carbon, phenolic, etc. Examples of other binder phases included are bismaleimide, phenolic, vinyl ester, etc. In addition, it includes both two-phase and three-phase syntactic foams with a line demarcating the division between them.

An Ashby plot of published syntactic foam data dating back to the 1970s is shown in Fig. 9–15. It illustrates the wide range in compressive strength that syntactic foams cover. In general the two-phase syntactic foams have higher specific compression strength than three-phase syntactic foams. Fig. 9–16 shows a graph of the specific modulus of syntactic foams versus density. In general when optimized for compressive properties, the specific compressive modulus of

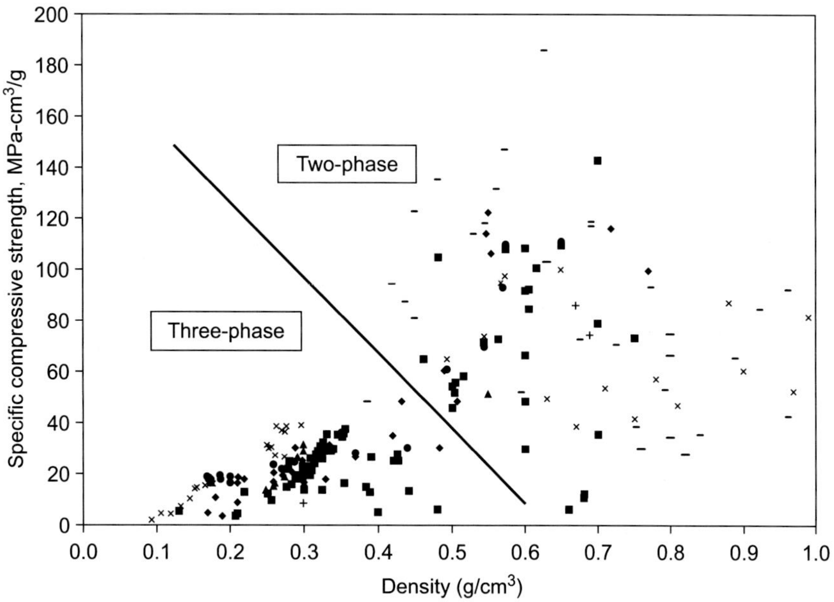

FIGURE 9–15 Specific compressive strength of syntactic foam versus density.

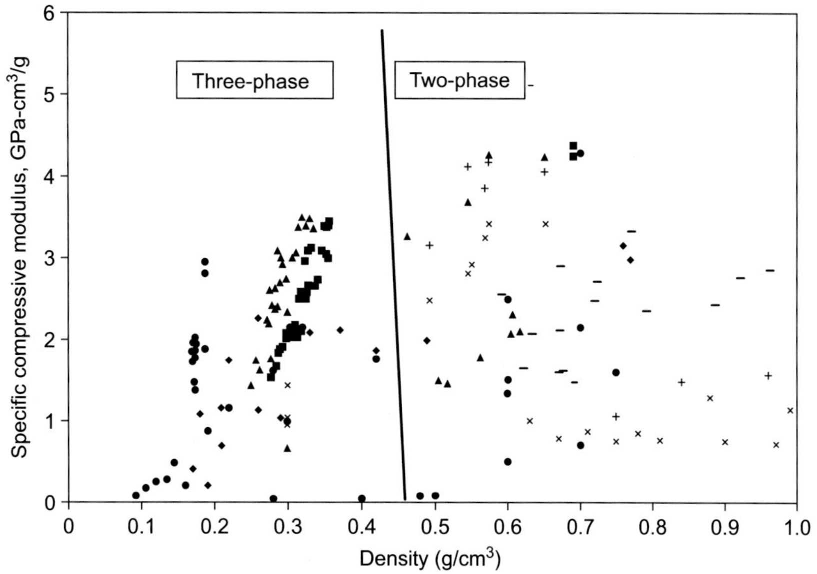

FIGURE 9–16 Specific compressive modulus of syntactic foams.

two-phase and three-phase syntactic foams is approximately equivalent in performance.

9.4.2 Compressive stress–strain relationship

When we deform a foam sample in compression, the following common features are observed:

1. Density changes with deformation.
2. Plastic deformation may occur which involves successive collapse of cells in localized bands.
3. Stress drops and becomes almost constant when only some noncollapsed cells are left.

An example of the stress–strain behavior of a crushable, three-phase, syntactic foam is shown in Fig. 9–17. The microballoons undergo crushing during compression. This figure shows compressive stress versus apparent strain of a glass microballoon/BMI. After an initial elastic compression, the microspheres in the foam start fracturing and the stress–strain curve starts to decrease, showing characteristic dips corresponding to the isolated crushing of microballoons. Eventually, enough microballoons are broken along a section that exfoliation or spalling occurs, which is indicated by the dip in the curve. Continued loading in compression leads to a gradual decrease in the stress because of the reduced cross section due to spalling. Eventually, the compressive loading of the fractured material that formed the walls of the microballoons will occur, followed by a rise in stress–strain curve due to densification.

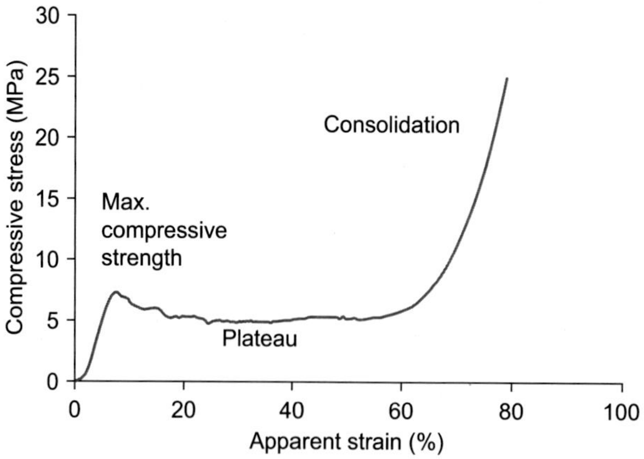

FIGURE 9–17 Stress-apparent strain curve for a glass microsphere syntactic foam showing an initial linear elastic loading segment, maximum compressive strength, plateau region, and final consolidation of the foam.

The strength of two-phase syntactic foam composites will decrease with the increasing content of hollow microspheres. This is due to two factors. A decreased load bearing cross-sectional area and a stress concentration effect of the microporosity. One can combine these effects in the following form (Okuno & Woodhams, 1974). The strength of the foam can be written as:

$$\sigma_1 = k\sigma_m(1 - V_v), \tag{9.20}$$

where σ_1 is the strength of the foam, σ_m is the strength of the matrix or binder phase, V_v is the volume fraction of the porosity or voids, and k is the stress concentration factor. This expression gives a reasonable approximation when the strength of the microballoon is negligible compared to that of the matrix surrounding it and applies to two-constituent syntactic foams. The reinforcing effect of these microballoons can be expressed as:

$$\sigma_2 = (V_{\text{mb}} - V_v)\sigma_{\text{mb},} \tag{9.21}$$

where σ_2 is the strength contribution due to the microballoons, V_{mb} is the volume fraction of the material that forms the microballoons, V_v in the volume fraction of the voids, σ_{mb} is the strength contributed by the material that forms the microballoons. Combining Eqs. (9.19) and (9.20), we can write:

$$\sigma_c = \sigma_1 + \sigma_2 = k\sigma_m(1 - V_v) + (V_{\text{mb}} V_v)\sigma_{\text{mb},} \tag{9.22}$$

$$\sigma_c/\sigma_m = k - (k - V_{\text{mb}}\sigma_{\text{mb}}/\sigma_{m)})V_v. \tag{9.23}$$

This expression indicates the expected trend of decreasing strength with increasing void volume fraction in the composite foam. Okuno and Woodhams (1974) observed reasonably good agreement with experimental results when an appropriate choice was made of the stress concentration factor k.

In three-phase composite foams, shown in Fig. 9–2, the compressive strength and modulus increase with increasing volume content of microballoons. In these foams, where very little polymer binder is present, the strength of the microballoon can contribute to the syntactic foam strength. In a two-phase syntactic foam, increasing microballoon content results in a decrease in density by the addition of microballoons. Unlike the two-phase, in a three-phase syntactic foam an increase in the volume fraction of microballoons will result in a decrease in the interstitial voids and increase in mass due to the increase in total cell walls. See Section 9.1.3 for a detailed discussion.

A number of expressions are available in the literature giving the elastic constants of a foam in terms of the elastic constants and volume fractions of the constituents. For syntactic foams, the following expressions are due to Nielsen (1983):

Shear modulus of the foam,

$$G_F = \left(\frac{1 + AB\phi_2}{1 - B\psi\phi_2}\right) G_1, \tag{9.24}$$

where

$$A = \frac{(7 - 5v_1)}{(8 - 10v_1)}; B = \frac{\left(\frac{G_H}{G_1} - 1\right)}{\left(\frac{G_H}{G_1} + A\right)}; \psi \approx 1 + \left(\frac{(1 - \varphi_2)}{\varphi_m^2}\right)\varphi_2.$$

In these expressions, G is shear modulus, a and b are the inner and outer radii of the microballoon, respectively, ν is Poisson's ratio, ϕ_m is the maximum packing factor that can be achieved, and ϕ_2 is the volume fraction of microballoons. The maximum packing factor, ϕ_m, is about 62.5% (McGeary, 1961). The subscripts H, F, and 1 indicate the hollow sphere, the foam, and the polymer matrix, respectively. The shear modulus can be converted to Young's modulus, E, by the following equation valid for isotropic materials:

$$E = 2G(1 - v). \tag{9.25}$$

In general, such equations due to Nielsen or others show a decrease in the modulus or stiffness of the foam with the increasing amount of microspheres, assuming a constant wall thickness of the microspheres.

9.5 Thermal properties

Foams or cellular solids are commonly used as insulation materials. The presence of porosity in a material reduces its thermal conductivity. A common example is a refractory brick containing a large amount of porosity, commonly used as insulation in high temperature furnaces. Yet another good example is the use of syntactic foams to insulate oil-drilling pipes; see Chapter 10, Applications. Cork is a natural material with very low thermal conductivity because it has a large number of tiny cells or voids. There is no heat convection within the tiny cells; it is primarily the solid cell wall that is responsible for material heat transfer via conduction. Aerogels are open cell foams with extremely low density and very low thermal conductivity. They are excellent candidates for insulation. Fig. 9–18 is a graph comparing traditional thermally conductive and insulative materials to that of materials with nanoscale porosity such as silica aerogel and carbon nanotubes.

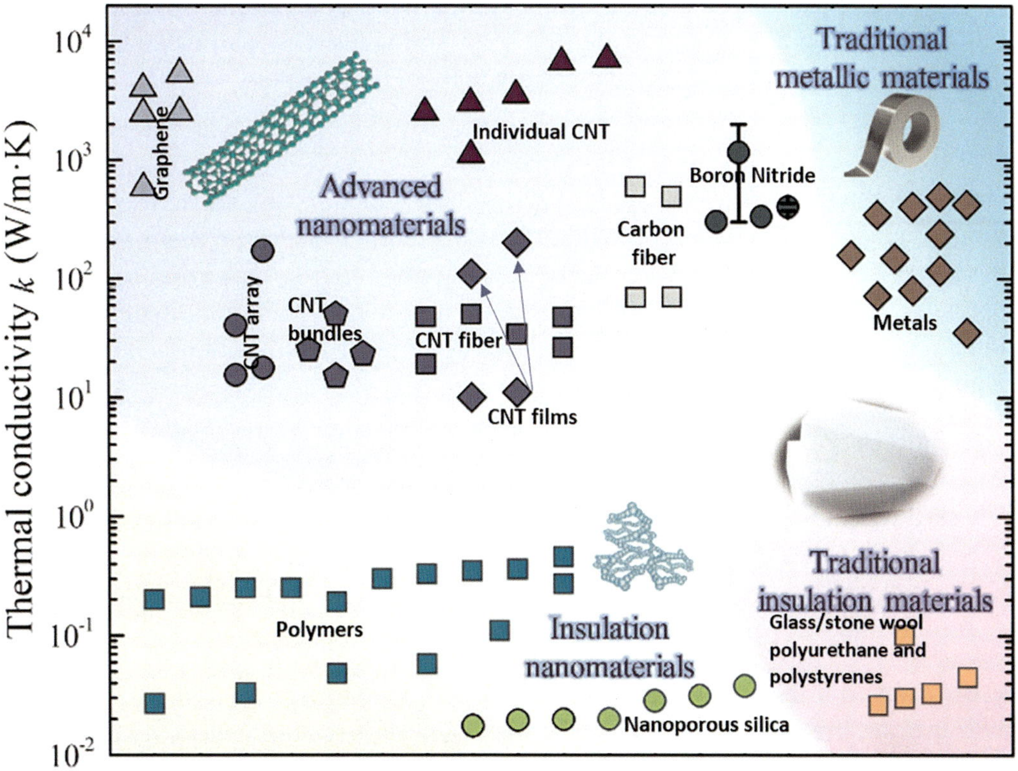

FIGURE 9–18 Thermal conductivities of materials with nanometer scale porosity compared to traditional insulative and conductive materials. *Source: Adapted from Qiu, L., Zhu, N., Feng, Y., Michaelides, E.E., Żyła, G., Jing, D., . . . Mahian, O. (2020). A review of recent advances in thermophysical properties at the nanoscale: From solid state to colloids https://doi.org/10.1016/j.physrep.2019.12.001.*

Thermal conductivity (k) measures the rate of heat transfer through a material. It has the units of W/mK. Thermal conductivity in heat transfer is analogous to diffusivity in mass transfer. Thermal energy can be transferred by electrons (in metals) and/or by phonons (in ceramics).

Recall that a foam consists of at least two distinct components, whereas a composite foam has more than two components. The thermal conductivity of a foam is a function of the thermal conductivities of its individual constituents and its internal structure, that is, the position and alignment of the individual constituents. Strictly speaking, because of variability in the internal structure of composite foams, the concept of a thermal conductivity as a material property is not valid. We will use the notation of k_f to indicate the *equivalent thermal conductivity* of a foam or cellular material. In what follows we provide some useful expressions for thermal properties of foams.

As a first approximation, the thermal conductivity of a cellular solid can be estimated by treating it as a composite of a solid and voids. One can use a simple rule-of-mixtures to describe the thermal conductivity of foam as:

$$k_f = k_s V_s + k_v V_v, \tag{9.26}$$

where k is the thermal conductivity, V the volume fraction, the subscripts, f, s, and v designate foam, solid material making the walls of the cells in the foam and the void or the gas enclosed in the void space. This equation can be extended to composite foams by accounting for the volume fraction and the thermal conductivity of additional constituents. There are the following implicit assumptions in the above relationship:

1. The pores are of uniform size.
2. No convection and radiation heat transfer.
3. The physical properties of the cell wall solid and the enclosed gas in the void remain constant throughout the temperature range.
4. The cell wall solid and the enclosed gas in the void are in local thermal equilibrium.

In reality, the thermal conductivity of a foam depends on the conductivity of the solid making the walls of the cells and the radiation between cells. One can reduce the radiation contribution by making the cells smaller.

Fourier's law states that the rate of heat conduction through a material is directly proportional to the temperature gradient across the material. The second law of thermodynamics tells us that heat flows from the higher temperature surface, T_2, to the lower temperature surface, T_1. The rate of heat conduction, q, through the sample of thickness, z, is given by,

$$q = -k\frac{dT}{dz} = -k\frac{T_1 - T_2}{z}. \tag{9.27}$$

We briefly describe a thermal conductivity measurement technique called the guarded heat flowmeter technique, which is based on Fourier's law of heat conduction. Fig. 9–19 shows the schematic of a guarded heat flowmeter with a sample with two opposite planar surfaces. The steady state temperatures on the surfaces are T_1 and T_2, with $T_2 > T_1$. The heat flux transducer measures the heat transferred to the sample to maintain the temperature, T_2, on the lower plate. A cylindrical guard heater surrounds the sample assembly. The guard heater is maintained at a temperature close to the mean sample temperature to avoid radial heat losses.

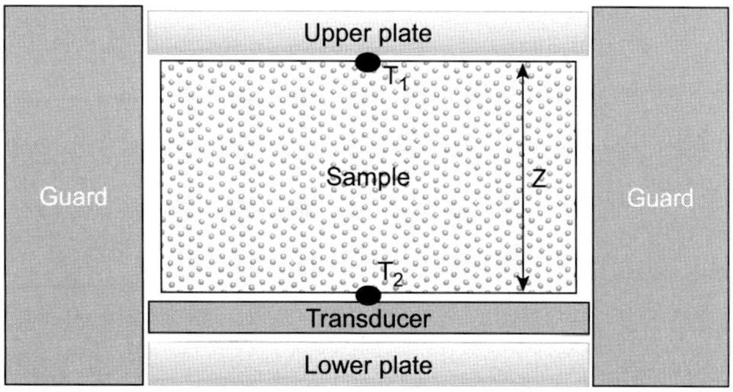

FIGURE 9–19 Schematic of a guarded heat flowmeter commonly used to determine equivalent thermal conductivity. In this experimental setup, the transducer measures the heat flow into the sample to maintain a constant temperature profile through the z-direction of the sample.

The main mechanism of heat transfer through foams is conduction, which occurs through the solid and gaseous phases. Typically, the thermal conductivity of the gaseous phase is very low ($\sim$0.025 W/mK at 298 K), thus the gaseous phase provides the most resistance to heat flow by conduction. Most of the models for thermal conductivity of composites are based on thermal conduction (Boomsma & Poulikakos, 2001; Cheng & Vachon, 1969; Shabde, Hoo, & Gladysz, 2006).

If cells in a ceramic foam are very small, then heat transfer by convection in gas enclosed in the void can be neglected (Ashby, 2005). Radiative heat transfer in foams is also negligible. Thus, the heat transfer in a cellular solid occurs mainly by conduction through the struts (in the open cell foam) and through the enclosed gas in the cell. In the case of closed cell foam, the heat conduction will occur through the walls of the cells. Following Ashby, we assume that one third of struts lie parallel to each axis. Then one can write for the conductivity of foam, k_f, as:

$$k_f = \frac{1}{3}\left(\frac{\rho_f}{\rho_s}\right)k_s + \left[1 - \left(\frac{\rho_f}{\rho_s}\right)\right]k_g, \tag{9.28}$$

where ρ is the density and k is the conductivity, and the subscripts f, s, and g indicate foam, solid of the cell walls, and the enclosed gas, respectively. The above equation describes well the thermal conductivity of foam at low densities. It does not work at high foam densities. At higher foam densities, one needs the following modified version:

$$k_f = \frac{1}{3}\left(\frac{\rho_f}{\rho_s} + 2\left(\frac{\rho_f}{\rho_s}\right)^{3/2}\right)k_s + \left[1 - \left(\frac{\rho_f}{\rho_s}\right)\right]k_g. \tag{9.29}$$

Because of the modification, Eq. (9.29) will give the correct value for thermal conductivity up to a fully dense solid; at $\rho = \rho_s$, $k_f = k_s$.

Boomsma and Poulikakos (2001) modeled the thermal conductivity of an aluminum foam. In their model, they used an idealized three dimensional, tetrakaidecahedron cell geometry of foam. For an aluminum foam with 95% porosity in vacuum, the three dimensional model predicted an equivalent thermal conductivity of 3.82 W/mK. There was not much effect of any change in the thermal conductivity based on the type of gas enclosed in the void space. In other words, in spite of the large content of porosity, the solid phase controls the equivalent thermal conductivity. Another example of this is that of a carbon based open cell foam which has a relatively low thermal conductivity; 1–2 W/mK. However, if the carbon in the foam is graphitized then the thermal conductivity of the foam can increase to as high as 150 W/mK, depending on the density of the foam (Klett, 1999).

We provide a summary of other expressions for thermal conductivity. Cheng and Vachon (1969) developed a resistance in series model analogous to electrical flow:

$$k_f = \frac{2}{\sqrt{C(k_d + k_c)}} \times \arctan\left(\frac{B}{2}\frac{\sqrt{C(k_d + k_c)}}{k_c + B(k_d - k_c)}\right) + \frac{1-B}{k_c}, \tag{9.30}$$

where $B = \sqrt{\frac{3V_d}{2}}$, $C = \frac{4}{B}$, k_c is the thermal conductivity of the continuous phase, k_d is the thermal conductivity of the discontinuous phase, and V_d is the volume fraction of the discontinuous phase.

Self-consistent field theory predicts the equivalent thermal conductivity of a composite based on the thermal conductivities on the individual constituents.

Eucken (1932) gives the following expression for the thermal conductivity of foam, k_f:

$$k_f = k_s\left(\frac{1 + V_g\left(\frac{1-Q_1}{1-2Q_1}\right)}{1 - V_g\left(\frac{1-Q_1}{1+2Q_1}\right)}\right), \tag{9.31}$$

where $Q_1 = \frac{k_s}{k_g}$, V is the volume fraction and k is the conductivity, and the subscripts f, s, and g indicate foam, solid of the cell walls, and the enclosed gas, respectively.

According to Leach (1993), using the self-consistent field theory, the equivalent thermal conductivity of a foam can be expressed as:

$$k_f = k_g + \frac{2}{3}\left(\frac{\rho_f}{\rho_s}\right)k_s, \tag{9.32}$$

where ρ is the density and k is the conductivity, and the subscripts f, s, and g indicate foam, solid of the cell walls, and the enclosed gas, respectively.

Radiative thermal conductivity can become significant at high temperatures and as cell dimensions become large. For a foam, one can calculate the radiative thermal conductivity using the following expression:

$$k_r = \frac{16kT^3}{3\beta}, \tag{9.33}$$

where κ is Boltzmann constant ($=1.381 \times 10^{-23}$ J/K), T is the absolute temperature, and β can be calculated from (Tao, Hsu, Chang, Hsu, & Lin, 2001):

$$\beta = 42.038\rho_f + 121.55, \tag{9.34}$$

where ρ_f is the density of the foam.

Shabde et al. (2006) used three models (Cheng & Vachon, 1969; Eucken, 1932; Leach, 1993) to predict the thermal conductivity of a three-phase syntactic foam. Both conductive and radiative thermal conductivities were used in these models. In a three-phase syntactic foam, there were two types of voids; reinforced voids (within the microballoon walls) and unreinforced interstitial-type voids. See Fig. 9–2 for the microstructure. The following equation due to Hashin (1968) was used to determine the equivalent thermal conductivity of the microballoon phase:

$$k_{\mathrm{mb}} = k_s\left(1 + \frac{V_g}{\frac{k_s}{k_s - k_g} + \frac{V_g}{3}}\right), \tag{9.35}$$

where k is the conductivity, V is the volume fraction, and the subscripts *mb*, *s*, and *g* indicate microballoon, solid of the cell walls, and the enclosed gas, respectively.

The modeling results were compared to experimental equivalent thermal conductivity values obtained using guarded heat flowmeter. This work found that self-consistent field theory models (Eucken, 1932; Leach, 1993) better reflected the experimental data. At higher temperatures, the self-consistent field theory model by Eucken (1932) had the best agreement.

For an extension of the self-consistent field theory models to two-phase syntactic foams, the reader should consult Felske (2004).

9.6 Finite element analysis (FEA)

Finite element methods (FEM) or FEA have become very popular in a whole host of fields. The field of foams or cellular solids is no exception. A detailed description of FEM techniques to analyze the foam behavior is beyond the scope of this text. In what follows, we provide a brief introduction to the reader and give references to dig deeper. There are many commercial software packages available to simulate the behavior of foams, especially the mechanical behavior of polymeric foams. For a review, the reader is referred to Srivastava and Srivastava (2014).

In a simple-minded analytical problem, one can use a differential equation to describe the problem; this is a continuous problem. In FEA, one breaks this continuous problem into many discrete smaller problems, or *finite elements*, so it is easier to solve. Simulation using FEA allows one to test and optimize a design before a component is actually built. FEA can identify areas of high stress and/or strain, test sensitivity and mechanical response to defect amount, size, shape, and position in a structure, and simulate in-service mechanical and thermal behavior of cellular materials.

In order to effectively use a commercially available FEA software, one must first have a conceptual understanding of the problem. This involves knowing what one wants to simulate; will it be a simplified ASTM-type sample or a more complex part? If modeling the microstructure, how big should the model or representative volume be? Whether simple or complex, one looks for ways of simplifying the problem. Simplifications such as recognizing symmetry can turn a 3D analysis into 2D, which will greatly decrease the computational time required for the simulation.

Once the geometry is decided and the problem defined, one must identify the relevant equations that apply to the problem. Most commercial software has been written for solid mechanics problems. One then needs to input the material property data, proper boundary conditions, and other inputs into the software. Quite often, material properties are included in a database for common metals, polymers, ceramics, and composites. For other unique materials, one must estimate these properties or experimentally determine them. In the case of foams, if using the discrete microstructure as an input, care should be taken to adequately mesh around the pores in the foam and to have enough elements in areas with very thin matrix ligaments.

The most important step in the FEA, before the simulation can run, is called segmentation or meshing, that is, breaking the geometry into discrete parts or finite elements. Meshing involves the use of elements (segments) and nodes. Nodes are points in space and usually placed where two or more elements intersect. Nodes are where unknown quantities are estimated. The elements carry the material properties that they represent. It is important to identify when a coarse mesh can be used and how to minimize the use of a refined mesh. Many times the mesh in the vicinity of important microstructural features, such as voids, needs to be refined, whereas areas or volumes representing homogeneous materials can be represented by a coarser mesh. Fig. 9–20 is an example of using both coarse and refined mesh in a tin-based solder (Padilla, Jakkali, Jiang, & Chawla, 2012).

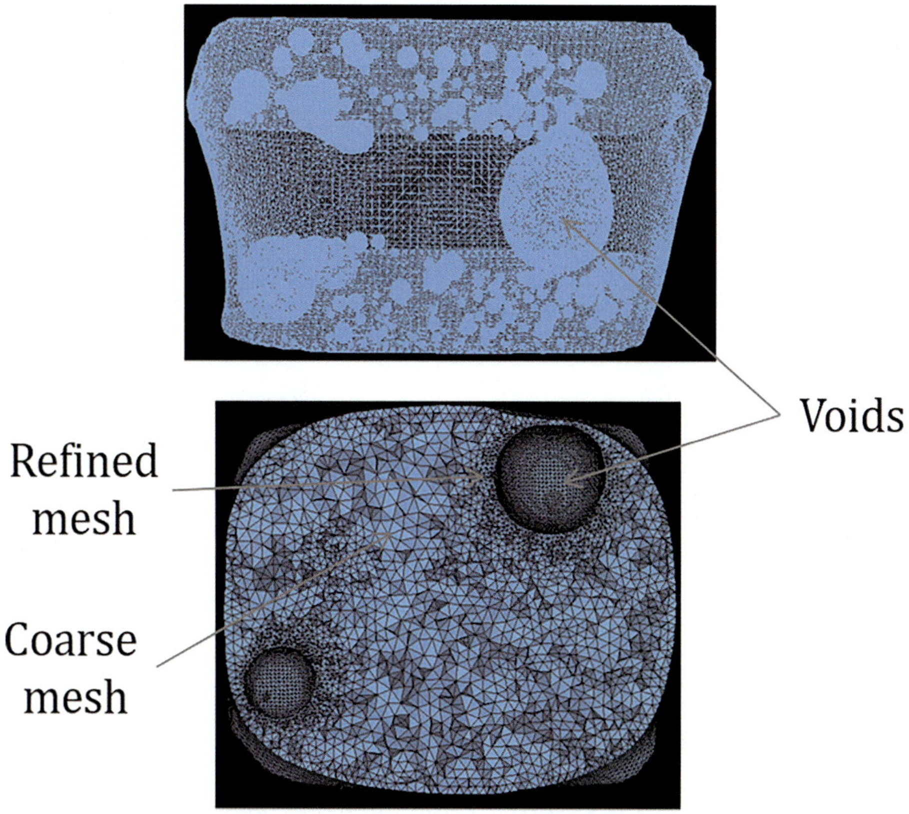

FIGURE 9–20 3D-meshed model of a tin-based solder containing voids. *Source: Padilla, E., Jakkali, V., Jiang, L., & Chawla, N. (2012). Quantifying the effect of porosity on the evolution of deformation and damage in Sn-based solder joints by X-ray microtomography and microstructure-based finite element modeling. Acta Materialia, 60(9), 4017–4026.*
Note that the mesh near the voids is refined or smaller while the matrix has a coarser mesh. The refined mesh significantly increases the computational time required to perform a simulation.

The mesh near a void is refined while the mesh in the matrix is coarse. A refined mesh will result in greater computational time but will allow more sensitivity and provide more insight into the material response near that feature.

Fig. 9–21 compares the results of an FEA simulation and the test of a tin-based solder with several large voids being deformed in shear (Padilla et al., 2012). The simulation (top) is a map of the strain within the joint while undergoing shear. The areas in red indicate regions of high strain and regions in blue indicate low strain. Note the high strain regions near the voids. The SEM micrograph below the FEM picture indicates the crack formation in the material near the large void. This actual cracking correlates very well with FEA simulation.

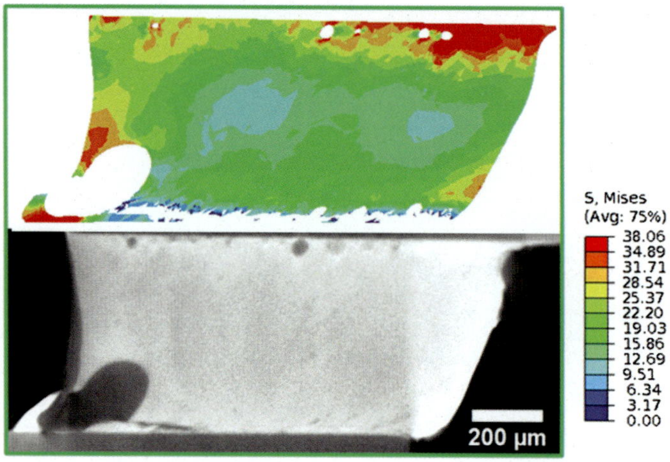

FIGURE 9.21 2D sections across the thickness of the tin-based solder joint tested in shear. The crack propagation path predicted by the FE model (top) correlates well with the actual crack propagation path (bottom) Note in both figures, the crack extends from the large void at the bottom left of corner of the joint. *Source: Padilla, E., Jakkali, V., Jiang, L., & Chawla, N. (2012). Quantifying the effect of porosity on the evolution of deformation and damage in Sn-based solder joints by X-ray microtomography and microstructure-based finite element modeling. Acta Materialia, 60(9), 4017–4026.*

9.7 Geopolymer foams

The term "geopolymer" was first used in 1979 to describe mineral polymers synthesized by *geosynthesis*, that is, the chemical reaction of aluminosilicate precursors with alkali polysilicates, yielding Al—O—Si bonds (Davidovits & Cordi, 1979). The widening attention on geopolymers is largely driven by environmental factors. It was initially envisioned as a replacement for Portland cement because a geopolymer has a low CO_2 embodied material. Further bolstering the positive environmental impacts, cure reactions of geopolymers have been shown to be relatively insensitive to solid fillers from industrial waste streams such as fly ash, glass waste, and slags when used as the dispersed phase for geopolymer composite. Beyond Portland cement, porous geopolymers are now being considered for construction applications such as thermal and acoustic insulation, and fire resistance.

Common forms of geopolymer foam are as bulk materials as well as discrete particles. Voids are introduced by many of the conventional methods discussed in Chapters 6, Techniques for introducing intentional voids into bulk materials, and 7, Techniques for introducing intentional voids into particles and fibers. These methods include physical and chemical blowing agents, sacrificial placeholders, emulsion templating or simply vigorous mixing to entrain a gas phase. Novais, Pullar, and Labrincha (2020) provide an extensive review of mechanical and thermal properties of geopolymer foams created by many of the techniques mentioned above. With such a wide range of techniques for creating voids the resulting properties also vary broadly. Geopolymers with void contents ranging from 32% to 89% by volume have been studied. The resulting compressive strengths and thermal conductivities are plotted in Fig. 9–22 and Fig. 9–23, respectively (Novais et al., 2020).

Geopolymeric structures have been successfully made by additive manufacturing techniques. Franchin and Colombo (2015) used an inverse replication technique to create hierarchical porosity in geopolymeric structures. Extrusion deposition was used to first create a polylactic acid (PLA) sacrificial framework. The geopolymer with a chemical blowing agent was then infiltrated throughout the 3D-printed framework. The geopolymer was subsequently foamed and cured and the PLA framework was then removed. The final structure was that of a geopolymer having micrometer scale stochastic porosity via a chemical blowing agent and digitally designed nonstochastic millimeter-scale porosity. The total porosity in these structures ranged from 66% to 71% by volume. More recently, Franchin et al. have reported the direct ink writing of geopolymer composite structures directly without the

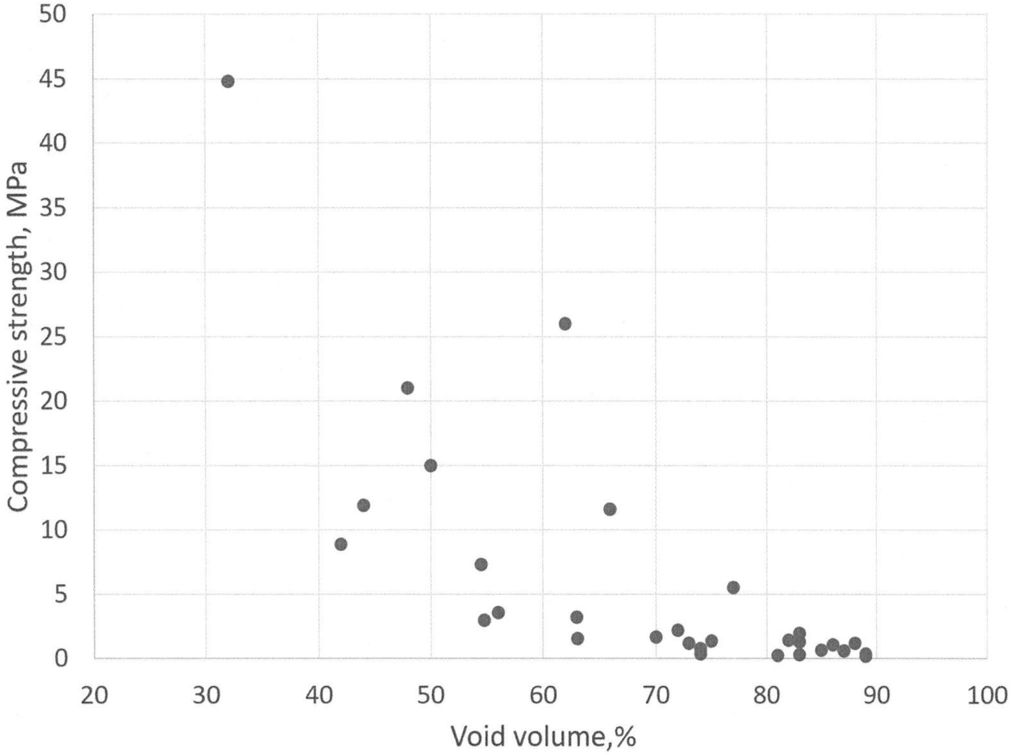

FIGURE 9.22 Compressive strength of geopolymer foams as a function of void content. *Source: Adapted from Novais, R.M., Pullar, R.C., & Labrincha, J.A. (2020). Geopolymer foams: An overview of recent advancements https://doi.org/10.1016/j.pmatsci.2019.100621.*

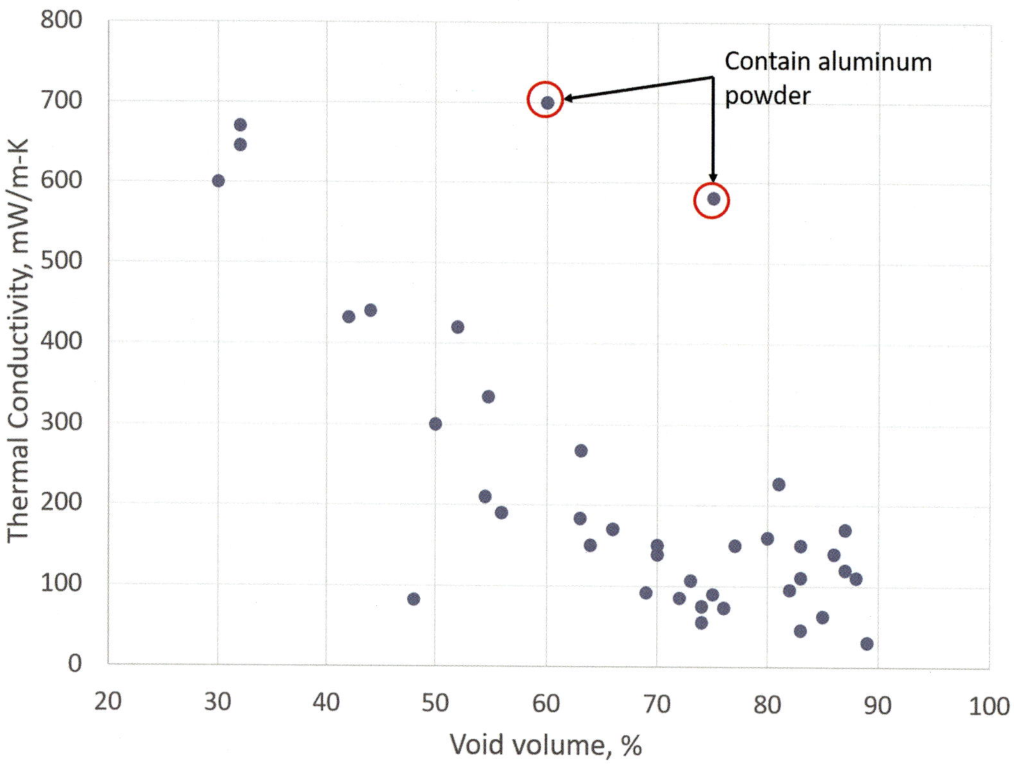

FIGURE 9–23 Thermal conductivity of geopolymer foams as a function of void content.
Source: Adapted from Novais, R.M., Pullar, R.C., & Labrincha, J.A. (2020). Geopolymer foams: An overview of recent advancements https://doi.org/10.1016/j.pmatsci.2019.100621.

need for infusion into a sacrificial framework (Franchin et al., 2017). Fig. 9–24A shows the direct ink writing structures with various span distances to create the 3D-printed frameworks with controllable void content. Fig. 9–24B shows the compressive strength as a function of total porosity for the different frameworks. The empirical data were fitted to the Ryshkewitch model (Ryshkewitch, 1953) for predicting compressive strength (σ):

$$\sigma = A \cdot e^{-B \cdot p}, \tag{9.36}$$

where A and B are experimental parameters and p is the total volume percent of void in the structure.

9.8 Metallic foams

Metallic foams have been studied extensively. Gibson and Ashby (Ashby, 2006; Ashby et al., 2000; Gibson & Ashby, 1999) have developed the most widely

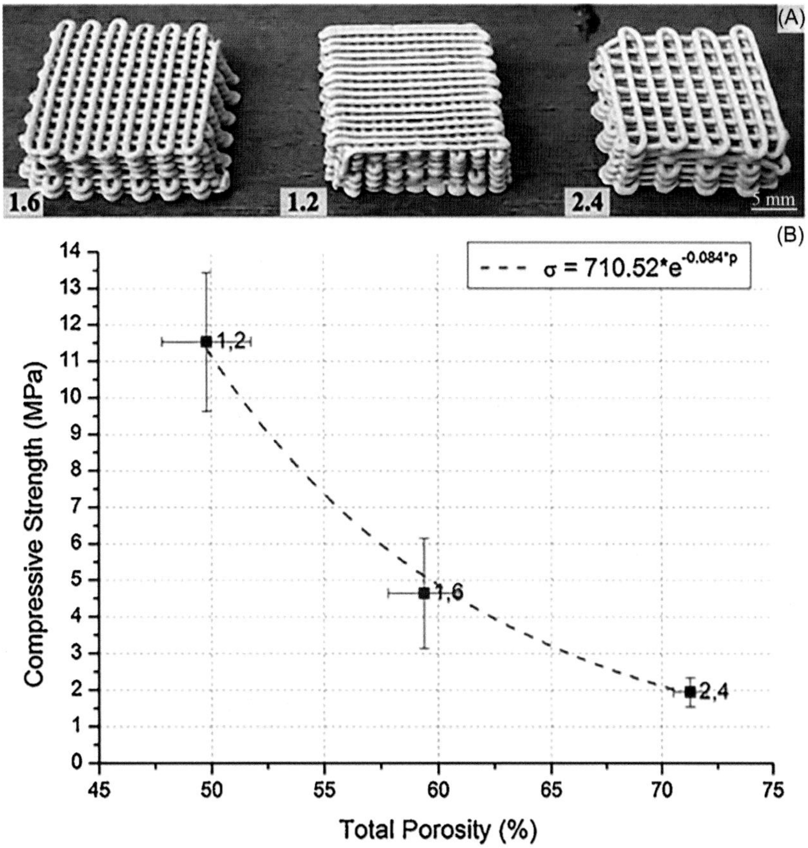

FIGURE 9–24 Direct ink writing of geopolymer foam (A) structures with various span distances and (B) the resulting compressive strength fitted to the Ryshkewitch model (exponential dependence).

accepted models that predict mechanical properties. The equations presented below are based on the relative density, that is, the foam density divided by the fully dense material, ρ_f/ρ_s. All other effect such as cell size, cell geometry, etc., are captured in multiplicative coefficient, *C*, with guidance given for the typical ranges for *C* in each of the equations. For clarity, we have included numerical subscripts to the *C* term to emphasize that the value for a particular foam may have different values depending on the mechanical property being calculated. We also present the equations for open and closed cell metallic foam separately.

Mechanical properties of open cell metallic foams can be predicted by the following equations. The equation for the Young's modulus, E_f, for a metallic foam depends on the relative density ρ_f/ρ_s, the Young's modulus on to fully dense material, E_s, and a coefficient, C_1. The relative modulus is proportional to the relative density squared:

$$\frac{E_f}{E_s} = C_1 \cdot \left(\frac{\rho_f}{\rho_s}\right)^2, \tag{9.37}$$

where the range of C_1 for predicting modulus is 0.1−4.

The equation for predicting the compressive yield strength of a metallic foam ($\sigma_{c,f}$) is similar; the relative compressive yield strength, $\sigma_{c,f}/\sigma_{c,s}$, is proportional to the relative density raised to the *3/2* power.

$$\frac{\sigma_{c,f}}{\sigma_{c,s}} = C_2 \cdot \left(\frac{\rho_f}{\rho_s}\right)^{\frac{3}{2}}, \tag{9.38}$$

where the range of C_2 for predicting compressive yield strength is 0.1−1.

Although the relative density term is not shown explicitly in the following tensile strength equation $\sigma_{c,f}$, it is directly proportional to $\sigma_{c,f}$. By inspection of Eq. (9.38), we can see that $\sigma_{c,f}$ is proportional to the relative density raised to the *3/2* power.

$$\sigma_{t,f} = C_3 \cdot \sigma_{c,f}, \tag{9.39}$$

where the range of C_3 for predicting tensile strength is 1.1−1.4.

Again, although the relative density term is not shown explicitly in the following equation, the shear modulus of a metallic foam, G_f, is directly proportional to E_f. By inspection of Eq. (9.37), G_f is proportional to the relative density squared.

$$G_f = \frac{3}{8} E_f. \tag{9.40}$$

The densification strain for the metallic foam, $\varepsilon_{D,f}$, is calculated by the following equation:

$$\varepsilon_{Df} = C_4 \cdot \left[C_5 \cdot \left(\frac{\rho_f}{\rho_s}\right) + 0.4 \cdot \left(\frac{\rho_f}{\rho_s}\right)^3\right], \tag{9.41}$$

where C_4 is in the range of 0.9−1.0 and C_5 is of 1.0−1.4

Equation for closed cell metallic foams are similar to open cell metallic foams but they have an additional linear term. The equations for closed cell foams with the coefficient ranges are given below. The Young's modulus for a metallic closed cell foam is calculated by the following equation:

$$\frac{E_f}{E_s} = C_6 \cdot \left[\mathit{0}.5 \cdot \left(\frac{\rho_f}{\rho_s}\right)^2 + 0.3 \cdot \left(\frac{\rho_f}{\rho_s}\right)\right], \tag{9.42}$$

where C_6 is in the range of 0.1–1.0.

The compressive yield strength for a metallic closed cell foam can be calculated from the following equation:

$$\frac{\sigma_{c,f}}{\sigma_{c,s}} = C_7 \cdot \left[0.5 \cdot \left(\frac{\rho_f}{\rho_s}\right)^{\frac{2}{3}} + 0.3 \cdot \left(\frac{\rho_f}{\rho_s}\right)\right], \tag{9.43}$$

where C_7 is in the range of 0.1–1.0.

The tensile strength for a metallic closed cell foam is calculated by the following equation:

$$\sigma_t = C_8 \cdot \sigma_{c,f}, \tag{9.44}$$

where C_8 is in the range of 1.1–1.4.

The shear modulus for a metallic closed foam can be calculated by the following equation:

$$G_f = \frac{3}{8} E_f. \tag{9.45}$$

The densification strain for a metallic closed foam can be calculated by the following equation:

$$\varepsilon_{D,f} = C_9 \cdot \left[C_{10} \cdot \left(\frac{\rho_f}{\rho_s}\right) + 0.4 \cdot \left(\frac{\rho_f}{\rho_s}\right)^3\right], \tag{9.46}$$

where C_9 is in the range of 0.9–1.0 and C_{10} is of 1.0–1.4.

Eqs. (9.37)–(9.46) provide good predictions for mechanical properties for metallic foams with some caveats. Properties of metallic foams will sometimes fall outside the predicted limits set by the coefficient range provided in each equation. Some types of metal foam that commonly fall outside of these limits are heat-treated metals, metals with high anisotropy and foams made from the bonding/sintering of thin-walled hollow spheres (Smith, Szyniszewski, Hajjar, Schafer, & Arwade, 2012). Fig. 9–25 is a comparison of metal foams processed in various ways to the open cell Gibson and Ashby (Open G&A) and closed cell Gibson and Ashby (Closed G&A) models for (A) normalized compressive yield strength and (B) normalized Young's modulus. The processes of making the foams are sintering of hollow spheres (HS), injection molding, composite of powder metal and hollow spheres (Composite HS), powder metallurgy (PM), and polymer/metal precursor (Precursor). To arrive at the limits, for example on the Closed G&A for normalized compressive yield, the value of coefficient in Eq. (9.38) was taken as the minimum ($C_2 = 0.1$), the maximum ($C_2 = 1.0$) and midpoint ($C_2 = 0.55$) (Smith et al., 2012).

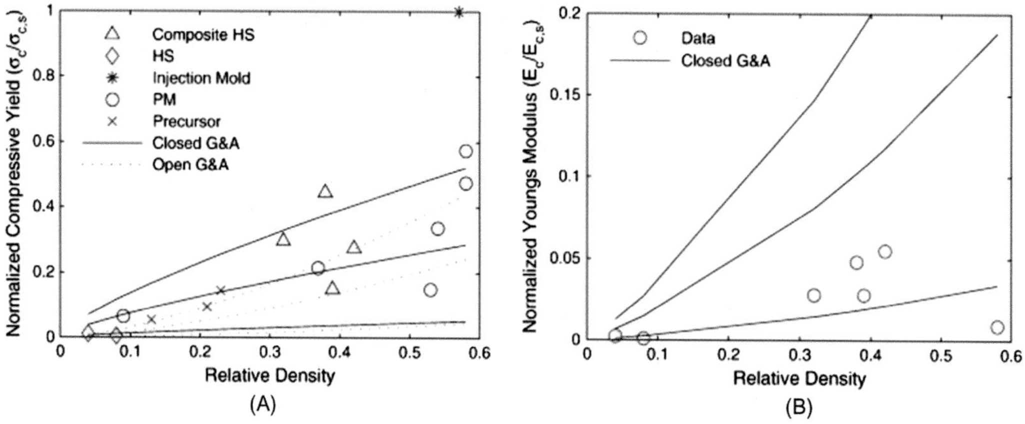

FIGURE 9–25 Comparison of available experimental (A) compressive and (B) Young's modulus data with Gibson & Ashby closed (G&A Closed) and open (G&A Open) cell models. In each graph, the limits on the Gibson & Ashby expressions are set by taking the coefficient equal to minimum ($C_2 = 0.1$), midpoint ($C_2 = 0.55$), and maximum ($C_2 = 1.0$). The various foam processing methods include sintering of hollow spheres (HS), injection molding, composite of powder metal and hollow spheres (Composite HS), powder metallurgy (PM), and polymer/metal precursor (Precursor).

References

Ashby, M. (2006). The properties of foams and lattices. *Philosophical Transactions of the Royal Society A: Mathematical, Physical and Engineering Sciences, 364*(1838), 15–30.

Ashby, M. F. (2005). Cellular solids—Scaling of properties. In M. Scheffler, & P. Colombo (Eds.), *Cellular ceramics: Structure, manufacturing, properties and applications* (pp. 1–17). Weinheim, Germany: Wiley-VCH Verlag GmbH KGaA.

Ashby, M. F., Evans, T., Fleck, N. A., Hutchinson, J., Wadley, H., & Gibson, L. (2000). *Metal foams: A design guide*. Elsevier, Cambridge, USA.

Ashby, M. F., & Medalist, R. M. (1983). The mechanical properties of cellular solids. *Metallurgical Transactions A, 14*(9), 1755–1769.

Benderly, D., Rezek, Y., Zafran, J., & Gorni, D. (2004). Effect of composition on the fracture toughness and flexural strength of syntactic foams. *Polymer Composites, 25*(2), 229–236.

Boomsma, K., & Poulikakos, D. (2001). On the effective thermal conductivity of a three-dimensionally structured fluid-saturated metal foam. *International Journal of Heat and Mass Transfer, 44*(4), 827–836.

Cheng, S., & Vachon, R. (1969). The prediction of the thermal conductivity of two and three-phase solid heterogeneous mixtures. *International Journal of Heat and Mass Transfer, 12*(3), 249–264.

Christodoulou, I., & Tan, P.J. (2013). Crack initiation and fracture toughness of random Voronoi honeycombs. *Engineering Fracture Mechanics, 104*, 140–161. Available from: https://doi.org/10.1016/j.engfracmech.2013.03.017.

Davidovits, J., & Cordi, S. (1979). Synthesis of new high temperature geo-polymers for reinforced plastics/composites. *Proceedings of the* SPE PACTEC *79*, Society of Plastic Engineers, Brookfield Center, 151–154.

Deshpande, V., & Fleck, N. (2000). Isotropic constitutive models for metallic foams. *Journal of the Mechanics and Physics of Solids, 48*(6), 1253–1283.

Duan, Y., Du, B., Zhao, X., Hou, N., Shi, X., Hou, B., & Li, Y. (2019). The cell regularity effects on the compressive responses of additively manufactured Voronoi foams. *International Journal of Mechanical Sciences,* 105151.

Eucken, A. (1932). Thermal conductivity of ceramic refractory materials; calculation from thermal conductivity of constituents. *Ceramic Abstracts, 11,* 576.

Felske, J. (2004). Effective thermal conductivity of composite spheres in a continuous medium with contact resistance. *International Journal of Heat and Mass Transfer, 47*(14), 3453–3461.

Franchin, G., & Colombo, P. (2015). Porous geopolymer components through inverse replica of 3D printed sacrificial templates. *Journal of Ceramic Science and Technology, 6*(2), 105–111.

Franchin, G., Scanferla, P., Zeffiro, L., Elsayed, H., Baliello, A., Giacomello, G., . . . Colombo, P. (2017). Direct ink writing of geopolymeric inks. *Journal of the European Ceramic Society, 37*(6), 2481–2489.

Gibson, L. J., & Ashby, M. F. (1997). *Cellular solids: Structure and properties.* Cambridge: Cambridge University Press.

Gibson, L. J., & Ashby, M. F. (1999). *Cellular solids: Structure and properties.* Cambridge university press.

Gladysz, G., Perry, B., Mceachen, G., & Lula, J. (2006). Three-phase syntactic foams: Structure-property relationships. *Journal of Materials Science, 41*(13), 4085–4092.

Grand View Research. (2019). *Polymer foam market size, share & trends analysis report by type (polyurethane, polystyrene, polyolefin, melamine, phenolic, PVC), by application, by region, and segment forecasts, 2019–2025* (Marketing Report no. 978-1-68038-932-6). San Francisco, CA: Grand View Research.

Gupta, N., Gupta, S. K., & Mueller, B. J. (2008). Analysis of a functionally graded particulate composite under flexural loading conditions. *Materials Science and Engineering: A, 485*(1), 439–447.

Gupta, N., & Nagorny, R. (2006). Tensile properties of glass microballoon-epoxy resin syntactic foams. *Journal of Applied Polymer Science, 102*(2), 1254–1261.

Gupta, N., & Woldesenbet, E. (2005). Characterization of flexural properties of syntactic foam core sandwich composites and effect of density variation. *Journal of Composite Materials, 39*(24), 2197–2212.

Gupta, N., Woldesenbet, E., & Mensah, P. (2004). Compression properties of syntactic foams: Effect of cenosphere radius ratio and specimen aspect ratio. *Composites Part A: Applied Science and Manufacturing, 35*(1), 103–111.

Gupta, N., Ye, R., & Porfiri, M. (2010). Comparison of tensile and compressive characteristics of vinyl ester/glass microballoon syntactic foams. *Composites Part B: Engineering., 41*(3), 236–245.

Hashin, Z. (1968). Assessment of the self consistent scheme approximation: conductivity of particulate composites. *Journal of Composite Materials, 2*(3), 284–300.

Huang, Y., Vaikhanski, L., & Nutt, S. R. (2006). 3D long fiber-reinforced syntactic foam based on hollow polymeric microspheres. *Composites Part A: Applied Science and Manufacturing, 37*(3), 488–496.

Klempner, D., & Frisch, K. C. (1991). *Handbook of polymeric foams and foam technology*. Munich: Hanser.

Klett, J. (1999). High thermal conductivity mesophase pitch-derived graphitic foams. *Composites in Manufacturing, 14*(4), 1–5.

Koopman, M., Chawla, K., Carlisle, K., & Gladysz, G. (2006). Microstructural failure modes in three-phase glass syntactic foams. *Journal of Materials Science, 41*(13), 4009–4014.

Leach, A. (1993). The thermal conductivity of foams. I. Models for heat conduction. *Journal of Physics D: Applied Physics, 26*(5), 733.

Lemaitre, J., & Lippmann, H. (1996). *A course on damage mechanics*. Berlin: Springer.

Ma, Z., Zhang, G., Yang, Q., Shi, X., & Shi, A. (2014). Fabrication of microcellular polycarbonate foams with unimodal or bimodal cell-size distributions using supercritical carbon dioxide as a blowing agent. *Journal of Cellular Plastics, 50*(1), 55–79.

Maharsia, R., Gupta, N., & Jerro, H. D. (2006). Investigation of flexural strength properties of rubber and nanoclay reinforced hybrid syntactic foams. *Materials Science and Engineering: A, 417* (1), 249–258.

Maharsia, R. R., & Jerro, H. D. (2007). Enhancing tensile strength and toughness in syntactic foams through nanoclay reinforcement. *Materials Science and Engineering: A, 454*, 416–422.

Markets and Markets. (2018). *Ceramic foams market by type (silicon carbide, aluminum oxide, zirconium oxide), application (molten metal filtration, thermal & acoustic insulation, automotive exhaust filters), end-use industry (foundry, automotive), and region—global forecast to 2023* (Marketing Analysis no. CH 6287). Northbrook, IL: Markets and Markets.

Markets and Markets. (2019). *Metal foam market by material (aluminum, copper, nickel), application (anti-intrusion bars, heat exchangers, sound insulation), end-use industry (automotive, construction & infrastructure, industrial), and region—global forecast to 2024.* (Marketing Analysis no. CH 5349). Northbrook, IL: Markets and Markets.

Martínez, J., Dumas, J., & Lefebvre, S. (2016). Procedural Voronoi foams for additive manufacturing. *ACM Transactions on Graphics, 35*(4), 44.

McGeary, R. K. (1961). Mechanical packing of spherical particles. *Journal of the American Ceramic Society, 44*(10), 513–522.

Miller, D., & Kumar, V. (2011). Microcellular and nanocellular solid-state polyetherimide (PEI) foams using sub-critical carbon dioxide II. Tensile and impact properties. *Polymer, 52*(13), 2910–2919. Available from: https://doi.org/10.1016/j.polymer.2011.04.049.

Nielsen, L. E. (1983). Elastic modulus of syntactic foams. *Journal of Polymer Science Polymer Physics Edition, 21*(8), 1567–1568.

Nistor, A., Topiar, M., Sovova, H., & Kosek, J. (2017). Effect of organic co-blowing agents on the morphology of CO2 blown microcellular polystyrene foams. *Journal of Supercritical Fluids, 130*, 30–39.

Notario, B., Pinto, J., & Rodriguez-Perez, M.A., (2016). Nanoporous polymeric materials: A new class of materials with enhanced properties. *Progress in Materials Science*, 78–79, 93–139. Available from: https://doi.org/10.1016/j.pmatsci.2016.02.002.

Novais, R.M., Pullar, R.C., & Labrincha, J.A. (2020). Geopolymer foams: An overview of recent advancements. *Progress in Materials Science*, 109, 100621. Available from: https://doi.org/10.1016/j.pmatsci.2019.100621.

Okuno, K., & Woodhams, R. (1974). Mechanical properties and characterization of phenolic resin syntactic foams. *Journal of Cellular Plastics, 10*(5), 237–244.

Padilla, E., Jakkali, V., Jiang, L., & Chawla, N. (2012). Quantifying the effect of porosity on the evolution of deformation and damage in Sn-based solder joints by X-ray microtomography and microstructure-based finite element modeling. *Acta Materialia, 60*(9), 4017–4026.

Qiu, L., Zhu, N., Feng, Y., Michaelides, E.E., Żyła, G., Jing, D., ... Mahian, O. (2020). A review of recent advances in thermophysical properties at the nanoscale: From solid state to colloids. *Physics Reports, 843*, 1–81. Available from: https://doi.org/10.1016/j.physrep.2019.12.001.

Ryshkewitch, E. (1953). Compression strength of porous sintered alumina and zirconia: 9th communication to ceramography. *Journal of the American Ceramic Society, 36*(2), 65–68.

Shabde, V., Hoo, K., & Gladysz, G. (2006). Experimental determination of the thermal conductivity of three-phase syntactic foams. *Journal of Materials Science, 41*(13), 4061–4073.

Shutov, F. A. (2004). *Cellular structure and properties of foamed polymers. Handbook of polymeric foams and foam technology* (pp. 17–53). Munich: Hanser.

Smith, B. H., Szyniszewski, S., Hajjar, J. F., Schafer, B. W., & Arwade, S. R. (2012). Steel foam for structures: A review of applications, manufacturing and material properties. *Journal of Constructional Steel Research, 71*(0), 1–10. Available from: https://doi.org/10.1016/j.jcsr.2011.10.028.

Srivastava, V., & Srivastava, R. (2014). On the polymeric foams: Modeling and properties. *Journal of Materials Science, 49*(7), 2681–2692.

Strååt, M.; Chmutin, I.; Boldizar, A. (2008). Dielectric properties of polyethylene foams at medium and high frequencies. *Proceedings of Polymer Processing Society (PPS-24)* (No. S15-770).

Tagliavia, G., Porfiri, M., & Gupta, N. (2012). Influence of moisture absorption on flexural properties of syntactic foams. *Composites Part B: Engineering, 43*(2), 115–123.

Tao, W., Hsu, H., Chang, C., Hsu, C., & Lin, Y. (2001). Measurement and prediction of thermal conductivity of open cell rigid polyurethane foam. *Journal of Cellular Plastics, 37*(4), 310–332.

Wouterson, E. M., Boey, F. Y., Hu, X., & Wong, S. (2007). Effect of fiber reinforcement on the tensile, fracture and thermal properties of syntactic foam. *Polymer, 48*(11), 3183–3191.

Zhang, L., & Ma, J. (2010). Effect of coupling agent on mechanical properties of hollow carbon microsphere/phenolic resin syntactic foam. *Composites Science and Technology., 70*(8), 1265–1271.

10 Applications

10.1 Introduction

Voids are ubiquitous in all natural and engineered materials. Different terms have been used to designate this void volume. This absence of material, which is what a void is, brings along with it very interesting and broad functionalities to the materials containing them. A few terms that we have used to name this "empty space" are: void, vacancy, porosity, free volume, cell, cell structure, and hollow structures.

Just as widespread as voids are the applications that exploit them. When most of us think of this topic, foams and more specifically, polymer-based foams immediately come to mind because they surround us in our daily lives. Car and furniture cushions, sponges, packaging material, insulation for coolers and construction, carpet backing, acoustic tiles, shoe insoles (Klempner & Frisch, 1991), etc., all are polymer foams. One cannot minimize the material design and testing that goes into these seemingly mundane applications. In most applications, the foams are highly engineered and have microstructures designed specifically to meet stringent design requirements. The materials and products require extensive testing and evaluation of properties, such as compression, fatigue, thermal conductivity, heat capacity, etc. In addition to the characteristics of foams in our daily life, even more highly engineered foams go into protecting life and equipment in harsh temperature and pressure environments as well as biomedical and drug delivery applications.

Analogous to conventional or nominally dense materials, there are foams other than polymer-based materials; for example, metal and ceramic, and composite-based foams. Ceramic foams have found widespread application in filtration. Most notably in filters used in low carbon dioxide emission diesel engine exhausts to remove particulate matter (Ohji, 2013). Fig. 10–1 shows the applications of ceramic foams as a function of void size. The functionalities that these applications exploit are materials with high temperature capability, tolerance of aggressive chemicals, high surface area, catalytic characteristics, and biocompatibility.

The applications of metal-based foams can be plotted in relation to the type of porosity, open versus closed cell. Fig. 10–2 plots these functional and structural applications. Metal foams have found widespread application as impact and energy absorbers, for example, as crumple zone materials.

Voids in Materials. DOI: https://doi.org/10.1016/B978-0-12-819282-5.00010-7

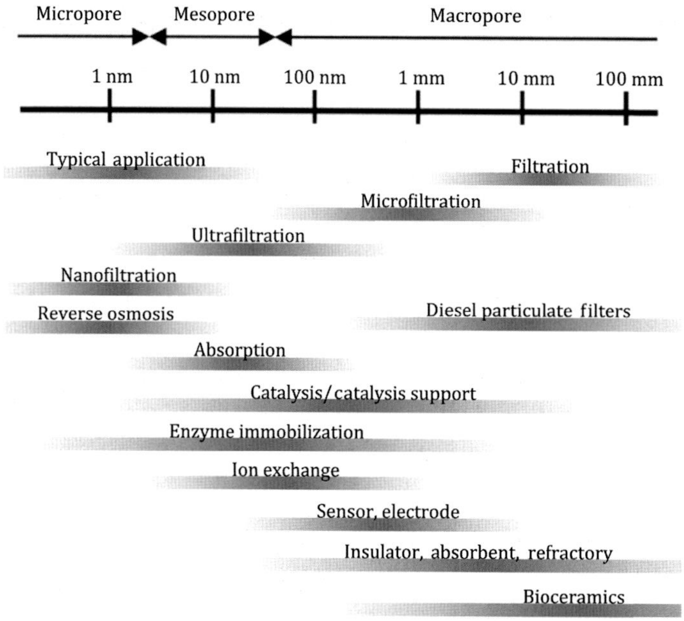

FIGURE 10–1 Applications of ceramic foams as a function of porosity size. *Source: Adapted from Ohji T. (2013) Chapter 11.2.2—Porous ceramic materials. In Somiya S. (Ed.) Handbook of advanced ceramics. 2nd ed. Academic Press: Oxford, pp. 1131–1148.*

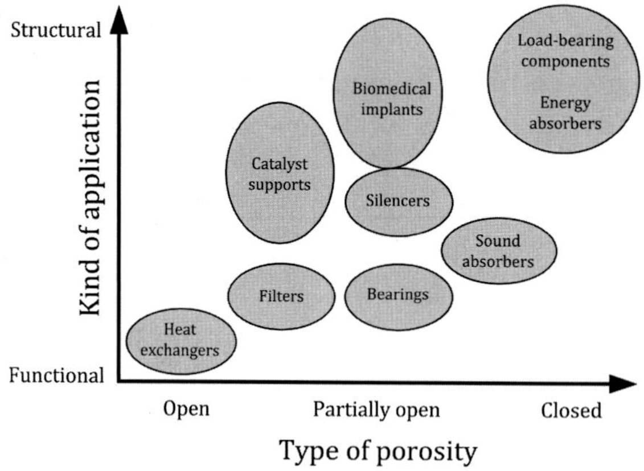

FIGURE 10–2 Application of metal foams as a function of the type of porosity. *Source: Adapted from: Banhart J. (2001) Manufacture, characterisation and application of cellular metals and metal foams. Progress in Materials Science 46 (6) 559–632.*

Another important topic that we have stressed repeatedly is hierarchical porosity; that is the engineering or control of voids simultaneously at multiple length scales. One of the important and rapidly growing areas of research involving hierarchical porosity is that of biomaterials for implantation into a body. Biomaterial were initially just biocompatible.

Now, because of hierarchical porosity and the ability to select materials, biomaterials are *bioactive and biosorbable, in addition to being biocompatible.*

In addition to foams, we discuss other applications of voids in materials. These applications may not be obvious but take advantage of voids at some level for functionality. As such we present functionality in this chapter and present examples from across the length scales.

10.2 Syntactic foams

10.2.1 Deep-sea buoyancy

In this section, we wish to highlight a unique application that involves the use of hollow structures, namely, deep-sea exploration. Buoyancy has played a huge part in man's ability to visit and explore the deepest parts of the ocean. In the 1960s deep-sea buoyancy went through an evolution; a step change that allowed for improved design of subsea vehicles. It was the transition from gasoline to syntactic foams as buoyancy for deep-sea applications. There are many types of underwater vehicles designed to operate deep in the ocean. Two human occupied vehicles have carried three people to the very bottom of the ocean; a place called Challenger Deep of the Marianas Trench off the coast of Guam. The most recent of which, used syntactic foam extensively in its design.

In the early 1960s the use of gasoline buoyancy was coming to an end, but it did so in a historic fashion. On January 26, 1960 a human occupied vehicle (HOV) called Trieste descended to the farthest depths of the ocean—a place called Challenger Deep, the deepest part of the Marianas Trench, see Fig. 10–3. Two men, Jacques Piccard and Don Walsh, took the bathyscaphe down 11 km to the very bottom of the ocean. It stayed there for only 20 min. It was not until 52 years later that another human would visit.

Trieste was basically a blimp or dirigible designed for the water. Buoyancy for the Trieste was provided by 128,700 L (34,000 gallons) of gasoline along with iron shot or pellets. Because gasoline is compressible, Treiste would lose buoyancy as it descended. The iron shot would slowly empty from ballast hoppers to compensate for the loss of buoyancy. Once at the bottom, the iron pellets would empty from the ballast hoppers at a faster rate for ascent back to the surface. Although Trieste II was built in 1964, it was clear that the end was coming for this type of buoyancy.

Displacing the Trieste bathyscaphe with its complicated buoyancy system was the Alvin class submersibles which used a new type of buoyancy agent, a *syntactic foam*. A syntactic foam is a composite foam consisting of a hollow filler material dispersed in a binder phase. Meant for deep-sea applications, this hollow phase in the syntactic foam was in the form of hollow glass microspheres (HGMS) or glass microballoons (GMB), see Fig. 10–4, and the binder phase was an epoxy matrix. This GMB average diameter is 50–75 μm.

FIGURE 10–3 Bathyscaphe Trieste being lowered into the water. For reference the observation sphere below the large tank is 2.4 m in diameter with 8.9 cm thick steel walls. *Source: Public domain, Art collection, U.S. Naval History and Heritage Command website.*

The emergence of syntactic foam as a buoyancy material is inextricably linked to the large-scale manufacturing of GMBs. The two men responsible for the industrial scale production of GMBs were Emerson and Cuming. They founded the company, Emerson & Cuming in 1948 near Boston, MA. Starting in the 1950s they produced GMBs for specialty electronic, thermal, and aerospace applications. With the launch of Alvin in 1964 the deep-sea applications of GMBs and syntactic foams quickly expanded. Today these materials are the mainstay for deep-sea buoyancy for offshore oil and gas, scientific exploration, and naval applications.

In the 1960s the buoyancy on the Alvin-class submersibles did not have the depth capability of Trieste; however, they were much more maneuverable and amenable for search-and-recovery and scientific exploration. In the following decades, the depth capability of syntactic foams improved to the point that syntactic foams can now reach ocean bottom

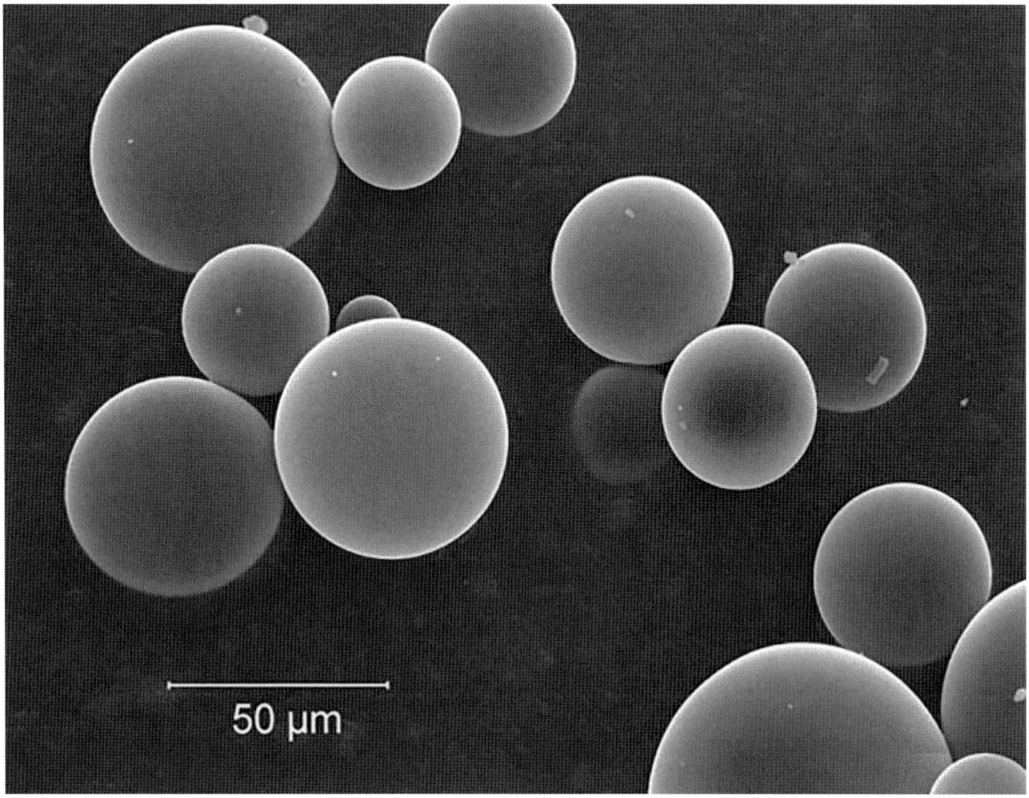

FIGURE 10–4 Glass microballoons (GMBs) have a generally spherical morphology with variation in size (SEM).

while providing more buoyancy than gasoline. The second HOV to reach Challenger Deep made significant use of syntactic foams, not only for buoyancy but as a structural material. James Cameron in his vehicle, Deepsea Challenger, made the 11 km descent on March 12, 2012. He spent a total of 3 h at 11 km. About 70% by volume of Deepsea Challenger was GMB/epoxy syntactic foam, see Fig. 10–5.

The greatest challenge, of course, operating in the deepest regions of earth's oceans known as the hadal zone, is the large hydrostatic pressure from the seawater. At 11 km, the hydrostatic pressure is approximately 116 MPa (1140 atm). Compare this to the pressure at sea level of only 0.101 MPa (1 atm), which is the pressure most systems on earth are required to function in. Any component going to this depth must either be strong enough to withstand the hydrostatic pressure or be inside a protective enclosure (pressure vessel/housing) at 0.101 MPa—which means this enclosure must withstand the hydrostatic pressure.

Syntactic foam contains *reinforced voids* and are the key to creating a light but strong foam capable of withstanding the high hydrostatic pressures experienced at these ocean

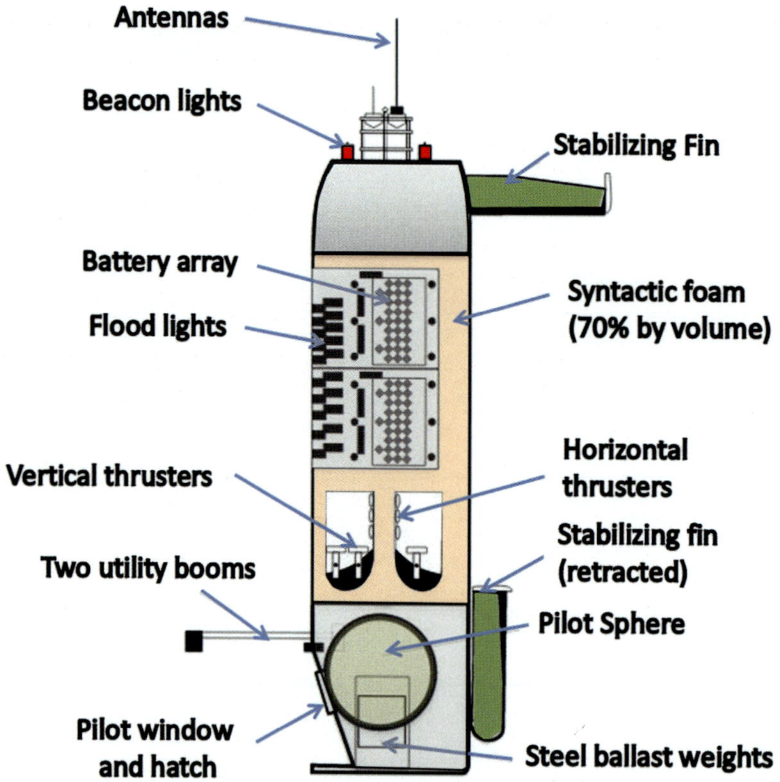

FIGURE 10–5 Schematic of Deepsea Challenger.

depths. The syntactic foams are significantly different from the common single-phase foams such as those found in furniture cushions and insulated coolers. The design of the rigid syntactic foams for buoyancy requires a balance among competing design criteria, minimizing density of the foam, and maximizing resistance to hydrostatic pressure at ocean depth. The strength of the syntactic foam is derived from three primary sources; the GMBs, resin, and interface between them. The strength of the GMB increases as the ratio of the wall thickness to diameter *(t/d)* increases. But as *t/d* increases so does the density, thus reducing buoyancy. The strength of the binder phase as well as the interface need to be considered as they both reinforce the strength of the GMB. In the case of Deepsea Challenger submersible, where the syntactic foam is structural and needs to handle more complex loading than just hydrostatic pressure, a more detailed analysis of the strains was needed to design an appropriate syntactic foam.

Also, the lifetime performance of any buoyancy agent will depend on the pressure cycle(s) it experiences. Here arrives a critical question: Will syntactic foams be continuously cycled as with an Alvin type submersible; descending and ascending hundreds of times a year? Or will it be at a single depth for a decade or more as are many applications in offshore oil and gas production? Physical aging (free volume), water ingress, and polymer and glass degradation are other factors that need to be considered.

While the Alvin was an HOV, there is another class of exploration vehicle called the hybrid remotely operated vehicle (HROV). An HROV, called Nereus, was designed and built by the Woods Hole Oceanographic Institution (WHOI) in Woods Hole, MA, with the goal of exploring the Earth's deepest trenches at the bottom of the oceans, see Fig. 10–6. The word Nereus comes from a mythical Greek god with a fish tail and a man's torso. The main point was to use the vehicle for exploration under extreme environmental conditions, namely, the crushing hydrostatic pressure at great ocean depths. The hydrostatic pressure due to water increases about 1 atm (0.1 MPa) per 10 m depth. Nereus was designed to operate like two types of vehicles, hence the word "hybrid." The first mode was to move and explore without any operator input, as an autonomous underwater vehicle (AUV). The second mode of operation was tethered to a surface ship and being controlled by a person not inside of the vehicle, that is, as a remotely operated vehicle, ROV. The tether supplies power and has wires that carry the signals to control Nereus's movement and instrumentation.

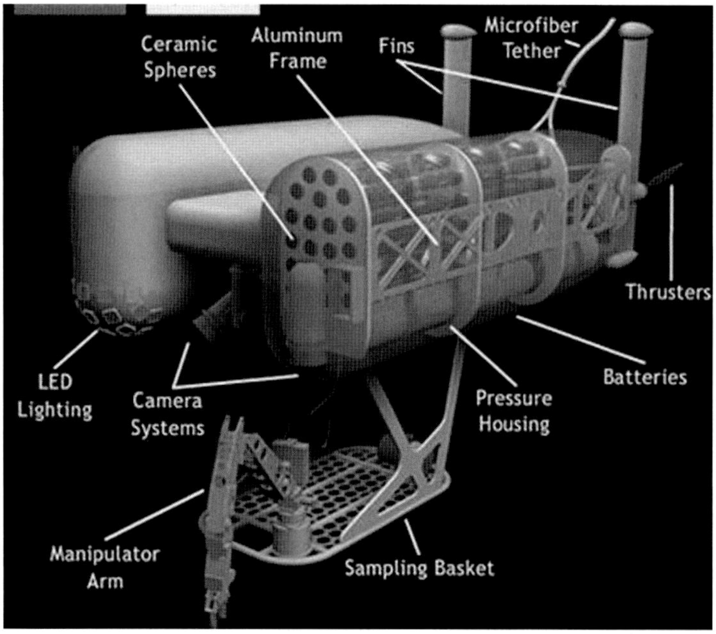

FIGURE 10–6 Hybrid Remotely Operated Vehicle (HROV) Nereus. It uses 1600 ceramic buoyancy macrospheres and nine pressure housings designed to contain sensitive electronics at atmospheric pressure (0.1 MPa or 1 atm) even at 11 km below the ocean surface. *Courtesy: Woods Hole Oceanographic Institution.*

We discussed the processing of the ceramic buoyancy spheres used on Nereus in Chapter 7, Techniques for introducing intentional voids into particles and fibers. What follows is a brief description of the conditions under which Nereus worked, the design tradeoffs pertaining to voids spaces, and an implosion that occurred on May 9, 2014 about 10 km below the surface, according to the expedition leader Machado (2014).

Nereus was designed to operate in the deepest regions of earth's oceans known as the hadal zone (down to 11 km). The greatest challenge, of course, at these depths is the large hydrostatic pressure from the seawater. At 11 km, the hydrostatic pressure is around 116 MPa (1140 atm). Compare this to the pressure at sea level of only 0.101 MPa (1 atm), which is the pressure most systems on earth are required to function in. Any component on Nereus must either be strong enough to withstand the immense hydrostatic pressure or be inside a protective enclosure (pressure vessel/housing) at 0.101 MPa. Nereus had nine ceramic pressure housings, which protected its electronics and batteries, all of which had the potential to implode. In addition to the pressure housings for electronics, Nereus also used hollow ceramic macrospheres (10 cm in diameter) for buoyancy. There were typically approximately 1600 of these spheres aboard Nereus.

The design challenge was to make strong housings that were also light weight (a priority while trying to provide enough flotation). This led to the use of ceramic materials in place of housings made entirely of titanium. Ceramic materials are low weight and very strong in compression. As we all know, ceramics are very brittle, that is, susceptible to catastrophic failure.

According to Machado (2014), while exploring the Kermadec Trench on May 9, 2014, there occurred an implosion. While the system could tolerate the loss of certain individual pressure housings, the energy released during an implosion generated a pressure wave that most likely proved too much for other nearby housings to withstand, leading to additional *sympathetic* implosions. According to Machado, a risk/probability analysis was conducted after the incident, which showed that the pressure housing containing a camera which included a glass viewport was one of the most likely culprit for the initial implosion.

10.2.2 Hollow composite macrospheres and composite syntactic foams

Hollow spheres represent another material type used in very high volumes in the deep-sea buoyancy applications. Typically, diameters of these hollow macrospheres range from 3 to 50 mm. These spheres are made via rotational molding around a low density, sacrificial core, such as polystyrene; see Fig. 10–7. The high-strength shells are formed by building up consecutive layers of a thermosetting resin, typically epoxy, and chopped fiber reinforcement.

A layer is cured before the next layer is started and the process continues until the final density is reached. There is a strong correlation between density and hydrostatic strength.

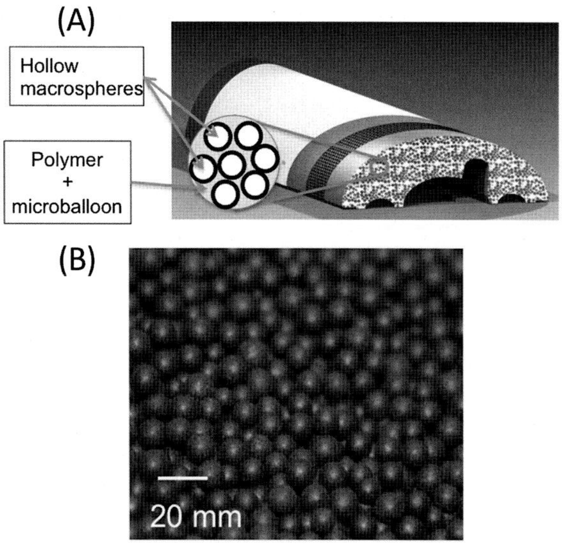

FIGURE 10–7 (A) A schematic cutaway of a drill riser buoyancy module. The dark spots on the front face represent the hollow macrospheres. The macrospheres are embedded in a two-phase syntactic foam matrix and are typically glass microballoons (GMB) and a thermosetting polymer (inset). (B) Carbon fiber/thermosetting polymer hollow macrosphere made by rotational molding. These hollow spheres are designed to have low density and high hydrostatic compressiive strength. They are used extensively in subsea buoyancy for the offshore oil and gas market. *Courtesy © Trelleborg Offshore, Skelmersdale, UK.*

The hydrostatic crush strength of a sphere is dependent upon its sphericity, sphere diameter, and the ratio of the sphere diameter to shell wall thickness. For deep-sea use, the hollow macrospheres are then embedded in a syntactic foam matrix (thermosetting polymer with HGMS) and are called composite syntactic foams. Composite syntactic foams are used for buoyancy in drill riser buoyancy modules, distributed buoyancy, and some subsea vehicles. When we sum up the volume occupied by hollow macro- and microspheres used in these buoyancy modules, we end up with as much as 80%–85% by volume of hollow particles (empty space + the shell). The density of these modules is optimized for maximum resistance to failure under hydrostatic conditions.

10.2.3 Deep-sea thermal insulation

Syntactic foams and hollow microspheres are important fields of research with expanding applications (Gladysz & Chawla, 2002). Fig. 10–8 shows the microstructure of a two-phase polymer-based syntactic foam with GMB commonly used in deep-sea insulation. Two-phase, GMB-based syntactic foams are, by far, the most used syntactic foams for deep-sea applications. The material design of deep-sea insulation for crude oil pipelines is challenging. It is worth pointing out that in offshore drilling crude exits the well

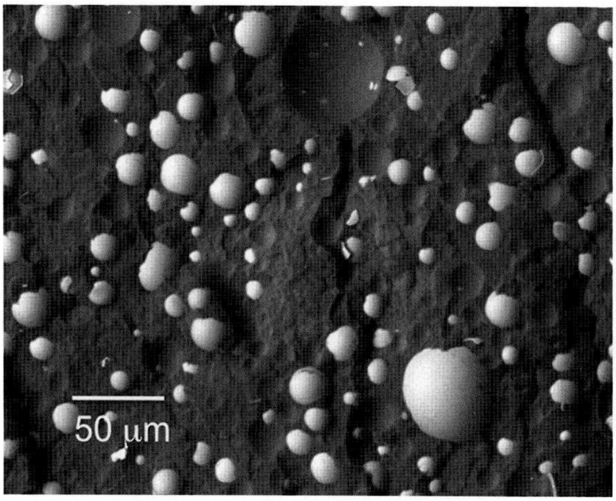

FIGURE 10–8 A polymer-based two-phase syntactic foam containing hollow glass microspheres. SEM.

at the ocean-floor hot, sometimes at 100°C and even hotter at deeper depths. Crude must retain this heat to prevent waxes from forming on the walls of the pipe as it flows to surface of the ocean. We should remind the reader that the temperature of the water at 1.5 km deep in the ocean is close to 0°C. Wax formations in the crude can restrict and even cut off the flow rate of crude that a pipe can deliver. The design of the insulation requires balancing of competing design criteria, minimizing thermal conductivity of the insulation, and maximizing resistance to hydrostatic pressure at ocean depth. One needs to keep the density of the GMB low, which minimizes the thermal conductivity while maximizing the density of the GMBs which, in turn, maximizes the resistance to hydrostatic pressures. Also, the lifetime of this insulation under these conditions (high temperature, high hydrostatic pressure) needs to be a decade or more. Physical aging, water ingress, and polymer and glass degradation also need to be considered.

10.2.4 Syntactic foams and explosive formulations

One of the largest applications, by volume, of hollow microspheres is their use in chemical explosives. The voids play an important part in tailoring the properties of explosive formulations. These voids are considered inert additives and the addition of voids is very important to the accurate targeting of density. The density, in turn, determines the detonation velocity, which controls the detonation pressure. Detonation pressure is a very important characteristic for any explosive. It is a measure of the shock wave energy. Since the shock wave energy is affected by density and detonation velocity, it is altered by controlling the void volume fraction of hollow microspheres in the explosive.

We can analyze the detonation velocity of an explosive composition in terms of a composite material consisting of explosive and voids. The detonation velocity of such a composite, D_{mix}, is a function of density and can be calculated very precisely by applying the rule-of-mixtures (Chawla, 2019) to the composite consisting of the explosive(s) and voids:

$$D_{mix} = \sum D_i V_i, \tag{10.1}$$

where D_{mix} is detonation velocity of the mixture, D is characteristic detonation velocity, V is volume fraction, and the subscript i indicates the component, in this case the explosive and void. Therefore, for a binary mixture containing just one explosive compound and one type of void, the equation becomes

$$D_{mix} = D_{exp} V_{exp} + D_v V_v, \tag{10.2}$$

where the subscripts v and exp indicate the *void* and *explosive*, respectively.

Detonation velocities are well-documented for various explosives and fillers. The characteristic detonation velocity for voids, D_v, is 1.5 km/s (Cooper & Cooper, 1996); substituting this into Eq. (10.2) gives

$$D_{mix} = D_{exp} V_{exp} + 1.5\,(1 - V_{exp}). \tag{10.3}$$

Since the volume fraction of void and explosive must equal 1, we use $V_v = (1 - V_{exp})$.

Eq. (10.3) can be simplified as:

$$D_{mix} = D_{exp} V_{exp} + 1.5 - 1.5 V_{exp}. \tag{10.4}$$

Rearranging:

$$D_{mix} = 1.5 + D_{exp} V_{exp} - 1.5 V_{exp}, \tag{10.5}$$

or

$$D_{mix} = 1.5 + V_{exp}(D_{exp} - 1.5). \tag{10.6}$$

Eq. (10.6) very accurately predicts the detonation velocity when compared with experimental data, as shown in Fig. 10–9. This figure compares values of a binary of pentaerythritol tetranitrate (PETN) and void from a volume fraction from 1.0 to essentially 0.0. The experimental data and the solid trend line are in good agreement.

Once the detonation velocity is known, as provided above, the detonation pressure P_{C-J} (GPa) can be estimated by the following (Cooper & Cooper, 1996):

$$P_{C-J} = \frac{\rho_{mix} D_{mix}^2}{\gamma - 1}, \tag{10.7}$$

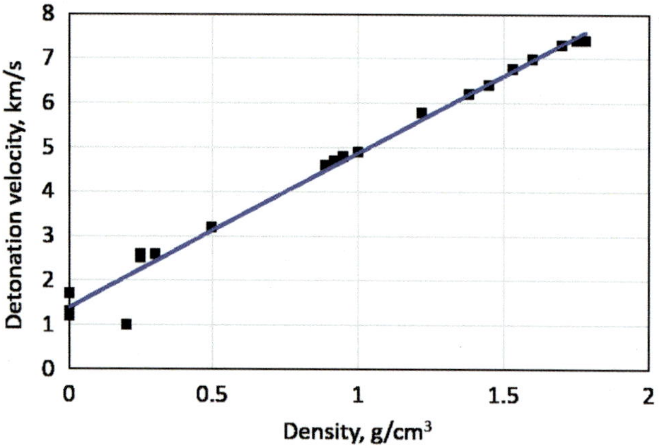

FIGURE 10–9 Detonation velocity of a pentaerythritol tetranitrate (PETN) and void mixture as a function of density. This illustrates the ability to tailor properties of explosives by the introduction of voids. *Source: Adapted from Cooper P.W. (1996) Explosives engineering. VCH: New York.*

where ρ_{mix} is the density of the explosive void material in g/cm^3, γ is the ratio of specific heat of product gases to the reactants, and D_{mix} is given by Eq. (10.6).

10.2.5 Other application

The use of hollow microspheres in a syntactic foam involves many material design options, the most basic being that of having reinforced (hollow microspheres) and/ or unreinforced (interstitial porosity) voids. Note that this aspect is not available in single-phase blown foams. When considering options of reinforced voids, one must take into account volume fraction of microballoons; shell material (glass, carbon, phenolic, etc. or a mixture of these); porous or dense shells; and microballoon size, shape, and distributions (i.e., monomodal, bimodal, or multimodal). One can consider many options for the unreinforced void as well, open or closed cell, volume fraction, and size/distribution. Reinforced and unreinforced voids can be varied independently or can have a gradient within a material. For example, one can have a density gradient in a syntactic foam (Gupta & Ricci, 2006), with higher void volume fraction near the surface, which gradually decreases toward the center.

Applications of syntactic foams stem from the high specific strength (strength/ density) and stiffness (modulus/density), as well as low thermal conductivity and low coefficient of thermal expansion. These applications include (Gupta, Zeltmann, Shunmugasamy, & Pinisetty, 2014):

1. Marine, primarily deep-sea buoyancy and insulation
2. Aerospace, aircraft, and spacecraft

3. Sports equipment
4. Synthetic wood for furniture
5. Plug assist material for vacuum forming
6. Materials used for composite tool layup
7. Radio equipment
8. Blast and fire protection.

10.3 Aerospace

Weight is critical in all aircraft and spacecraft parts. Parts typically require low-density materials, often designed with intentional voids to minimize weight. Lessening the weight reduces the energy burden for a vehicle to take flight or be lifted into orbit or to escape the pull of Earth's gravity. When one thinks of aerospace applications, lightweight, strong, and stiff materials meant for structural applications most likely come to mind. Although lightweight structural materials are very important, the meaning of the term *aerospace applications* and what materials technologies are included have evolved and expanded. Some examples of this include micro-electro-mechanical-systems (MEMS), thermal protection systems, sustainable materials, heat transfer and power systems, sensors, filtration, antistatic, etc. Aerospace materials can be extended to those materials designed for processing in space and even harvesting materials from natural resources on planets other than Earth for manufacturing materials. In this regard, it is worth pointing out that National Aeronautics & Space Administration (NASA) has launched the 3D-Printed Habitat Challenge seeking to develop materials and technology for manufacturing habitats on Mars while using the resources indigenous to Mars. In this section we will cover some of the aerospace applications and important materials with intentional voids.

10.3.1 Carbon nanotubes (CNT)

The first multiwalled carbon nanotubes (MWNT), a kind of CNTs, were synthesized and identified in the early 1990s (Iijima, 1991; Iijima & Ichihashi, 1993). A singlewalled carbon nanotube (SWCNT) is a hollow seamless cylinder made from a single layer of graphene, as shown in Fig. 7.16. One can form a *double-walled carbon nanotube* (DWCNT) by the addition of another graphene layer of a concentric seamless tube on the top of an SWCNT. Further successive additions of graphene layers result in *multiwalled carbon nanotubes*. All of these different types of CNTs find applications in aerospace.

As broad as the term *aerospace material* has now become, so too have the applications of CNTs as a material in *aerospace science* since the initial discovery in 1991. Because of superior mechanical, thermal, and electrical properties, the use of CNTs has led to lighter, stronger composite materials and expanded usage of them. Applications are across the gamut of systems in commercial and military aircraft, helicopters, unmanned aerial vehicles, satellites, and space launch vehicles. These same properties have led to many different advances beyond structural applications in the broader field of aerospace science. As such CNTs are being utilized in energy storage (Islam et al., 2016), coatings (Verma, Anoop, Sasidhara Rao, Sharma, & Uma Rani, 2018), conductive adhesives (Rosca & Hoa, 2011) absorption of light over a broadband wavelength range (Jin et al., 2020), lightning strike protection (Xia et al., 2020), deicing (Chu, Zhang, Liu, & Leng, 2014; Wang, Tay, Sun, Liang, & Yang, 2019), and many others.

In addition to aerospace, CNTs have found applications in a number of composite materials including sporting goods, wind-turbine blades, antifouling marine coatings, and electromagnetic shielding (De Volder, Tawfick, Baughman, & Hart, 2013). CNTs have been added to polyacrylonitrile fibers, a precursor material for carbon fibers (Chae, Choi, Minus, & Kumar, 2009). Upon carbonization, the addition of CNTs allowed for stable carbon fiber formation with diameters down to 1 μm, see Fig. 10–10, and significant increase in strength and modulus. Applications of CNTs will continue to expand in the future. New and expanded uses are expected in commercial and military aircraft, unmanned aerial vehicles (UAVs), micro air vehicles (MAVs), space vehicles, satellites, aerostats, and the space elevator (Gohardani, Elola, & Elizetxea, 2014).

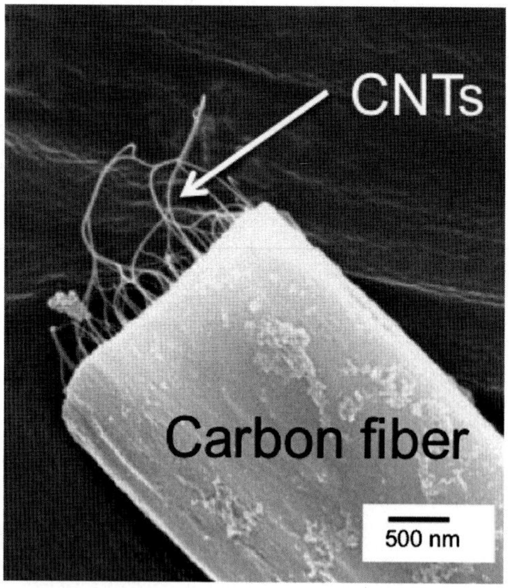

FIGURE 10–10 Carbon nanotubes (CNTs) in a 1 μm carbon fiber. CNTs are added to the polyacrylonitrile precursor before spinning. CNTs allow for a stable formation of a smaller diameter fiber than that of conventionally processed fibers without CNTs additions. The carbon fibers with CNTs also have superior mechanical properties. SEM. *Source: Adapted from Chae H.G.; Choi Y.H.; Minus M.L.; Kumar S. (2009) Carbon nanotube reinforced small diameter polyacrylonitrile based carbon fiber. Composites Science and Technology 69 (3–4) 406–413.*

10.3.2 Honeycombs

Everybody knows what a honeycomb is. It is the engineered structure that bees make of beeswax to store honey. It turns out that the hexagonal form of stacking results in the most compact structure. It can be easily shown that hexagonal packing will result in 90.6% packing efficiency. In engineering, we call any structure resembling honeycomb made by bees a honeycomb.

Honeycombs, traditionally, are hexagonal closed cell columnar structures leading to very rigid properties in the out-of-plane direction and weak properties in the inplane directions. Composite structures in aircrafts are commonly formed by bonding honeycomb cores with face sheets or skins in both primary (wings) and secondary (floors and bulkheads) structural components, see Fig. 10–11. The objective is similar to that of an I-beam, namely, the core provides the shear resistance and enables enhanced stiffness by keeping the facing sheets apart. The attractive features of honeycomb core materials and their uses in sandwich composites are low density, high strength, and stiffness at a relatively low cost. In addition to these advantages, the honeycomb structures can add other functionalities such as improved thermal insulation, sound abatement, fire resistance, and vibration damping properties (Black, 2003).

Since the 1960s, polymer composite-based laminates using honeycomb cores have been used extensively in aerospace and commercial aircraft applications. Fig. 10–12 highlights the extensive use of honeycomb cores on the Airbus A380–800. The components using honeycomb on the A380 aircraft include loadbearing floorboards, control surfaces on the wings and tail sections, and external engine parts. The first aerospace application of a laminate structure with honeycomb core dates back to 1915 when Hugo Junkers patented the sandwich structure using a honeycomb core with metallic facing sheets. This sandwich concept would eventually lead to the replacement of fabric covered structures with metal laminates in aircraft construction. The most common

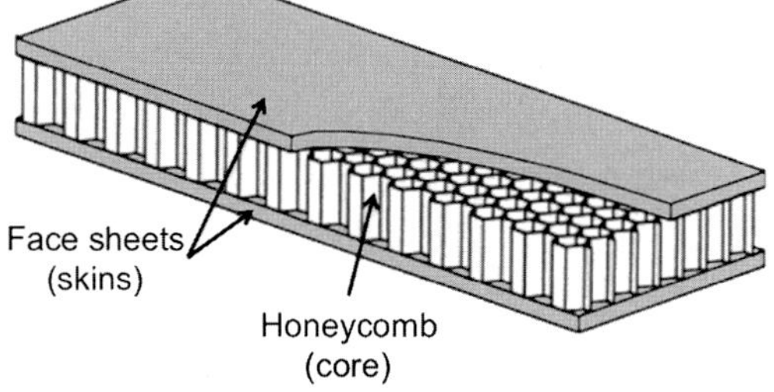

FIGURE 10–11 A panel made from a honeycomb core and face sheets. *Source: Adapted from Kee Paik J.; Thayamballi A.K.; Sung Kim G. (1999) The strength characteristics of aluminum honeycomb sandwich panels. Thin-Walled Structures 35 (3) 205–231.*

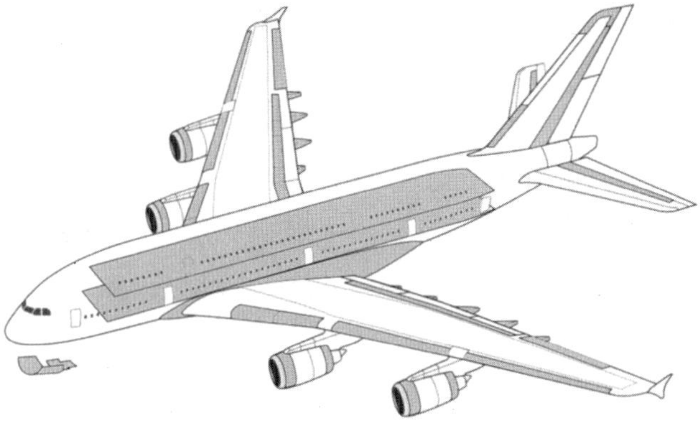

FIGURE 10–12 The Airbus A380–800 utilizes many composite materials.
The portions highlighted indicate the components made from sandwich panels having a honeycomb core.

aerospace honeycomb cores are made from aluminum facings and aramid (Kevlar, Twaron, Nomex, etc.) honeycombs.

One drawback of honeycomb is the anisotropic behavior stemming from the columnar, closed cell void structure (Xiong et al., 2014b). Attempts to create a three-dimensional honeycomb with interconnected porosity have been made to address this shortcoming and add additional functionality (Xiong et al., 2014a, 2014b). Xiong et al. found that more complex stresses can be accommodated by pyramidal and "egg" or three-dimensional geometries, see Fig. 10–13.

10.3.3 Thermal protection systems and heat shields

Orion is NASA's spacecraft designed to take humans farther than ever into space. With such forward looking and grand plans, it is ironic that NASA has gone back to the 1960s formulations for the heat-shield material. NASA is using ablative materials for the heat shield in contrast to the tiles used on the Space Shuttles.

These ablative heat shields are not reusable, whereas tiles used on the Space Shuttles were reusable. The ablative heat shield functioning is in distinct contrast to the porous Space Shuttle tiles that were thermally stable during the heating that occurred during reentry into the atmosphere. Although the black Space Shuttle tiles are part of Orion's thermal management system making up the back shell, the material exposed to the highest temperatures will be an ablative material that dates to the Apollo program.

Ablation is the process of removing the surface of a material by vaporizing it. Composite and multiphase syntactic foams are used extensively as *ablative materials*. These materials are found in rocket nozzles to protect internal structures from hot combustion gases and as heat shields for planetary entry. During atmospheric entry of a spacecraft, extreme ablative conditions are present; these include high velocity, high

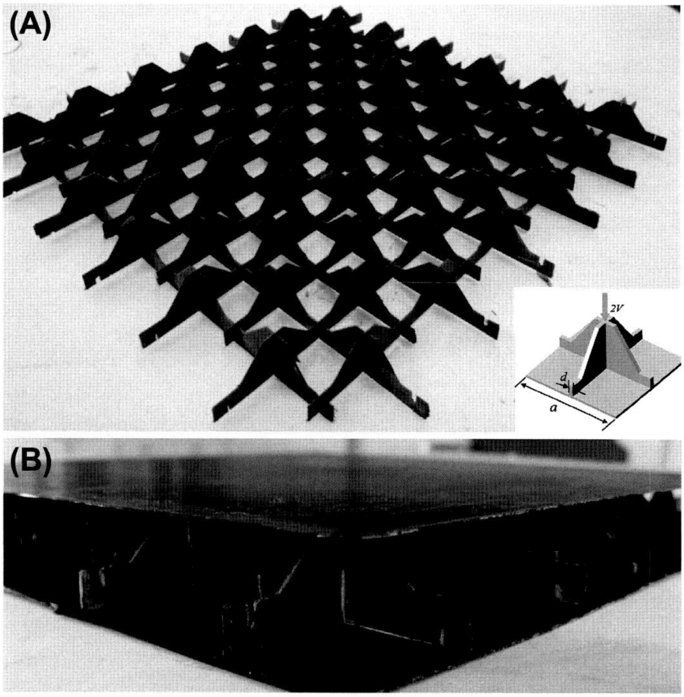

FIGURE 10–13 An example of a (A) three-dimensional honeycomb and (B) sandwich composite with three-dimensional honeycomb and carbon fiber composite skins. *Source: Adapted from Xiong J.; Ma L.; Stocchi A.; Yang J.; Wu L.; Pan S. (2014a) Bending response of carbon fiber composite sandwich beams with three dimensional honeycomb cores. Composite Structures 108 234–242.*

temperature potentially in an oxidizing environment, and erosion from friction with the atmospheric gases. The important material properties for ablative resistance are low density, low thermal conductivity, high temperature resistance, formation of a stable char, and high char shear strength.

Fig. 10–14A shows a scanning electron micrograph of a typical ablative material. Voids play an important role in the functioning of an ablator. As mentioned above, space travel requires low-density materials to minimize weight, thus reducing the energy requirement to lift a vehicle either into orbit or to escape the pull of Earth's gravity. Designing voids into materials reduces density. Other important properties that voids bring to ablative materials are low thermal conductivity and low coefficient of thermal expansion. The ablative material in Fig. 10–14A is a multiphase syntactic foam containing both reinforced and unreinforced voids as well as a discontinuous fibrous phase. Orion's four-phase syntactic foam heat shield is composed of:

1. hollow particles (reinforced voids)—hollow phenolic microspheres,
2. binder phase—novolac epoxy,

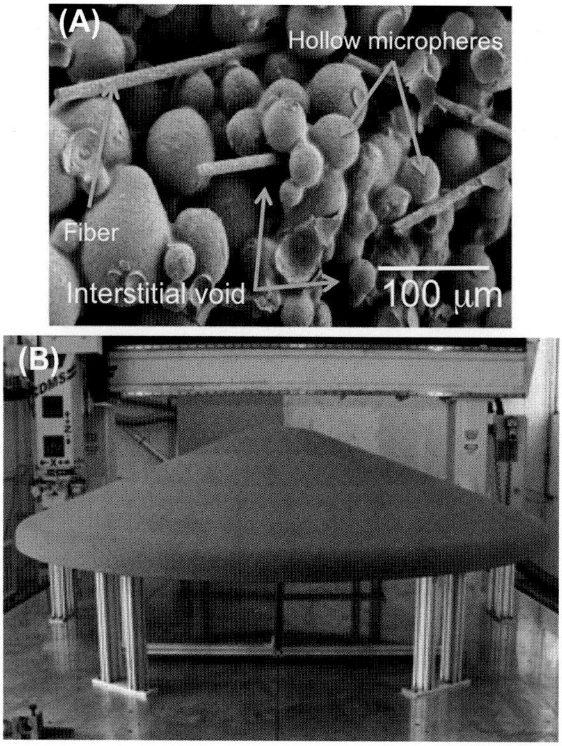

FIGURE 10–14 (A) Microstructure of a silicone-based ablative material. SEM. Note the presence of reinforced and unreinforced voids. The silicone binder, not labeled, is only present in a small volume fraction. This binder phase coats the fibers and hollow spheres, bonds them together, and chars upon entry into the planetary atmosphere. (B) Picture of a finished silicone-based heat shield, ~3 m in diameter. *Micrograph and picture courtesy: ARA Ablatives Laboratory, Centennial, Colorado, United States.*

3. unreinforced interstitial voids—engineered porosity in the novolac epoxy, and
4. glass fibers.

The hollow particles used in the Orion formulation are hollow phenolic spheres with an average 75 μm, the binder phase is a novolac epoxy, there are interstitial voids *between* hollow phenolic particles, and finally the glass fibers. Providing additional structural support to the syntactic foam is a honeycomb made from glass fiber with a phenolic matrix. Each cell of the honeycomb is filled with the syntactic foam material described earlier. This heat-shield formulation is called AVCOAT® and has a density of 0.51 g/cm^3 (32 lb/ft^3). The honeycomb adds compressive strength and helps maintain the structural integrity of the syntactic foam during service. The unreinforced void is the interstitial void between the filler materials. One of the filler materials is the reinforced void, a hollow microsphere, and other one is the fiber. Common binder phase materials for ablative materials are phenolics and silicones because of their high char yield. Fig. 10−14B shows a finished silicone based heat shield, approximately 3 m in diameter.

The assessment of the heat shield, and more generally the thermal protection system, is one of the critical tests conducted on Orion spacecraft. Specifically, a flight objective as set by NASA and Lockheed Martin reads:

"Demonstrate thermal protection system performance during a high-energy return, when Orion will travel near 20,000 mph (32,000 kph), generating 4,000 degrees Fahrenheit (2,200 degrees Celsius) on its heat shield and 3,150 degrees Fahrenheit (1,730 degrees Celsius) on its backshell."

Considering the high temperature and velocity, testing and modeling (Sawant, Rao, Harpale, Chew, & Levin, 2019) of heat-shield materials is a nontrivial job. One test involving an arc jet is shown in Fig. 10−15A. Another example involves an array of solar collectors, collecting and focusing solar light on a sample mounted on a testing tower. This testing was done at Sandia National Laboratory's Solar Tower Facility (located in Albuquerque, New Mexico, United States); see Fig. 10-15B. Both the tests are designed to simulate the conditions of planetary entry. Such tests can be used to assess new formulations of ablative materials. The image captured in the inset in Fig. 10-15B was taken just 1 s into the test and shows the blow off of the ablator surface. Fig. 10-15C is a photograph of a heat shield after the test. The test shows the ablated surface layers and the exposed underlying honeycomb structure. Typically during processing, the syntactic foam is packed and cured into the hollows of the honeycomb. The honeycomb adds compressive strength and helps to maintain the structural integrity of the syntactic/composite foam during service.

10.3.4 Silica aerogel for a comet dust collector

In Chapter 6, Techniques for Introducing Intentional Voids into Materials, we discussed the ability of foams to absorb impact energy when hit with a projectile. That discussion involved relatively dense foams, large projectiles traveling at relatively low speeds. In this section, we wish to discuss a special and unusual example. This example involves the use of an aerogel, which is a special kind of foam with 98%−99% porosity. Aerogels were discovered in 1931 by Samuel Kistler. An aerogel is an extremely porous, therefore extremely low-density material made by removing the liquid from a gel. Kistler reported on silica, alumina, nickel tartarate, stannic oxide, tungsten oxide, gelatin agar, nitrocellulose and egg albumin aerogels (Kistler, 1931). Aerogels are the lightest- weight, lowest mass solid known, and they are ideal for capturing tiny particles in space.

In this example, we describe a comet dust collection apparatus that used an aerogel component to absorb projectiles smaller than a grain of sand, moving at a relative velocity of 21,960 km/h (13,650 miles/h). Obviously, both the dust grain and the collector are moving through space. The relative velocity of an object (dust grain)

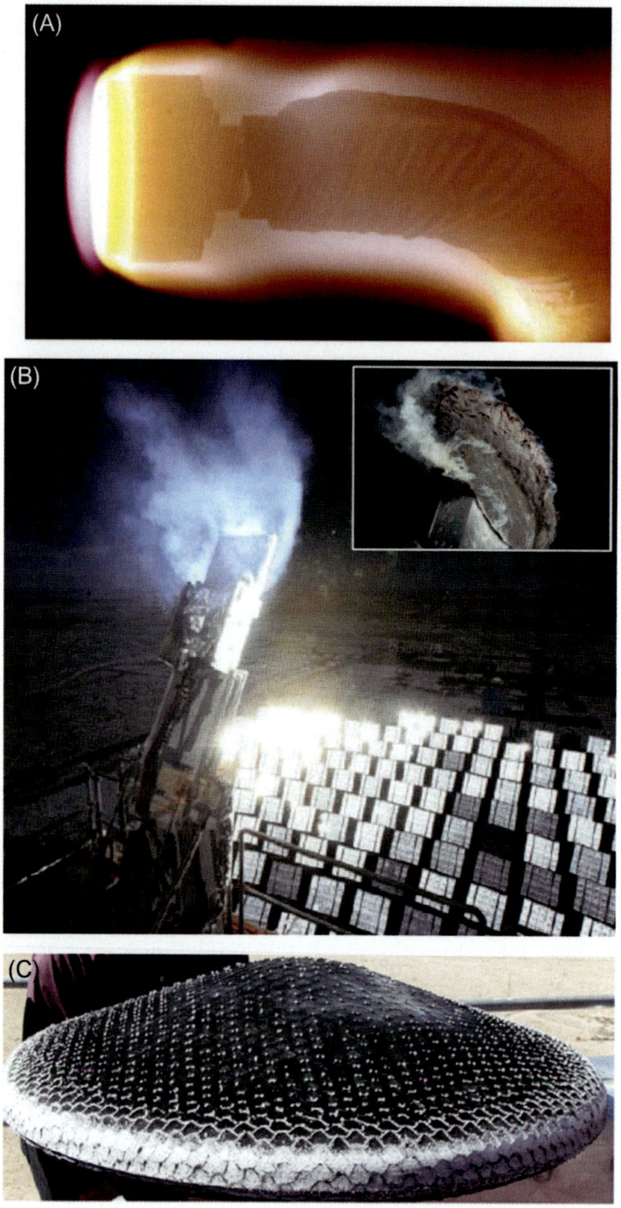

FIGURE 10–15 (A) Arc jet and (B) solar tower tests of a phenolic-based ablative material (Mark Thiessen/National Geographic Creative). The image captured in the inset in (B) was taken just 1 s into the test. (C) The posttest condition from the solar tower test. Note the presence of honeycomb in the heat shield. *Courtesy: ARA Ablatives Laboratory, Centennial, Colorado, United States.*

with respect to another (collector) is how fast the grain appears to move from the perspective of the (stationary) collector. Particles collected ranged from a few nanometers to approximately 100 μm and density ranged from a nominally dense material at 3 g/cm^3 to porous material at <1 g/cm^3 (Kearsley et al., 2008). It is worth relating the full story to show the importance of aerogel.

In 1999 NASA launched Stardust, a space probe that would enter the dust cloud of a comet, collect samples of the dust grains and return those samples to Earth. Stardust would maneuver through the solar system for 5 years before encountering the comet Wild 2 on January 2, 2004. On board Stardust was a scientific instrument called the aerogel collection grid, Fig. 10–16. It was deployed once inside the dust

FIGURE 10–16 The Stardust aerogel dust collection instrument used to capture comet dust grains in the dust cloud of comet Wild 2. Each compartment in the collection grid was filled with a silica aerogel. The density of the aerogel was graded from 0.01–0.05 g/cm^3. The grid was deployed once inside of the dust cloud and the function of the aerogel was to absorb the kinetic energy and capture the comet grains. *Smithsonian Institute https://airandspace.si.edu/collection-objects/stardust-capsule/nasm_A20080417000.*

cloud of the comet to capture the tiny grains of comet dust. The tiny grains would impact with and become embedded in the aerogel. The aerogel collection grid was safely returned to Earth on January 15, 2006 when the capsule gently touched down in Utah, United States.

The aerogel material used was silica aerogel, one of the most common aerogel materials. We described the processing of silica aerogel in Chapter 6, Techniques for Introducing Intentional Voids into Materials. Within the Stardust probe's aluminum collection grid, there were 132 compartments, see Fig. 10–16, these these compartments were filled to a depth of 3 cm with silica (SiO_2) aerogel. Within that 3 cm depth, the aerogel had a density gradient ranging from 0.01 to 0.05 g/cm^3. The lowest density was near the surface where the grains first impacted and the gradient progressively got denser through the 3 cm thickness. The density gradient was designed to gradually slow the grain speed, thus minimizing damage and stopping it in less than a microsecond ($<10^{-6}$ s). It is interesting to compare the characteristics of fully dense silica glass and silica aerogel used in the Stardust probe, which we do in Table 10–1. Because the tiny grains move at such a high relative velocity, the temperature of the aerogel would rise to ~10,000 K (Leroux et al., 2008) as a grain enters the aerogel. As a result, the aerogel melts and encases the grain in a dense silica glass phase. Significant research has been done on Earth using olivine, $(Mg,Fe)_2SiO_4$ to simulate comet grains to the characterize the damage that occurs during such an impact. The simulated grains were launched at 26,000 km/h into silica aerogel (Barrett, Zolensky, & Bernhard, 1993; Zolensky et al., 2008). Under these conditions damage occurred to both the grains and aerogel. This damage included thermal ablation, fragmentation, thermal effects on microstructure, and mixing of the grain/aerogel (Leroux, 2012).

Table 10–1 Comparison of important characteristics of silica aerogel and fully dense silica glass.

Property	Silica aerogel	Silica glass
Density, kg/m^3	5–200	2300
Specific surface area, m^2/g	500–800	0.1
Refractive index at 632.8 nm	1.002–1.046	1.514–1.644
Optical transmittance at 632.8 nm, %	90	99
Coefficient of thermal expansion (20°C–80°C), K^{-1}	2×10^{-6}	10×10^{-6}
Thermal Conductivity, W/(m·K)	0.016–0.03	1.2
Sound velocity, m/s	70–1300	5000–6000
Acoustic Impedance, kg/m^2/s	10^4	10^7
Electrical resistivity, ohm-cm	1×10^{18}	1×10^{15}
Dielectric constant at 3–40 GHz	1.008–2.27	4.0–6.75

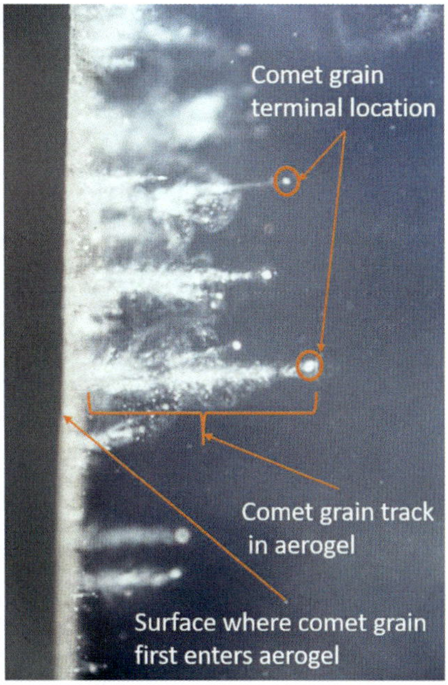

FIGURE 10–17 Image of Wild 2 grains impacts with aerogel from the Stardust collection grid. The grains of comet dust enter on the left of the figure. The kinetic energy of the grain is dissipated by creating damage "tracks" through the aerogel. The grain is also damaged and fragments are left throughout the track with the grain stopping at the end of the track. *After NASA image: https://www.nasa.gov/mission_pages/stardust/multimedia/pia03186.html.*

Let us describe the structure of the silica aerogel used in the Stardust probe and its interaction with the comet grains. The Stardust silica aerogel is made of long chains of nanometer-sized clusters of amorphous silica. The chains are interconnected, making a three-dimensional network with an open system of nanometer sized pores (Fricke & Tillotson, 1997). Fig. 10–17 is a micrograph showing a Stardust aerogel with tracks, made by comet grains. The white regions are the places where grains interacted with the silica aerogel. Grains were recovered from these tracks and chemical analysis was performed on them. The interaction modified the aerogel structure; collapsing nanometer-size voids, compressing into a denser aerogel and causing localized melting of the aerogel to yield fully dense silica glass.

The damage that is done to the aerogel and the mechanisms employed to absorb the kinetic energy of low mass but very high velocity grain of comet dust is similar, in some ways, to compressing other traditional brittle foams. As the aerogel interacts with the grain, it is compressed into higher density regions. Friction with the aerogel increases the farther it penetrates into the aerogel because of the density gradient. The friction will cause the temperature to increase leading to localized melting of the aerogel, followed by

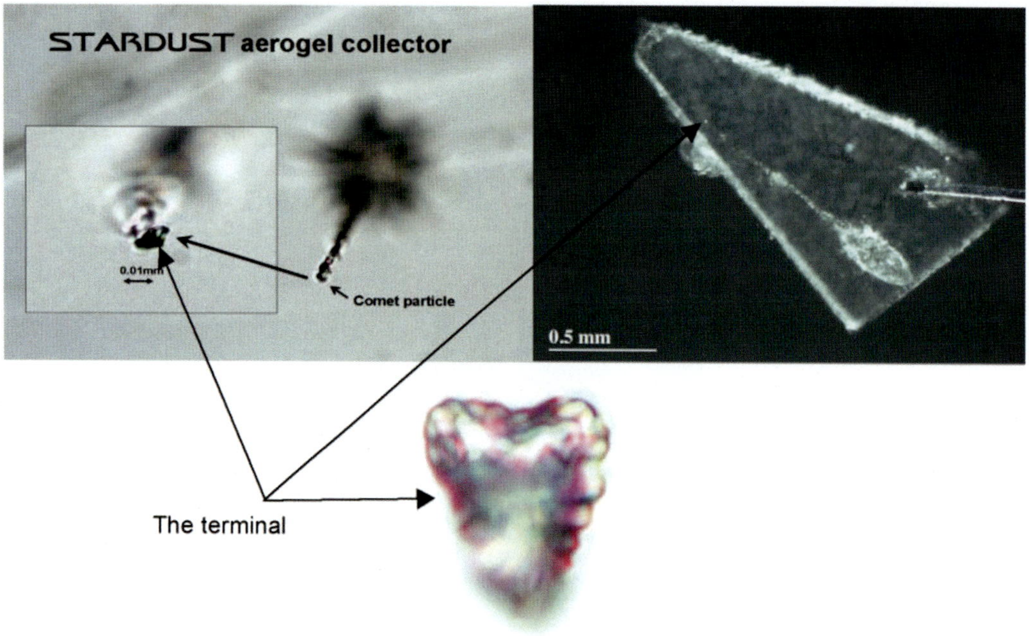

FIGURE 10–18 Image of a Wild 2 comet grain extracted from aerogel collector. It was returned to Earth as part of the Stardust program. *NASA https://www.nasa.gov/mission_pages/stardust/multimedia/stardust-20060221.html.*

solidification and encasing the grain in silica glass. Fig. 10–18 shows the location of a particle embedded in the aerogel and the final extracted particle.

10.3.5 Thermal barrier coatings (TBCs)

The porous structure in thermal barrier coatings, TBCs, is used to cool jet engine turbine blades; these are processing induced but designed porosity, see Fig. 10–19. The TBCs are designed to mitigate the extreme heat from the jet engine environment thus reducing the heat load on the structural nickel-based superalloys. The TBC protects the superalloy component and reduces temperature by 100°C–200°C. The porosity is an artifact of the deposition process used to make the coating. Electron beam chemical vapor deposition and plasma spray techniques are used to create these porous TBCs.

There are four main components, with unique functions in a jet turbine blade, see Fig. 10–20 (Darolia, 2013):

1. Topcoat layer: It provides thermal insulation.
2. Thermally grown oxide (TGO) layer: It provides bonding of TBC to bond coat and slows subsequent oxidation.

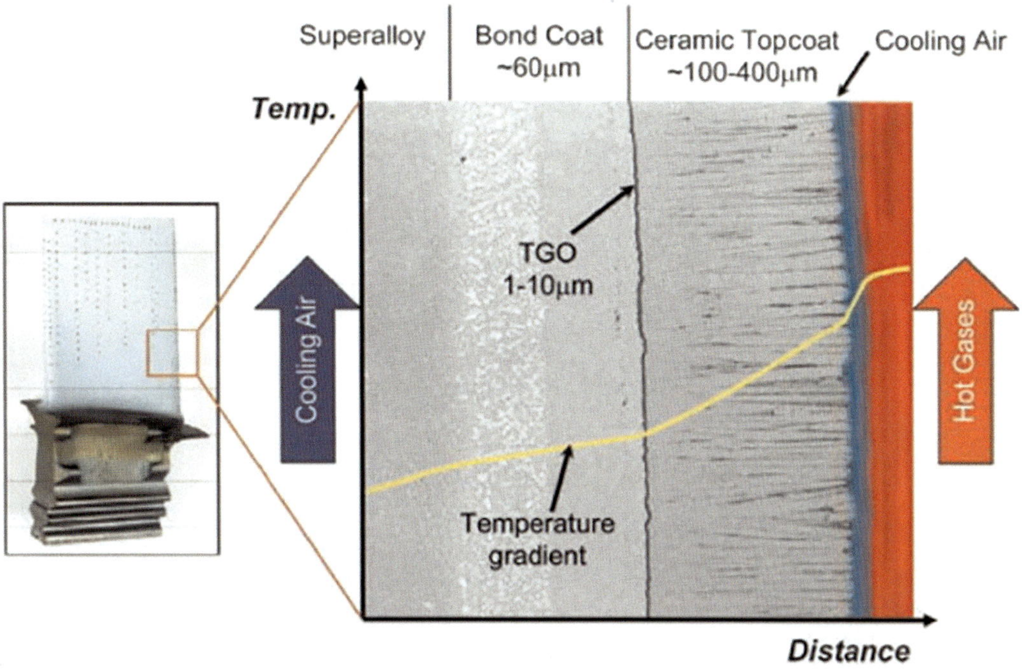

FIGURE 10–19 Schematic of the layers in a turbine blade. The porosity in the topcoat is essential to the functioning of the thermal barrier coating. The temperature drop through the topcoat can up to 200°C, as indicated by the yellow line. *Source: Adapted from Darolia, R. (2013). Thermal barrier coatings technology: Critical review, progress update, remaining challenges and prospects. International Materials Reviews, 58(6), 315.*

3. Bond-coat layer: It contains the source of elements to create TGO in oxidizing environment and provides oxidation protection.
4. Superalloy substrate: It carries the mechanical load. Each of these components has markedly different physical, thermal, and mechanical properties that are strongly affected by processing conditions.

During the processing each of the above components interact chemically and mechanically to create an optimum structure. This dynamic relationships between layers during processing and in-service aging controls the durability and lifetime of the TBC. As such, porosity in the topcoat, typically made from yttria stabilized zirconia (YSZ), is essential to the TBC function and the overall system functionality.

The porous columnar YSZ microstructure is a result of stresses that develop during processing caused by a coefficient of thermal expansion mismatch between the metal and ceramic. The porosity can range from 10% to 25% by volume and can be characterized as pores, gaps, and microcracks, see Fig. 10–20. The functionality this porosity brings

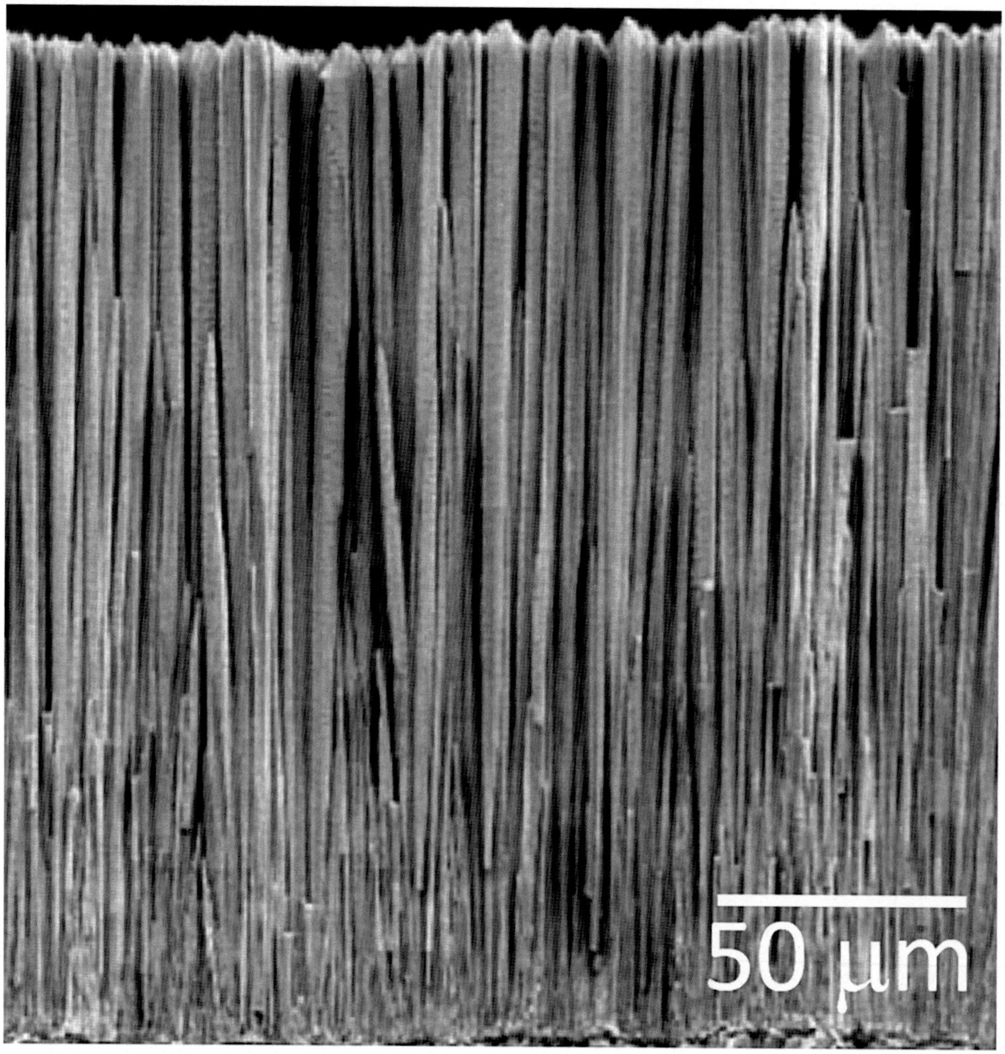

FIGURE 10–20 A expanded view of the yttria stabilized zirconia (YSZ) topcoat of a turbine blade. The porous columnarx structure is needed for proper functionality. This microstructure is an artifact of the high temperature processing of this multilayer system. *Source: Adapted from Darolia, R. (2013). Thermal barrier coatings technology: Critical review, progress update, remaining challenges and prospects.* International Materials Reviews, *58(6), 315.*

to the topcoat is strain tolerance and erosion resistance and also acts as a thermal expansion buffer between the substrate and top layer during service. In addition, the porosity reduces the thermal conductivity compared to a fully dense coating, as the boundaries and pores tend to lie parallel to the surface, that is, perpendicular to the temperature gradient (Darolia, 2013).

10.4 Energy

10.4.1 Lithium-ion battery

Lithium ion batteries are rechargeable batteries that are characterized by very high power densities. Such batteries have become very commonplace: from everyday electronic products such as cell phones to electric vehicles. What is not commonly appreciated is that voids play a very important role in such batteries. As this example will illustrate the void structure in a material, it does not always need to be spherical. Let us first briefly describe the main features of a lithium-ion battery and then point out the important role of voids in it.

There are four components in a lithium-ion cell: anode, cathode, separator, and the nonaqueous electrolyte. Different chemistries are used; the anode is graphite, the cathode is an oxide ($LiCoO_2$), and the alternating layers of anode and cathode are separated by a porous polymer separator, which is generally made of polypropylene (PP), polyethylene (PE), or a laminate of PP and PE. In all cases a critical feature of the separator is a controlled amount and uniform size of porosity in the separator.

The electrolyte consists of an organic solvent and dissolved lithium salt, it provides the media for Li-ion transport. Lithium ions move from the anode to the cathode during discharge and are intercalated into, that is, are inserted into, open spaces in the voids in the cathode. Li ions make the reverse journey during charging. A lithium-ion battery (or battery pack) is made from one or more individual cells packaged together with their associated protection electronics.

Cells are constructed by stacking alternating layers of electrodes such as in prismatic cells or by winding long strips of electrodes into a "jelly roll" configuration typical for cylindrical cells, see Fig. 10–21. Generally, cell form factors are classified as prismatic, cylindrical, and pouch cells (also known as polymer, soft-pack polymer, or lithium polymer).

A separator is nothing but a porous membrane that separates the anode and the cathode in a lithium-ion battery. It allows flow of ionic charge carriers but prevents electrical contact between the electrodes (Arora & Zhang, 2004). All separators contain pores or voids. They can be made of nonwoven fibers (e.g., cotton, polyester, nylon, or glass); films of PE, PP; or laminates of PP and PE. Lithiumbased batteries use nonaqueous electrolytes because of the reactivity of lithium with water. Most of these batteries use porous membranes made of polyolefins. There are two processes of making separators: dry and wet. In the dry process, a polyolefin resin is melted, extruded into a film and annealed, and subjected to a controlled tensile stretching to form pores ($\sim$40% by volume). Examples of the microstructure of such separators made by Celgard are shown in Fig. 10–22. Note the slitlike form of voids. The mechanical properties of the separator are obviously anisotropic. The wet process is

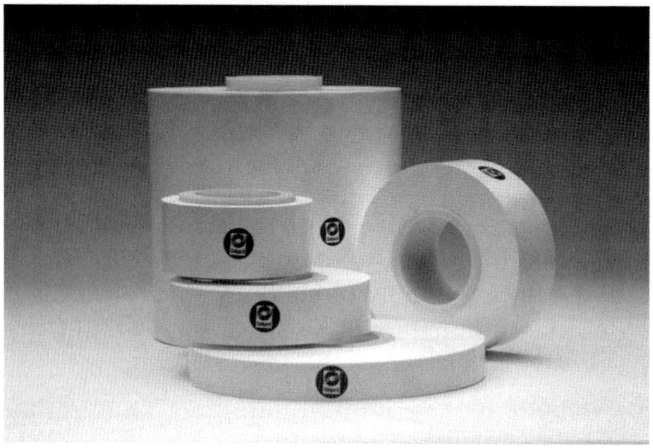

FIGURE 10–21 Polypropylene separators of various widths. *Courtesy: Celgard.*

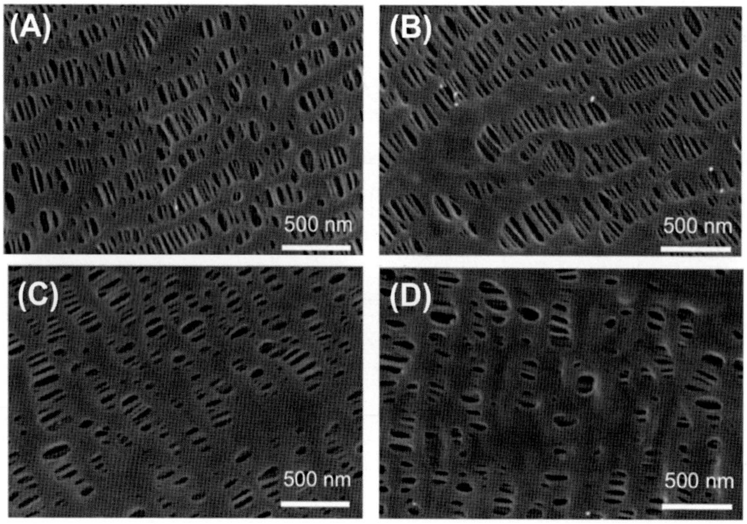

FIGURE 10–22 A characteristic of the Celgard material is that as stress on the material is increased there is a decrease in pore volume fraction.
This is illustrated in the micrographs where the stress state is (A) unstressed, (B) 5 MPa, (C) 10 MPa, and (D) 30 MPa (Peabody & Arnold, 2011). *Courtesy: Celgard.*

an example of a χ-induced syneresis technique of introducing voids discussed in Chapter 6. In the wet process, a hydrocarbon liquid is mixed with a polyolefin resin, followed by heating, melting, and extrusion into a sheet. The sheet is oriented and the hydrocarbon liquid is extracted with a volatile solvent. Typically, a controlled amount of porosity ($\sim$40% by volume) and submicrometer pore size are specified.

It is worth mentioning that pore closure can occur with increasing applied compressive stress. Fig. 10–22 shows the situation under different states of stress (A) unstressed, (B) 5 MPa, (C) 10 MPa, and (D) 30 MPa. With increasing applied stress, there is a decrease in the pore volume fraction (Peabody & Arnold, 2011).

10.4.2 Electrochemical energy storage with porous metals

There is significant research activity into energy storage systems. Improving electrochemical energy storage systems can open up the new possibilities for portable electronics, electric vehicles, and grid-scale energy storage (Zhang et al., 2019). As the world increasingly moves toward renewable energy, such as solar and wind power, there is a need to store this energy. Improving electrochemical energy storage (EES) is one key to the success of these new power generating technologies. The ability to store wind and solar energies is necessary so excess energy can be stored and used during down times, that is, at night or on non-windy days, for example. Common commercially available examples of EES are standard lead acid, Ni–Cd batteries and the more modern lithium-ion batteries. We already discussed the importance of porosity in the polypropylene separator component of the lithium-ion battery. In this section we will discuss the importance of porosity in metal components regularly used in EES devices as supports, current collectors, or active electrode materials.

Intrinsic properties of metal such as surface activity, chemical potential, and oxidation state are important to the functioning of an ESS; so too is the porosity. Porosity in metal offers great advantages over nonporous, planar metal components imparting larger surface area, superior space-time yield and enhanced mass transport (Arenas, Ponce de León, & Walsh, 2019). Fig. 10–23 shows micrographs of three-dimensional Cu foil current collectors all of which have open cell porosity introduced via the Kirkendall effect. The Kirkendall effect can be used to create intentional voids in materials and was discussed in Chapter 5. Fig. 10–23A–C used Sn, Zn, and Al, respective to form a diffusion couple with Cu. Fig. 10–23D is a schematic representation of the porous foil current collector. The open cell porosity can reduce current density, provide "cages" for dendrite, enhance structural stability, and hold high Coulombic efficiencies of after 200 cycles and life spans of more than 2000 h (Zhang et al., 2019).

Traditional methods of introducing voids into metals can, to only limited extent, control void size, size distribution but in most cases these techniques thus resulting in a stochastic cell structure; see Chapters 7 and 9. The Kirkendall effect to create porous structures, as in the previous example also produce a stochastic pore structure. Also discussed in Chapter 7 is the potential of additive manufacturing to create nonstochastic and hierarchical porosity. Using additive manufacturing to

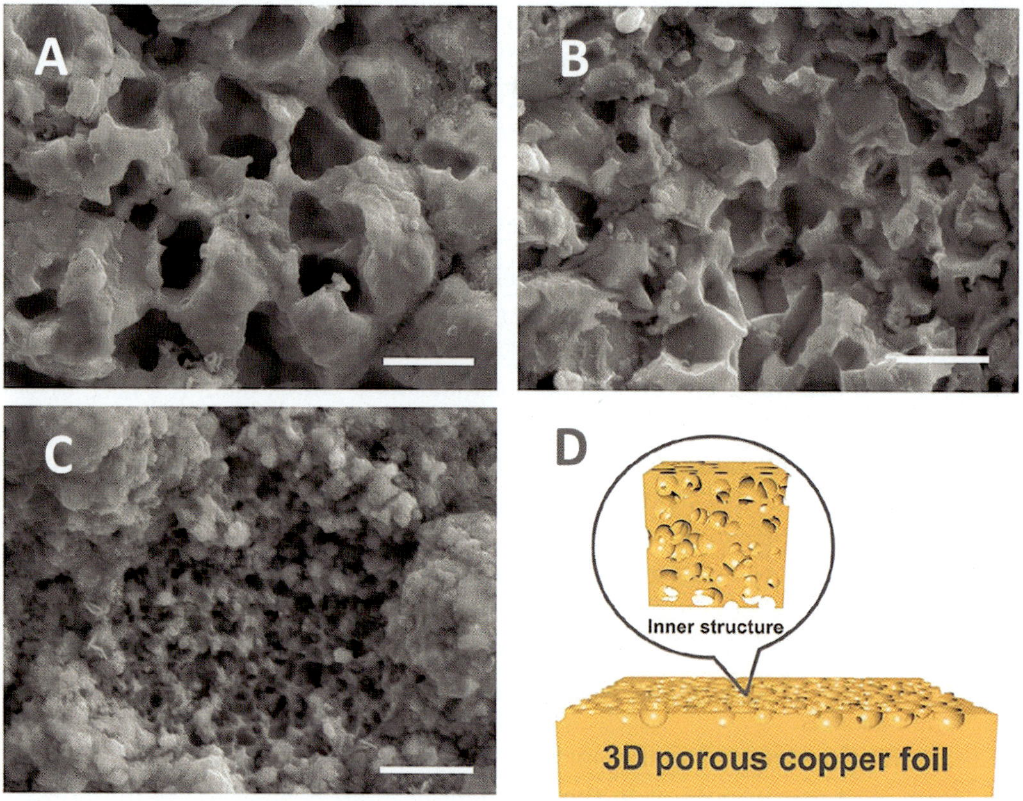

FIGURE 10–23 SEM images of three-dimensional porous copper current collectors made via the Kirkendall effect with (A) Cu-Sn, (B) Cu–Zn and (C) Cu–Al diffusion couple. (D) Schematic of the three-dimensional pore structure of the Cu current collector. The scale bar represents 2 μm in (A), (B), and (C). *Source: Adapted from Zhang, W., Jin, H., Xu, C., Zhao, S., Du, Y., & Zhang, J., Diffusion couples Cu-X (X5Sn, Zn, Al) derived 3D porous current collector for dendrite-free lithium metal battery, Journal of Power Sources,* ***440****, 2019, 227142. https://doi.org/10.1016/j.jpowsour.2019.227142.*

create complex porous architectures one can optimized structures to meet each individual EES application (Egorov & O'Dwyer, 2020).

The use of 3D printing to digitally design porosity and tailor pore structure, pore size and size distribution and periodicity, volumetric space filling and other factors open new avenues to optimize components for ESS applications (Arenas et al., 2019; Egorov & O'Dwyer, 2020). Fig. 10–24 shows a 3D-printed Ni/stainless steel electrode. Knowing the intrinsic properties of metals coupled with ability to tailor pore structures and the reproducibility inherent with 3D printing, one can more precisely distinguish and quantify geometric and intrinsic contributions in the performance of the EES (Egorov & O'Dwyer, 2020).

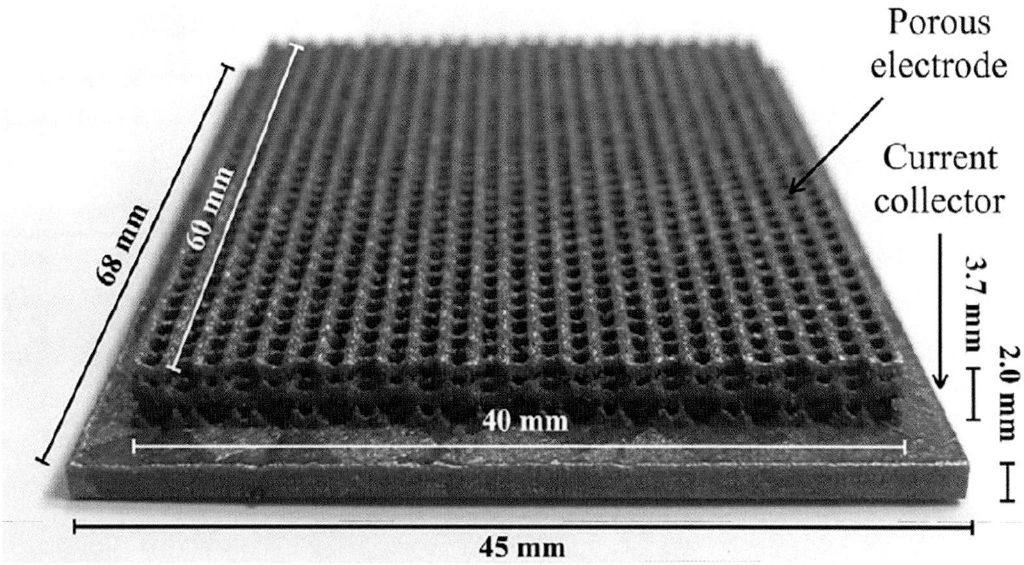

FIGURE 10–24 A 3D-printed porous Nickel-stainless steel electrode for an EES. *Source: Adapted from Arenas, L.F., Ponce de León, C., & Walsh, F.C. (2019). Three-dimensional porous metal electrodes: Fabrication, characterisation and use https://doi.org/10.1016/j.coelec.2019.02.002.*

10.4.3 Guest–host complexes

We discussed at length the role that free volume plays in the thermomechanical performance in polymers and the role of plasticizers in increasing the free volume and plasticity. These are classic studies on polymers that explain this behavior. Another important functionality that certain materials exhibit is accommodating or hosting a molecule in free volume. This class of materials is known as *guest–host complexes*. These are a very important class of materials that can be used in biomimetic membranes, nonlinear optics, and selective transport through liquid membranes. Examples of these guest–host complexes are clathrates and calixarenes. The best known clathrate is methane hydrate. This material has been given considerable attention both as a source of global warming and as a source for the cleanest burning hydrocarbon fuel. The chemical formula is $CH_4 \cdot n\ H_2O$, where $n \sim 6$ (Circone, Kirby, & Stern, 2005). This material has also garnered attention as an engineered material, again for the solid storage of natural gas (methane) within a cage of water crystals, see Fig. 10–25.

The drawback of the material is stability. This material is stable up to approximately 0°C, or a few degrees above 0°C while under pressure. Outside of these conditions the methane will be released. As a greenhouse gas, methane, is far more aggressive than CO_2, $\sim$21 times more damaging. This raises considerable concern about the release of

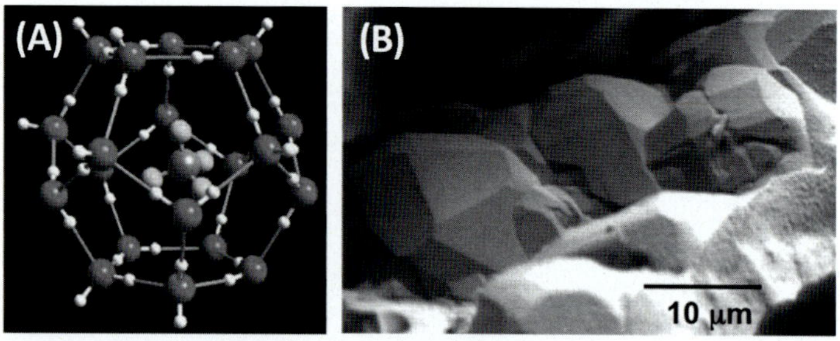

FIGURE 10–25 (A) A methane hydrate crystal consisting of a methane molecule in a cage of ice molecules. Large spheres represent oxygen, small spheres represent hydrogen in water, central spheres represent hydrogen in methane, and the sphere at the very center represents carbon. (B) A micrograph of methane hydrate crystals. SEM. *Images courtesy: US Geological Survey.*

FIGURE 10–26 Burning of methane hydrate. *Image courtesy: US Geological Survey.*

methane into the environment from natural deposits in arctic permafrost. Fig. 10–26 shows the release and burning of methane from the ice cage. This dramatic appearance has led to methane hydrate being referred to as "burning ice." In 2004, Mitsui in collaboration with Osaka University opened one of the first facilities to produce methane hydrate.

In addition to water cages, one can utilize the free volumes in organic solids, such as calixarenes, to host methane (Atwood, Barbour, Thallapally, & Wirsig, 2005).

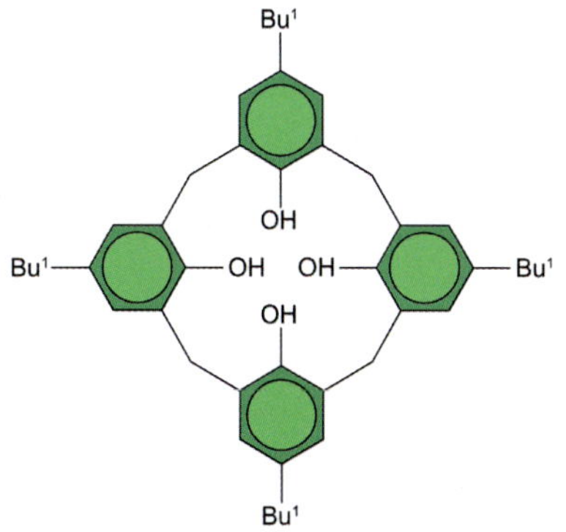

FIGURE 10–27 Chemical structure of p-tert-Butylcalix[4]arene used as a host to capture methane gas in the free volume. *Source: Adapted from Thallapally P.K.; Wirsig T.B.; Barbour L.J.; Atwood J.L. (2005) Crystal engineering of nonporous organic solids for methane sorption.* Chemical Communications *(35) 4420–4422.*

The structure shown in Fig. 10–27 has a free volume of ~235 Å^3 which can host two methane molecules (Thallapally, Wirsig, Barbour, & Atwood, 2005).

10.4.4 Solar power

Solar power is another technology along with the methane hydrate example described earlier as a power source. Intense research and development effort have gone into solar power, being a renewable energy source. The efficiency of solar cells is continuously being optimized to produce more power at a reduced cost. The degradation in efficiency in solar cells has been attributed to the formation of voids in the amorphous silicon wafers. This degradation, which can be as much as 15% during the first 1000 h of use, is known as the Staebler–Wronski effect. Unlike the crystalline silicon wafers, only the amorphous wafers demonstrate the Staebler–Wronski effect due to the disordered nature of the amorphous atomic structure (Fehr et al., 2014). In the beginning, processing induced intrinsic nanoscale voids are present in the amorphous silicon network. These are the sites for the formation of the small, 1–2 nm, voids. The formation of the smaller voids is caused by light-induced breaking of the hydrogen–silicon bonds. These 1–2 nm voids are now thought to be the cause of the Staebler–Wronski effect. The mechanism of this degradation is shown in Fig. 10–28.

10.5 Titania and photocatalysis

One cannot overstate the importance of titania (TiO_2). It is used as a white pigment for paints, but its reach into technology is broad and functionalities are based on its electronic and optical properties. Titania's applications are vast, from paints and

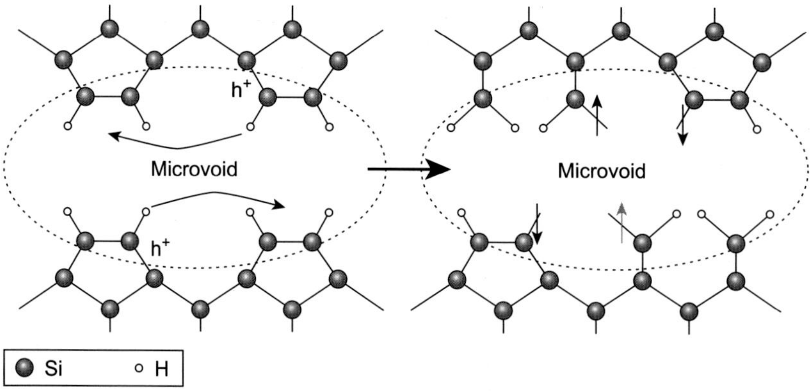

FIGURE 10–28 In the initial state (left), the internal surfaces of voids are saturated with hydrogen atoms so that no defects are observed. Light-induced breakage of atomic bonds causes defects (indicated by the vertical arrows on the right-hand side), which translates into reduced solar cell efficiency. *Source: Adapted from Fehr M.; Schnegg A.; Rech B.; Astakhov O.; Finger F.; Bittl R.; . . . Lips K. (2014) Metastable defect formation at microvoids identified as a source of light-induced degradation in a—Si:H.* Physical Review Letters *112 (6) 066403.*

cosmetics to photocatalysis for water purification (Koopman et al., 2009), solar cells, lithium-ion batteries, sensors, biomaterials, and catalyst supports (Diebold, 2003; Zhang, Elzatahry, Al-Deyab, & Zhao, 2012).

Many of these applications are based on functionalities that are derived from its pore structure. Fig. 10–29 shows nanometer-scale porosity in titania. This material and its pore structure have been used to split water (Fujishima & Honda, 1972) to create hydrogen for fuel cell use, see Fig. 10–30. There are three main steps in the process:

1. Absorption of photons by TiO_2 for the generation of electron–hole pairs,
2. Carrier separation and migration to the catalyst surface, and
3. Redox reactions producing H_2 and O_2.

Titania is a semiconductor. When it absorbs energy, for example from photons, it can create an electron–hole pair. The electron–hole pair is created when the absorbed energy is equal to the bandgap energy. The bandgap between the valence and conduction bands in titania depends on the crystal phase. The bandgap for rutile and anatase phases is 3.1 and 3.27 eV (Valencia, Marín, & Restrepo, 2010), respectively. To split water into H_2 and O_2, it takes 1.23 eV. Water molecules on active sites will split and form H_2 and O_2. It is interesting to note that atomic defects in the titania crystal structure are an undesirable void as they will facilitate the recombination of the electron–hole pair (Zhang et al., 2012), inhibiting photocatalysis. However, the process of splitting water in Fig. 10–30 is greatly enhanced by having a large surface area. Nanometer-scale surface porosity, as

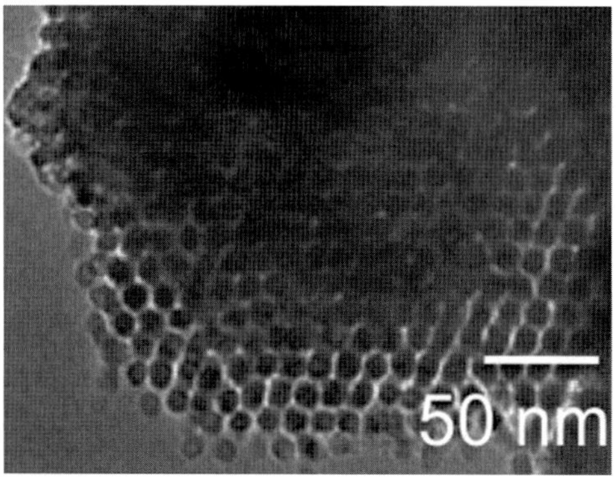

FIGURE 10–29 Micrograph showing nanometer-scale porosity in TiO_2. The combination of chemical functionality and pore structure leads to many interesting applications. *Source: Adapted from Zhang R.; Elzatahry A.A.; Al-Deyab S.S.; Zhao D. (2012) Mesoporous titania: From synthesis to application.* Nano Today *7 (4) 344–366.*

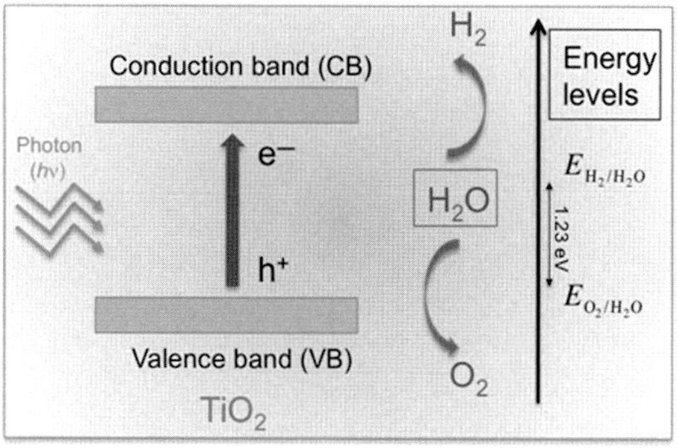

FIGURE 10–30 Mechanism of TiO_2 photocatalysis used inwater splitting for hydrogen production. *Source: Adapted from Zhang R.; Elzatahry A.A.; Al-Deyab S.S.; Zhao D. (2012) Mesoporous titania: From synthesis to application.* Nano Today *7 (4) 344–366.*

shown in Fig. 10–29, creates TiO_2 with a large surface area and provides many sites for the absorption of photons and the redox reaction.

10.6 Biomaterials and healthcare

10.6.1 Introduction

We dedicate a section of this chapter to reflect the important role that voids play in the biomaterials. When it comes to the effectiveness of biomaterials, the functionality is as much about the voids present as it is about the solid materials that are present.

Designing voids of specific sizes becomes an integral part of the successful repair of the body. Biomaterials are not just simple replacement "components" for damaged bone or tissue; they facilitate regeneration.

One of the important and rapidly growing applications involving control of voids simultaneously at multiple-length scales is that of biomaterials for implantation into a body. To put it another way, *in vivo* biomaterials with hierarchical porosity. The field has undergone a remarkable evolution since the 1960s and 1970s when the first generation of biomaterials was introduced; see Fig. 10–31 (Holzapfel et al., 2013). The first generation biomaterial was chosen mainly for chemical inertness, that is, *biocompatibility*. The materials were chosen so the body would not attack and reject the implant; but it also meant that the materials did not encourage healing and the growth of new tissue. An important example of this is the use of titanium. Subsequent generations of biomaterials, referred to as *bioactive materials*, are open to increasing interactions with the body and eventually being absorbed into the body, leaving only natural tissue. Thus, biomaterials are *bioactive and biodegradable/biosorpable, in addition to being biocompatible*. We use the term biomimetic to describe them. We define biomimetic as a

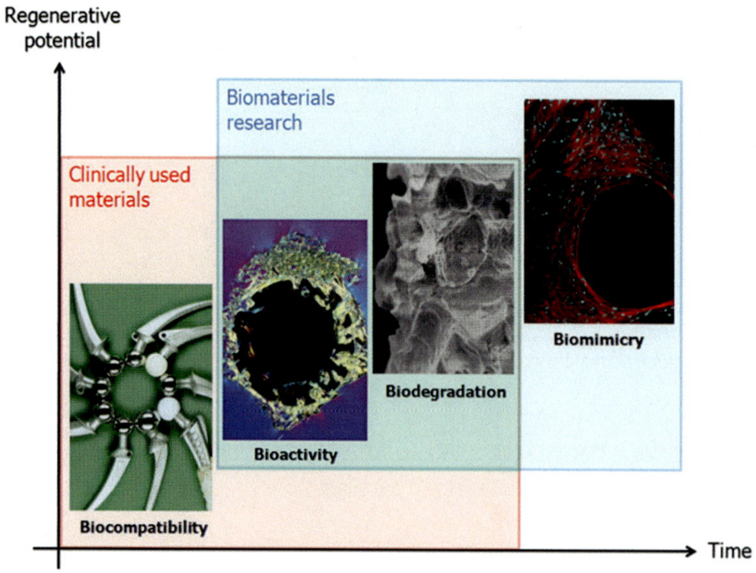

FIGURE 10–31 An illustration of the progression in the approach that biomaterials have taken. The first generation of biomaterials was focused on inertness, using materials such as titanium. The approach has changed to have the materials that are biodegradable, bioactive, and biocompatible, that is, biomimicry. Hierarchical porosity is an important feature of biomimetic materials as it enables tissue regeneration. *Source: Adapted from Holzapfel, B. M., Reichert, J. C., Schantz, J., Gbureck, U., … Rackwitz, L., Nöth, U., (Hutmacher, D. W. (2013)). How smart do biomaterials need to be? A translational science and clinical point of view. Advanced Drug Delivery Reviews, 65(4), 581-603.*

material or material process that replicates one in nature or biology. As shown in Fig. 10–31, hierarchical porosity plays an important role in the tissue-regeneration process by facilitating growth of cellular and extracellular material (ECM). ECM is a term used to describe all of the supporting materials required to allow the cells to grow and function.

10.6.2 Biomaterials scaffold

Biomaterials implanted in a body are in the form of a scaffold, that is, these materials provide the framework for the regrowth of new tissue. Functions of ECM can be, for example, structural and biochemical. To regenerate bone, for example, we need three specific sizes of porosity for optimum functionality. A material with designed porosity on several scales is said to have a *hierarchical porosity*. Each scale of porosity has essential interactions with the body that are required for successful tissue regeneration. It is important to have interconnected porosity (open cell) larger than 100 μm and, in certain cases, as large as 300 μm so that new bone can grow throughout the scaffold. The maximum cell size on this level is usually controlled by mechanical properties. If the cell size gets too large, the strength and fracture toughness are degraded to a point where the material is no longer suitable for scaffolding applications. Pores in the range $<10\ \mu m$ are important for intensifying adsorption of cell differentiation inducing factors and ion exchange (Holzapfel et al., 2013).

In addition, an increase in surface area is needed for the proliferation and differentiation of anchorage-dependent cells for tissue regeneration. Nanoscale texture and surface features facilitate interactions between host cells and the biomaterial. Surface features and properties determine the organization of adsorbed protein layers, which in turn determine specific cellular responses.

Throughout this book we have discussed the importance of porosity for successful implantation of materials in vivo and the growing impact of 3D printing is having on the design of materials and components. The major advantage of this intersection is the customization of an implant designed specifically for an individual patient (Wilcox et al., 2017). One example requiring specific pore structure and 3D printing uses porous metal implants in surgeries involving spinal repair. Stryker Corp. (Allendale, NJ, USA), has developed a 3D printed biocompatible, porous titanium alloy engineered for bone regeneration. Fig. 10–32 shows an anterior cervical cage used as a spinal fusion implant made with Stryker's proprietary Tritanium In-Growth Technology® (Choy et al., 2018).

The Tritanium In-Growth Technology is a general material design philosophy for biomaterial implants. The porosity design of Tritanium In-Growth Technology facilitates healing of the body via enabling regrowth of bone into the porous titanium

FIGURE 10–32 An example of 3D printed anterior cervical cage used in cervical spinal implant surgery. The Tritanium In-Growth Technology® is from Stryker Corp. It is a biocompatible titanium alloy. The portions that are porous have a random cell structure, R ~ 0, and specifically designed for facilitate cell attachment, proliferation, and bone regeneration. *Source: Choy, W. J., Parr, W., Phan, K., Walsh, W. R., & Mobbs, R. J. (2018). 3-dimensional printing for anterior cervical surgery: a review. Journal of spine surgery (Hong Kong), 4(4), 757–769. https://doi.org/10.21037/jss.2018.12.01*

alloy structure. The 3D printed pore structure is designed to mimic cancellous bone and thus is optimized to facilitate infiltration and regrowth of bone throughout the volume of the porosity. In this case the titanium alloy scaffold is not bioabsorbable but it allows the bone to regenerate around the titanium alloy struts. Other Stryker Corp. implant designs using this technology are curved posterior lumbar cage, posterior lumbar cage, and cementless total knee arthroplasty.

The porosity is digitally designed to be random, meaning the regularity, R, is low and approaches zero. Despite the randomized nature of the pore size distribution, the fact that it is digitally designed and 3D printed makes the pore structure reproducible in a manufacturing setting. The porous section of the anterior cervical cage, shown in Fig. 10–32, is 100% open cell porosity with a size distribution between 100 and 700 μm; the mean cell size is ~450 μm. The surface roughness is also a designed feature, which further optimizes the bone growth (Choy et al., 2018).

10.6.3 Nerve regeneration

There are surgical and engineering techniques used to repair nerve damage (Raza, Riaz, Anjum, & Shakeel, 2020). See Fig. 10–33 for a schematic of a nerve. This section will focus on those engineering techniques that involve biomaterials with designed porosity to repair severed axon. In particular, nerve guidance conduits have been shown to increase speeds of recovery as well as repair of severed nerves when regeneration greater than

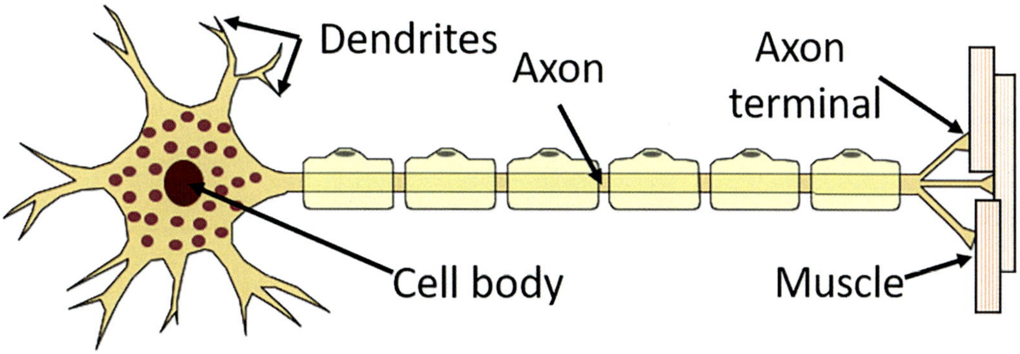

FIGURE 10–33 Schematic of the main components of a nerve. Much of the research for repair focuses on rejoining severed axon via porous biomaterials that facilitate axon regeneration.

5 mm is required. As in all biomaterials for in vivo application, compatibility, interaction with the body to facilitate nerve repair, regeneration, and eventual absorption in the body are essential to an effective material. Natural axon nerve repair is a very slow process, regenerating at ~ 1 mm/day and gaps larger the 5 mm typically cannot be regenerated naturally (Manoukian, Arul, Rudraiah, Kalajzic, & Kumbar, 2019).

As with the bone scaffold, voids are a necessary component in the biomaterial to promote healing to damaged nerve tissue. The nerve tissue is highly vascularized, it demands a highly porous scaffold capable of adequate cell penetration and adhesion in order to enable the infiltration and development of the nerve axons and blood vasculature within and/or through the scaffold network (Manoukian et al., 2019).

Natural polymers such as chitosan, derived from sea crustaceans, are currently being used for nerve regeneration. Chitosan is biocompatible, antimicrobial, and biodegradable, but in general has limited uses because of poor mechanical properties. The addition of a naturally occurring nanotube, halloysite, has been found to improve mechanical properties of the chitosan in addition to promoting axon regeneration, see Fig. 10–34. Halloysite is an aluminosilicate clay mineral having the composition, $Al_2(OH)_4Si_2O_5 \cdot nH_2O$. It is found naturally in the form of plates, spheroids, but nanotubes are the most common form. Applications for halloysite are nanocontainment for drug delivery, reinforcement in composite materials, catalyst supports, and are naturally biocompatible (Tharmavaram, Pandey, & Rawtani, 2018).

Manoukian et al. (2019) developed a porous 5 wt% halloysite/chitosan composite infused with the drug 4-Aminopyridine (4-AP) known to promote nerve regeneration. This porous composite material is used as a conduit material for nerve regeneration. The liquid 4-AP was drawn into the central cavity of the halloysite nanotube using a vacuum chamber. The porosity was introduced into the chitosan by a freeze-drying process.

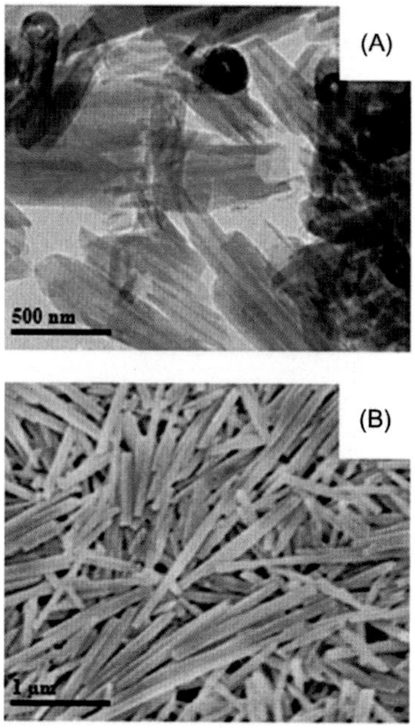

FIGURE 10–34 Characterization of Halloysite nanotubes through (A) TEM; (B) SEM; Halloysite. *Source: Adapted from Tharmavaram, M., Pandey, G., & Rawtani, D. (2018).* Surface modified halloysite nanotubes: A flexible interface for biological, environmental and catalytic applications *https://doi.org/10.1016/j.cis.2018.09.001.*

Chitosan was first dissolved in acetic acid and then the halloysite was introduced. This was followed by a freeze-drying process in a mold to form a conduit or tube. This was followed by crosslinking the chitosan with alkaline epichlorohydrin. The final structure has longitudinally aligned porosity 20–60 μm in length, as shown in Fig. 10–35. In applications that involve severed nerves, where regeneration is greater than 5 mm, conduits guide the nerve growth, see Fig. 10–36. Without the conduit, that is, relying on natural healing, the nerve will slowly regrow but misdirectional growth is common (Raza et al., 2020). As the chitosan dissolves and is absorbed into the body, the halloysite is exposed and the drug 4-AP is released. Promising result has been seen in vivo nerve repair in mice with 15 mm defects.

This example of nerve regenerating biomaterial illustrates many functionalities that hierarchical voids bring to materials. The central void of the nanometer-scale porosity of the halloysite nanotube was used as nanocontainment for drug delivery. The longitudinally aligned porosity in the chitosan helped direct the directional healing of the axon and provided a high surface area for tissue regeneration (Fig. 10–36).

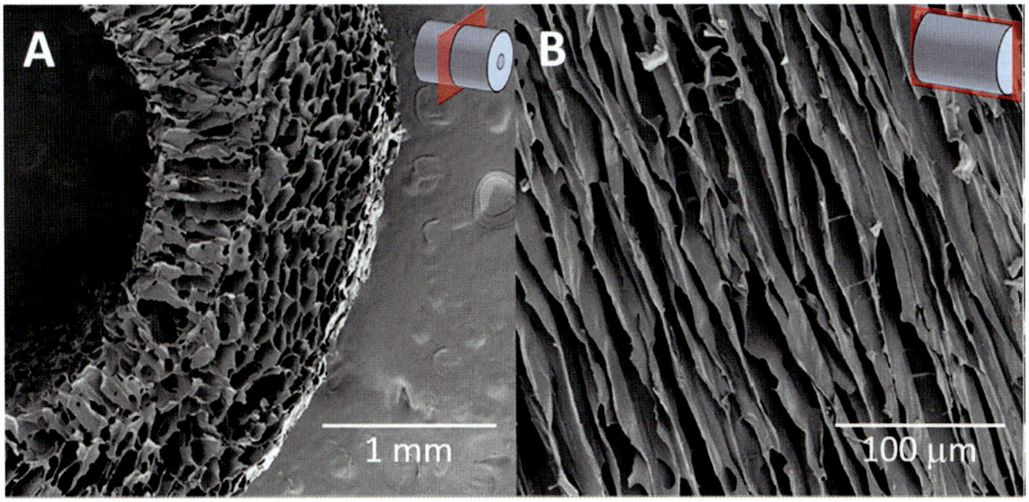

FIGURE 10–35 Porous halloysite/chitosan composite conduit (A) cross-section and (B) longitudinal view used as conduit for nerve regeneration. Note the longitudinal porosity that is an artifact of the freeze-drying process. *Source: Manoukian, O.S., Arul, M.R., Rudraiah, S., Kalajzic, I., & Kumbar, S.G., Aligned microchannel polymer-nanotube composites for peripheral nerve regeneration: Small molecule drug delivery, Journal of Controlled Release, 296, 2019, 54-67. https://doi.org/10.1016/j.jconrel.2019.01.013.*

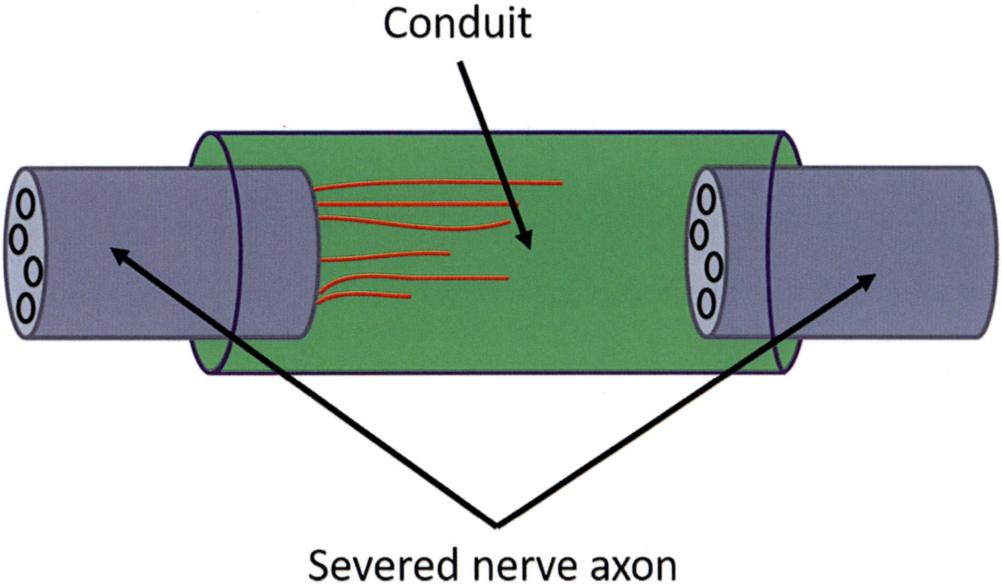

FIGURE 10–36 Illustration of a conduit facilitating the directional healing and regeneration of nerve axon. *Source: Adapted from Raza, C., Riaz, H.A., Anjum, R., & Shakeel, N.u.A. (2020). Repair strategies for injured peripheral nerve: Review https://doi.org/10.1016/j.lfs.2020.117308.*

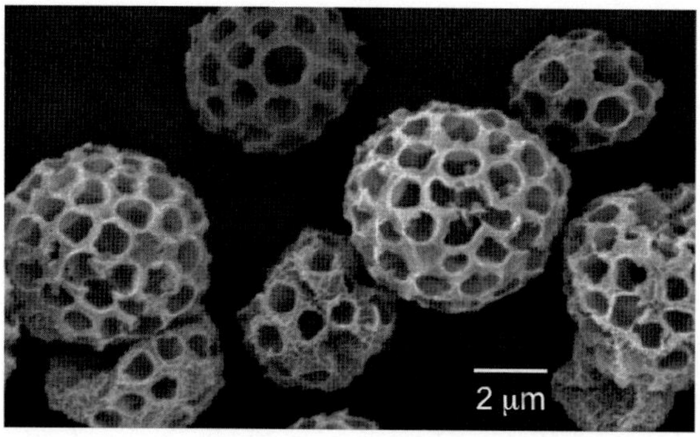

FIGURE 10–37 Microstructure of the hollow porous particles, PulmoSpheres™. SEM.

10.6.4 Drug delivery

PulmoSphere™ represents a nice example of a hollow porous particle used for drug delivery. Note that the PulmoSpheres are not the medicine per se; these are the vehicles for delivering the drug. These porous, hollow spheres can be formulated with different drugs. These are 3–5 μm in diameter. The shape and structure of PulmoSpheres give them the aerodynamic properties required for pulmonary drug delivery for use in dry-powder inhalers, nebulizers, etc., see Fig. 10–37. The void structure is designed to ensure that the initial surface area to volume ratio, as well as the evolving particle surface area (as the particle dissolves) will provide the proper dosage and drug release characteristics. The continuous phase contains any water soluble drug (e.g., Tobramycin). Tobramycin is a common antibiotic used for health conditions such as lung infections, asthma, and cystic fibrous.

10.7 Menger sponges

Menger sponges, sometimes referred to as Menger structures, were first described by the mathemetician Karl Menger in 1926. As the term sponge suggests, this is, a structure that contains voids. It is a nonstochastic foam that has a fractal characteristic, meaning no matter what the magnification, the structure retains the same statistical character or general pattern. Though, initially it was merely a mathematical representation, with the advent of additive manufacturing, reseachers are finding interesting applications for these structures. We describe briefly the Menger structures.

In order to construct a Menger structure, we begin with a solid cube, see Fig. 10–38A. This cube, a level zero Menger structure (M_0), is then divided into 27 equally sized, smaller cubes, and the center face cubes as well as the center cube are removed, see Fig. 10–38B. The structure, now made up of 20 cubes in given in Fig. 10–38B, is refered

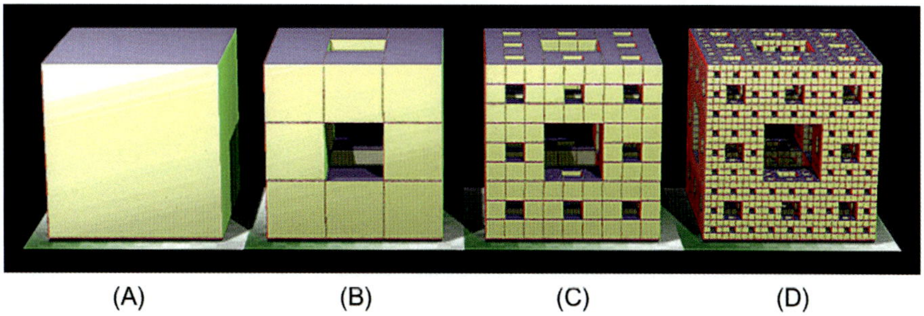

FIGURE 10–38 Illustration of a A) solid cube or M_0 Menger structure, B) M_1 structure, C) M_2 structure and D) M_3 structure. *Source: Public domain, Solkoll*

to as a level one Menger structure, M_1. A level two Menger structure, M_2, is shown in Fig. 10–38C and is made by dividing each of the 20 M_1 cubes into 27 equally sized cubes, followed by removal of the center face cubes as well as the center cubes. The M_3 structure, shown in Fig. 10–38D, was obtained by repeating this same procedure on the M_2 structure. From a mathematical standpoint, this procedure can go on indefintely, creating a fractal geometry or curve.

The use of Menger sponges in modeling has been used in light scattering (Sakoda et al., 2005), describing void structures for groundwater transport and water retention in porous media (Vita et al., 2012) and in acoustic applications, such as metamaterials, negative refraction, sound tunneling, and acoustic focusing (Liu et al., 2019). Structures have been 3D printed and tested for dielectric response (Kirihara et al., 2006), biomimetic architectures (Yang et al., 2017), and dynamic response to shockwaves (Dattelbaum et al., 2020).

Dattelbaum et al. (2020) created M_1, M_2, and M_3 Menger structures starting with a 1.7 mm × 1.7 mm × 1.7 mm M_0 cube. Depending on the Menger level, structures were made with cubic voids, dimensions of a side being 0.560 mm, 0.180 mm, and 0.063 mm. These structures were created via vat photopolymerization 3D printing process; using photons to polymerize an acrylate polymer. To characterize the shockwave mitigation characteriztics of each of the Menger structures, a gas gun driven plate traveling at a velocity of 0.3 km/s impacted the Menger structures. The materials response was collected using real-time shockwave propagation via time-resolved X-ray phase contrast imaging (PCI). The result for the M_3 structure with all three cubic void dimensions (0.56, 0.18, and 0.063 mm, respectively), is shown interacting with a 0.3 km/s shockwave in Fig. 10–39 using X-ray PCI with the shock wave traveling from right to left in each frame of Fig. 10–39B–H.

The first image in Fig. 10–39A shows the situaion prior to the shockwave, the time stamp is at an arbitrary time. The subsquent seven images, Fig. 10–39B–H, are time

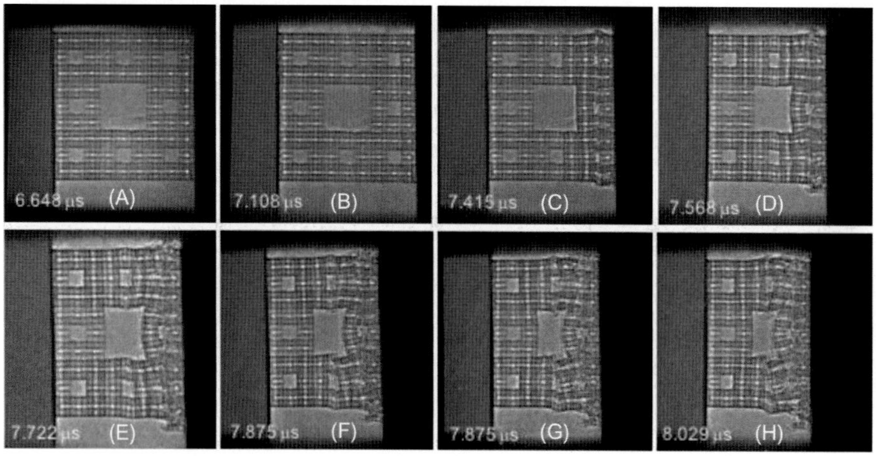

FIGURE 10–39 X-ray phase contract images of a shockwave interaction with an M_3 acrylate structure. A) pre-shock time stamp is an arbitrary time. B-H show the shockwave moving from right to left in each frame. The time stamps are relative to the time stamp in (A). The interaction with the M_3 cube shows plastic deformation of the free surfaces of the cubic voids and lateral displacement of acrylate normal to the shock propagation direction. *Source: Dattelbaum, D. M., Ionita, A., Patterson, B. M., Branch, B. A., & Kuettner, L., Shockwave dissipation by interface-dominated porous structures. AIP Advances, 10(7), 2020, 075016.*

stamped relative the the first one and show the deformation of the M_3 structure as the shock wave propagated through the cross-section of the cube. Dattelbaum et al. reported that the free surfaces of the cubic voids initially plastically deformed creating diamond shaped voids followed by collapse of the voids. As the shockwave propagated there was a lateral displacement of material normal to the shockwave propagation direction (Dattelbaum et al., 2020). The transmitted shockwave pressure of the M_3 structure was $\sim$0.04 GPa. This is approximately a 16$\times$ reduction compared to a nominally dense acrylate, with a transmitted pressure 0.65 GPa.

Evolution of heat is another way the Menger structure dissipates the energy of the shockwave. Fig. 10–40A shows the schematic representation of the M_1, M_2, and M_3 structures, respectively. Fig. 10–40B shows the finite element model of the temperature change of each Menger structures. Heating is localized to the void faces and is a result of plastic deformation occuring as the shock wave propagates. The temperature scale is shown in Fig. 10–40 is in Kelvin and indicates that greatest temperature rise of 170 K was seen in the M_3 structure. The degree of localized heating increased with fractal order. In complementary research, similar investigations were performed on stochastic and other nonstochastic foams structures (Branch et al., 2017, 2019).

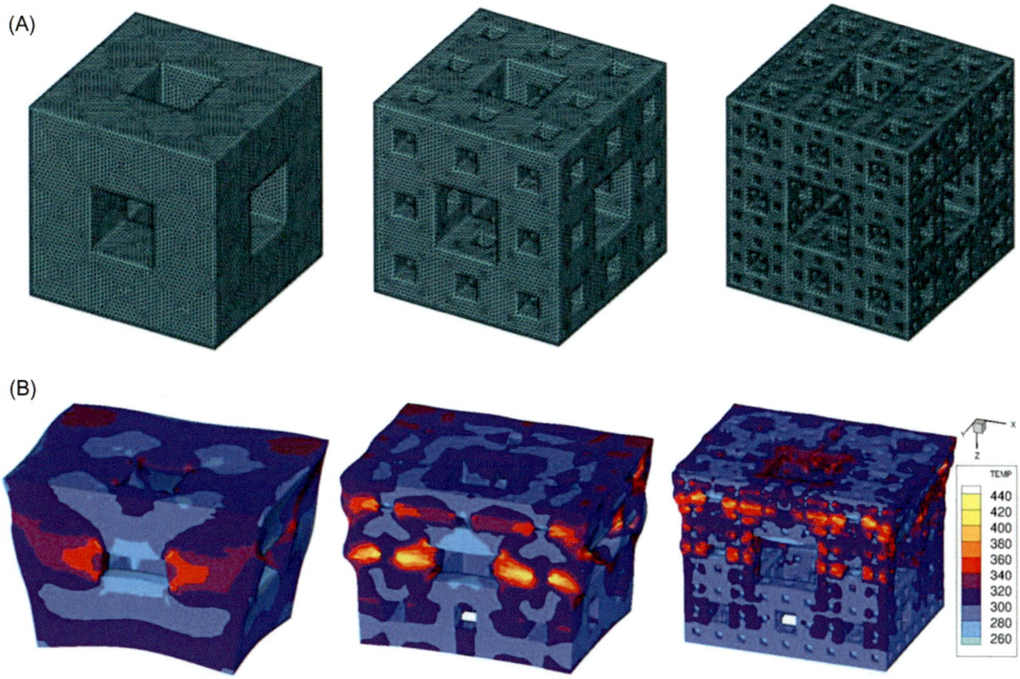

FIGURE 10–40 (A) Meshed finite element model of M_1, M_2, and M3 structures. (B) Temperature rise in the M_1, M_2, and M_3 structures subjected to a shockwave having an initial impact velocity on ~ 0.3 km/s on the top surface. The scale is in units of kelvin. The highest temperature change is 170 K and occurs in the M3 structure. Heat generation via plastic deformation is one mechanism to dissipate energy of the shockwave. *Source: Dattelbaum, D. M., Ionita, A., Patterson, B. M., Branch, B. A., & Kuettner, L., Shockwave dissipation by interface-dominated porous structures. AIP Advances, 10(7), 2020, 075016.*

References

Arenas, L.F., Ponce de León, C., & Walsh, F.C., Three-dimensional porous metal electrodes: Fabrication, characterisation and use, *Current Opinion in Electrochemistry*, **16**, 2019, 1-9. https://doi.org/10.1016/j.coelec.2019.02.002.

Arora, P., & Zhang, Z. (2004). Battery separators. *Chemical Reviews, 104*(10), 4419–4462.

Atwood, J. L., Barbour, L. J., Thallapally, P. K., & Wirsig, T. B. (2005). A crystalline organic substrate absorbs methane under STP conditions. *Chemical Communications, 1*, 51–53.

Banhart, J. (2001). Manufacture, characterisation and application of cellular metals and metal foams. *Progress in Materials Science, 46*(6), 559–632.

Barrett, R., Zolensky, M., & Bernhard, R. (1993). Mineralogy of chondritic interplanetary dust particle impact residues from LDEF. Paper presented at the *Lunar and Planetary Science Conference, 24* 65.

Black, S. (2003). Core materials: Getting to the core of composite laminates, October *Composites Technology, 9*(5), 24–29.

Branch, B., Ionita, A., Clements, B. E., Montgomery, D. S., Jensen, B. J., Patterson, B., . . . Dattelbaum, D. M. (2017). Controlling shockwave dynamics using architecture in periodic porous materials. *Journal of Applied Physics, 121*(13), 135102.

Branch, B., Ionita, A., Patterson, B. M., Schmalzer, A., Clements, B., Mueller, A., & Dattelbaum, D. M. (2019). A comparison of shockwave dynamics in stochastic and periodic porous polymer architectures. *Polymer, 160*, 325–337.

Chae, H. G., Choi, Y. H., Minus, M. L., & Kumar, S. (2009). Carbon nanotube reinforced small diameter polyacrylonitrile based carbon fiber. *Composites Science and Technology, 69*(3–4), 406–413.

Chawla, Krishan K. (2019). Micromechanics of composites. *Composite Materials*, 341–390.

Choy, W. J., Parr, W., Phan, K., Walsh, W. R., & Mobbs, R. J. (2018). 3-dimensional printing for anterior cervical surgery: A review. *Journal of Spine Surgery (Hong Kong), 4*(4), 757–769. Available from https://doi.org/10.21037/jss.2018.12.01.

Chu, H., Zhang, Z., Liu, Y., & Leng, J., Self-heating fiber reinforced polymer composite using meso/macropore carbon nanotube paper and its application in deicing, *Carbon,* **66**, 2014, 154-163. https://doi.org/10.1016/j.carbon.2013.08.053.

Circone, S., Kirby, S. H., & Stern, L. A. (2005). Direct measurement of methane hydrate composition along the hydrate equilibrium boundary. *Journal of Physical Chemistry B, 109*(19), 9468–9475.

Cooper, P. W., & Cooper, P. W. (1996). *Explosives engineering*. New York: VCH.

Darolia, R. (2013). Thermal barrier coatings technology: Critical review, progress update, remaining challenges and prospects. *International Materials Reviews, 58*(6), 315.

Dattelbaum, D. M., Ionita, A., Patterson, B. M., Branch, B. A., & Kuettner, L. (2020). Shockwave dissipation by interface-dominated porous structures. *AIP Advances, 10*(7), 075016.

De Volder, M. F. L., Tawfick, S. H., Baughman, R. H., & Hart, A. J. (2013). Carbon nanotubes: present and future commercial applications. *Science, 339*(6119), 535–539.

Diebold, U. (2003). The surface science of titanium dioxide. *Surface Science Reports, 48*(5–8), 53–229.

Egorov, V., & O'Dwyer, C. , Architected Porous Metals in Electrochemical Energy Storage, *Current Opinion in Electrochemistry*, **21**, 2020, 201-208. https://doi.org/10.1016/j.coelec.2020.02.011.

Fehr, M., Schnegg, A., Rech, B., Astakhov, O., Finger, F., Bittl, R., . . . Lips, K. (2014). Metastable defect formation at microvoids identified as a source of light-induced degradation in a—Si:H. *Physical Review Letters, 112*(6), 066403.

Fricke, J., & Tillotson, T. (1997). Aerogels: Production, characterization, and applications. *Thin Solid Films, 297*(1–2), 212–223. Available from https://doi.org/10.1016/S0040-6090(96)09441-2.

Fujishima, A., & Honda, K. (1972). Photolysis-decomposition of water at the surface of an irradiated semiconductor. *Nature, 238*(5385), 37–38.

Gladysz, G. M., & Chawla, K. K. (2002). *Composite foams. Encyclopedia of polymer science and technology*. New York: John Wiley & Sons.

Gohardani, O., Elola, M.C., & Elizetxea, C., Potential and prospective implementation of carbon nanotubes on next generation aircraft and space vehicles: A review of current and expected applications in aerospace sciences, *Progress in Aerospace Sciences*, **70**, 2014, 42-68. https://doi.org/10.1016/j.paerosci.2014.05.002.

Gupta, N., & Ricci, W. (2006). Comparison of compressive properties of layered syntactic foams having gradient in microballoon volume fraction and wall thickness. *Materials Science and Engineering: A, 427*(1), 331–342.

Gupta, N., Zeltmann, S. E., Shunmugasamy, V. C., & Pinisetty, D. (2014). Applications of polymer matrix syntactic foams. *JOM, 66*(2), 245–254.

Holzapfel, B. M., Reichert, J. C., Schantz, J., Gbureck, U., Rackwitz, L., Nöth, U., . . . Hutmacher, D. W. (2013). How smart do biomaterials need to be? A translational science and clinical point of view. *Advanced Drug Delivery Reviews, 65*(4), 581–603.

Iijima, Sumio (1991). Helical microtubules of graphitic carbon. *Nature, 254*(6348), 56–58.

Iijima, S., & Ichihashi, T. (1993). Single-shell carbon nanotubes of 1-nm diameter. *Nature, 363* (6430), 603–605.

Islam, M.S., Deng, Y., Tong, L., Faisal, S.N., Roy, A.K., Minett, A.I., & Gomes, V.G., Grafting carbon nanotubes directly onto carbon fibers for superior mechanical stability: Towards next generation aerospace composites and energy storage applications, *Carbon*, **96**, 2016, 701-710. https://doi.org/10.1016/j.carbon.2015.10.002.

Jin, Y., Zhang, T., Huang, Z., Zhao, J., Zhao, Y., Wang, Z., . . . Li, Q., Broadband omnidirectional perfect absorber based on carbon nanotube films, *Carbon*, **161**, 2020, 510-516. https://doi.org/10.1016/j.carbon.2020.01.106.

Kearsley, A. T., Borg, J., Graham, G. A., Burchell, M. J., Cole, M. J., Leroux, H., . . . Bland, P. A. (2008). Dust from comet wild 2: Interpreting particle size, shape, structure, and composition from impact features on the stardust aluminum foils. *Meteoritics & Planetary Science, 43*(1-2), 41–73.

Kee Paik, J., Thayamballi, A. K., & Sung Kim, G. (1999). The strength characteristics of aluminum honeycomb sandwich panels. *Thin-Walled Structures, 35*(3), 205–231.

Kistler, S. S. (1931). Coherent expanded aerogels and jellies. *Nature, 127*(3211), 741–741.

Klempner, D., & Frisch, K. (1991). *Polymeric foams*. New York: Hanser.

Koopman, M., Chawla, K., Ricci, W., Carlisle, K., Gladsyz, G., Lalor, M., . . . Gouadec, G. (2009). Titania-coated glass microballoons and cenospheres for environmental applications. *Journal of Materials Science, 44*(6), 1435–1441.

Kirihara, S., Takeda, M. W., Sakoda, K., Honda, K., & Miyamoto, Y. (2006). Strong localization of microwave in photonic fractals with Menger-sponge structure. *Journal of the European Ceramic Society, 26*(10–11), 1861–1864.

Leroux, H. (2012). Fine-grained material of 81P/wild 2 in interaction with the stardust aerogel. *Meteoritics & Planetary Science, 47*(4), 613–622.

Leroux, H., Rietmeijer, F. J., Velbel, M. A., Brearley, A. J., Jacob, D., Langenhorst, F., . . . Cordier, P. (2008). A TEM study of thermally modified comet 81P/wild 2 dust particles by interactions with the aerogel matrix during the stardust capture process. *Meteoritics & Planetary Science, 43* (1-2), 97–120.

Liu, Y., Xu, W., Chen, M., Pei, D., Yang, T., Jiang, H., & Wang, Y. (2019). Menger fractal structure with negative refraction and sound tunneling properties. *Materials Research Express, 6*(11), 116211.

Machado, C. (2014). *Email communication pertaining to the HROV Nereus*. July.

Manoukian, O.S., Arul, M.R., Rudraiah, S., Kalajzic, I., & Kumbar, S.G., Aligned microchannel polymer-nanotube composites for peripheral nerve regeneration: Small molecule drug delivery, *Journal of Controlled Release*, **296**, 2019, 54-67. https://doi.org/10.1016/j.jconrel.2019.01.013.

Ohji, T. (2013). Chapter 11.2.2—Porous ceramic materials. In S. Somiya (Ed.), *Handbook of advanced ceramics* (2nd ed.). Oxford: Academic Press, pp. 1131–1148.

Peabody, C., & Arnold, C. B. (2011). The role of mechanically induced separator creep in lithium-ion battery capacity fade. *Journal of Power Sources, 196*(19), 8147–8153.

Raza, C., Riaz, H.A., Anjum, R., & Shakeel, N. u. A., Repair strategies for injured peripheral nerve: Review, *Life Science*, **243**, 2020, 117308. https://doi.org/10.1016/j.lfs.2020.117308.

Rosca, I.D., & Hoa, S.V., Method for reducing contact resistivity of carbon nanotube-containing epoxy adhesives for aerospace applications, *Composites Science and Technology*, **71**(2), 2011, 95–100. https://doi.org/10.1016/j.compscitech.2010.10.016.

Sakoda, K., Kirihara, S., Miyamoto, Y., Takeda, M. W., & Honda, K. (2005). Light scattering and transmission spectra of the Menger sponge fractal. *Applied Physics B, 81*(2–3), 321–324.

Sawant, S.S., Rao, P., Harpale, A., Chew, H.B., & Levin, D.A., Multi-scale thermal response modeling of an AVCOAT-like thermal protection material, *International Journal of Heat and Mass Transfer*, **133**, 2019, 1176-1195. https://doi.org/10.1016/j.ijheatmasstransfer.2018.12.182.

Thallapally, P. K., Wirsig, T. B., Barbour, L. J., & Atwood, J. L. (2005). Crystal engineering of non-porous organic solids for methane sorption. *Chemical Communications, 35*, 4420–4422.

Tharmavaram, M., Pandey, G., &, D., Surface modified halloysite nanotubes: A flexible interface for biological, environmental and catalytic applications, *Advances in Colloid and Interface Science*, **261**, 2018, 82-101. https://doi.org/10.1016/j.cis.2018.09.001.

Tsou, P., Brownlee, D., Sandford, S., Hörz, F., & Zolensky, M. (2003). Wild 2 and interstellar sample collection and earth return. *Journal of Geophysical Research: Planets, 108*(E10).

Valencia, S., Marín, J. M., & Restrepo, G. (2010). Study of the bandgap of synthesized titanium dioxide nanoparticles using the sol–gel method and a hydrothermal treatment. *Materials Science Journal, 4*(1), 9–14.

Verma, P., Anoop, S., Sasidhara Rao, V., Sharma, A.K., & Uma Rani, R., Multiwalled carbon nanotube-poly vinyl alcohol nanocomposite multifunctional coatings on aerospace alloys, *Materials Today: Proceedings*, **5**(10), 2018, 21205-21216. https://doi.org/10.1016/j.matpr.2018.06.520.

Vita, M. C., De Bartolo, S., Fallico, C., & Veltri, M. (2012). Usage of infinitesimals in the Menger's Sponge model of porosity. *Applied Mathematics and Computation, 218*(16), 8187–8195.

Wang, F., Tay, T.E., Sun, Y., Liang, W., & Yang, B., Low-voltage and -surface energy SWCNT/poly (dimethylsiloxane) (PDMS) nanocomposite film: Surface wettability for passive anti-icing and surface-skin heating for active deicing, *Composites Science and Technology*, **184**, 2019, 107872. https://doi.org/10.1016/j.compscitech.2019.107872.

Wilcox, B., Mobbs, R. J., Wu, A. M., & Phan, K. (2017). Systematic review of 3D printing in spinal surgery: The current state of play. *Journal of Spine Surgery (Hong Kong), 3*(3), 433–443. Available from https://doi.org/10.21037/jss.2017.09.01.

Xia, Q., Mei, H., Zhang, Z., Liu, Y., Liu, Y., & Leng, J., Fabrication of the silver modified carbon nanotube film/carbon fiber reinforced polymer composite for the lightning strike protection application, Composites Part B: Engineering, 180, 2020, 107563. https://doi.org/10.1016/j.compositesb.2019.107563.

Xiong, J., Ma, L., Stocchi, A., Yang, J., Wu, L., & Pan, S. (2014a). Bending response of carbon fiber composite sandwich beams with three dimensional honeycomb cores. *Composite Structures, 108*, 234–242.

Xiong, J., Zhang, M., Stocchi, A., Hu, H., Ma, L., Wu, L., & Zhang, Z. (2014b). Mechanical behaviors of carbon fiber composite sandwich columns with three dimensional honeycomb cores under in-plane compression. *Composites Part B: Engineering, 60*, 350–358.

Yang, Y., Chen, Z., Song, X., Zhang, Z., Zhang, J., Shung, K. K., . . . Chen, Y. (2017). Biomimetic anisotropic reinforcement architectures by electrically assisted nanocomposite 3D printing. *Advanced materials,*
29(11), 1605750.

Zhang, R., Elzatahry, A. A., Al-Deyab, S. S., & Zhao, D. (2012). Mesoporous titania: From synthesis to application. *Nano Today, 7*(4), 344–366.

Zhang, W., Jin, H., Xu, C., Zhao, S., Du, Y., & Zhang, J., Diffusion couples Cu-X (X = Sn, Zn, Al) derived 3D porous current collector for dendrite-free lithium metal battery, *Journal of Power Sources*, **440**, 2019, 227142. https://doi.org/10.1016/j.jpowsour.2019.227142.

Zolensky, M., Nakamura-Messenger, K., Rietmeijer, F., Leroux, H., Mikouchi, T., Ohsumi, K., . . . Weisberg, M. (2008). Comparing Wild 2 particles to chondrites and IDPs. *Meteoritics & Planetary Science, 43*(1-2), 261–272.

Glossary

We define some terms as they relate to voids.

Aerogel a highly porous (extremely low density) material made by the removal of a liquid porogen from a gel.

Asymmetric particle transformation Kirkendall effect conversion mechanism for transforming a solid particle into a hollow particle with a nonuniform shell thickness.

Blowing agent a material that forms a gas and expands, forming a cell or cells in a material.

Bubble a gas stabilized void.

Buoyancy an upward force exerted by a fluid opposing the weight of a partially or fully immersed object.

Cell a void that is intentionally introduced into a material.

Chemical blowing agent the gas formed as a result of a reaction between starting materials or the gas formed as a result of a decomposition chemical reaction, resulting in a cellular structure.

Chemically selective weakness in solids a void formation technique that regioselectively targets portions of a molecule to be removed through chemical, mechanical, or physical stimuli.

Closed cell each void in a foam is completely surrounded by solid material.

Coble creep diffusional creep via migration of the vacancies through the grain boundary region.

Composite foam a multicomponent material with one or more intentional void phases present.

Conjugated microporous polymer a three-dimensional semiconducting polymer in which rigid aromatic groups are linked together, either directly or via double or triple bonds. Conjugation in this sense means that there are alternating single and double- or triple-bonds throughout the extended network.

Coquina A porous sedimentary rock, that is, it is made up of fragments of other preexisting rocks. It was used as a material of construction in Castella de San Marcus. It is a composite of fragmented marine shells, fossils and coral, limestone, sand, minerals, and clay, with significant void space between these solid components that have the ability to absorb the energy of projectiles, such as bullets and cannon balls.

Covalent organic frameworks (COFs) a class of crystalline porous polymers that allows the atomically precise integration of organic molecules into extended structures with periodic skeletons and ordered nanopores. The pores have polygonal shapes that can be designed to assume hexagonal, tetragonal, trigonal, rhombic, and Kagome (star-shaped) structures.

Covalent triazine frameworks a chemical structure of triazine that is similar to that of benzene; however, in triazine the six-member ring contains three nitrogen and three carbon atoms. The void size can be designed as small as $\sim$ 1/10 of a nanometer.

Excess free volume the volume difference between the volume at the equilibrium glass transition temperature, T_g^∞ and the volume at the glass transition temperature, T_g. It is the available volume that can be lost during the physical aging process.

Foam a material with a plurality of voids that are intentionally introduced into a material forming a cellular structure.

Free volume the space between polymer chain networks or inorganic networks in noncrystalline materials.

Frit a network of particles connected at contact points with interstitial porosity.

Gasar metallic foam a metallic foam made by a two-stage process. The first stage is creating a molten metal that is saturated with gas. The second stage is unidirectional solidification of the melt. As the melt cools and eventually solidifies, the dissolved gas becomes less soluble in the molten metal and comes out of solution as a series of bubbles at the solid/liquid interface.

Guest–host complexes a material functionality in which a smaller guest molecule becomes encapsulated within the structure of a larger host molecule.

Hard template a solid sacrificial placeholder material that, when removed from the matrix material, forms an intentional void.

Hierarchical porosity the simultaneous control of voids and functionality at multiple length scales.

Hole free volume a discontinuous component in free volume, which increases greatly with temperature.

Hollow material/particle a structure with a void having the same general geometry as the shell or skin layer.

Hollow, porous particle a particle that has a void with the same general geometry as the shell or skin layer but the shell is not fully dense.

Hydrogel gel made of hydrophilic polymers with crosslinked networks. They can absorb up to thousand times than their dry weight in water.

Intentional voids voids that are incorporated into a solid material by design.

Interstitial free volume a continuous part of free volume that results from oscillations of atoms and that persists and increases slightly as the temperature is raised. Also called oscillation free volume.

Interstitial void a void located between particles, fibers, or atoms in a crystal structure specifically in face centered cubic (FCC) or body centered cubic (BCC) crystals.

Intrinsic void typically thought of as defects; they appear because of the nature of the material, natural processes, processing limitations, and/or aging during service.

Kirkendall voids porosity that appears when two materials, A and B, forming a diffusion couple have significantly different diffusivities or diffusion coefficients.

Latent gas gas released by a solid or liquid additive when a certain processing conditions are satisfied. The additive can be a chemical or physical blowing agent.

Macropores pores greater than 50 nm in diameter as defined by the International Union of Pure and Applied Chemists (IUPAC).

Mechanical stretching mechanical straining of a material to produce a unique, slit-like pore structure in membranes (e.g., for lithium ion battery separators)

Menger sponge nonstochastic foam that has a fractal characteristic, meaning no matter what the magnification, the structure retains the same statistical character or general pattern. Specifically, the pattern is constructed dividing a cube into 27 equally sized, smaller cubes, and removing the center face cubes as well as the center cube.

Mesopores pores between 2 and 50 nm in diameter as defined by International Union of Pure and Applied Chemists (IUPAC).

Microballoon or hollow microspheres a hollow particle, generally in the micrometer size range.

Microcracks voids that take the form of a slit, that is, a flattened pore.

Micropores pores less than 2 nm in diameter as defined by International Union of Pure and Applied Chemists (IUPAC).

Mixing/aeration vigorous stirring of a gas into the liquid phase of a material or precursor material, thus causing entrainment of that gas.

Nabarro–Herring creep diffusional creep, which involves vacancy currents through the grain.

Nanoporous organic networks a class of materials consisting solely of the lighter elements in the periodic table such as B, C, N, and O. Such void spaces in a molecular network are intrinsic to the molecular structure.

Nanoporous polymers typically polymer materials (polyurethane, polyethylenimine, polymethylmethacrylate, etc.) have micrometer scale porosity. When the void size is reduced to nanometer scale we get nanoporous materials. This is typically done to enhance specific properties.

Nonstochastic structure in a cellular material, it consists of ideal repeat units of open cells in a foam.

Occupied volume it is the volume taken up by the polymer molecules and is considered impenetrable.

Octahedral void interstitial void between six spheres that are in contact. The centers of these six spheres are at the corners of a regular octahedron.

Open cell in a foam, part of the cell wall is missing, leaving voids that are interconnected with each other.

Physical blowing agent a material that undergoes vaporization to introduce gas into a material. The gas generated creates the voids and cellular structure in the material.

Pore or cell structure the arrangement and types of voids in a solid.

Porogen a general term used to describe any material that can be used to form a void.

Porous material/particle a material or particle having distributed voids throughout the entire volume.

Positron annihilation lifetime spectroscopy (PALS) a nondestructive spectroscopic technique that is used to study a variety material properties on an atomic scale. Specifically, PALS measures the elapsed time between implantation of positron into the material and the emission of annihilation radiation. PALS can be used for determining the void space or free volume in different materials.

r-type cavity voids in grain boundaries perpendicular to the tensile axis.

Radius ratio dimensionless number obtained by the size of the lattice hole (described by the radius of a hypothetical sphere, r_{hole}) than be contained within the void created by the surrounding atoms of radius r_{atom}.

$$\text{radius ratio} = \frac{r_{\text{hole}}}{r_{\text{atom}}}$$

Regularity (of pore structure) a dimensionless parameter defined by the ratio of δ/d_0. In a volume, V, made up of n number of cell, d_0 is the minimum distance between cell center points needed to fill a volume, V, having n identical tetrakaidecahedral cells and δ is the minimum distance between the center points of any two adjacent cells in volume V. A value of $R = 1$ is a nonstochastic foam (honeycomb) and a $R = 0$ has a completely random cell structure.

Reinforced void a void in a material formed by the introduction of a hollow material (e.g., microballoons, nanotubes).

Sandwich construction a structure made of a low density core and rigid face sheets or skins.

Schottky defect type of vacancy in an ionic crystal where cations and anions are missing from a crystal lattice. The cations and anions are missing in a stoichiometric ratio.

Single-phase foam a cellular material made by incorporating voids into a homogeneous material.

Sintered metal fibrous media a nonwoven fibrous cloth with a three-dimensional internal pore structure.

Soft templates liquid or gas sacrificial porogen material.

Solid freeform fabrication a technique of building finished three-dimensional parts, layer-by-layer, based on computer models. Also known as additive manufacturing.

Stochastic structure a cell structure having a random variation in cell size and shape.

Symmetric particle transformation Kirkendall effect conversion mechanism for transforming a solid particle into a hollow particle with a uniform shell thickness.

Syntactic foam a cellular material that contains at least one reinforced void phase.

Template-free formation of a hollow particle a process that does not require a sacrificial core, porogen, or blowing agent to form a void.

Template, sacrificial pore former or placeholder a material, usually a particle of specific size and distribution that creates a cellular structure. The particle is subsequently removed leaving a void in its place.

Tetrahedral void interstitial space between four spheres that are in contact. The centers of these four spheres are at the corners of a regular tetrahedron.

Unreinforced void a void contained in a material without a skin or shell.

Vacancy diffusion the flow of vacancies and atoms in opposite directions. Also called self-diffusion.

Void any internal volume in which the solid phase is absent. A general term used in the context of a solid, independent of length scale.

Void cluster growth of voids and eventual coalescence of voids.

Voronoi foam a structure with porosity having a designed regularity from completely ordered and repeatable cell structure to one that is completely random. Used in additive manufacturing to digitally design complex cell structures within a single part.

w-type cavity void formation at the grain boundary triple points.

Author Index

Note: Page numbers followed by "*f*" and "*t*" refer to figures and tables, respectively.

A

Abbasi, M. H., 107, 170*f*
Abdullah, M., 140–141
Adams, D. J., 64–65, 68*f*
Ahmad, M., 56
Akins, D. L., 162*t*
Albernaz, F. O., 3–4, 4*f*
Al-Deyab, S. S., 263–265, 265*f*, 266*f*
Alfrey, T., 49
Alivisatos, A. P., 81–82, 81*f*
Amaranan, S., 106
An, G., 155–156
Anastasio, M. A., 175–176, 176*f*
Anderson, C. D., 172
Anjum, R., 268–269, 272*f*
Anoop, S., 243–244
Apparao, G., 181
Appel, A. A., 175–176, 176*f*
Arakawa, C., 53
Arenas, L. F., 259–260
Arnold, C. B., 259, 260*f*
Arora, P., 123, 257–258
Arul, M. R., 268–269, 271*f*
Arunachalam, V. S., 25, 27–28
Arwade, S. R., 96, 225
Ashby, M., 206–207, 222–223
Ashby, M. F., 12, 96, 197, 203–206, 208, 215, 222–223
Astakhov, O., 265*f*
Atwood, J. L., 262–263, 264*f*
Axet, M. R., 58
Aydogmus, T., 106
Azarniya, A., 120–121
Aziz, M. J., 72

B

Babcsán, N., 99–100
Balandin, A. A., 48
Balluffi, R. W., 22
Bandosz, T. J., 162*t*
Banhart, J., 96, 99–100, 232*f*
Barbour, L. J., 262–263, 264*f*
Barhoum, A., 54
Barigou, M., 97, 98*f*
Barrett, R., 252
Barthlott, W., 130*f*
Basaran, C., 55–56
Baughman, R. H., 244
Beachley, V., 109*f*
Bechtold, T., 183*f*, 184, 184*t*
Becker, J., 168
Bekoz, N., 106
Benderly, D., 208
Benes, N. E., 156*f*
Bernardo, E., 117–119, 118*f*
Bernhard, R., 252
Bittencourt, C., 54, 80–82, 85
Bittl, R., 265*f*
Black, S., 245
Boccaccini, A. R., 102
Boey, F. Y., 208
Boey, F. Y. C., 157, 162*t*
Boldizar, A., 207
Bondi, A., 46
Boomsma, K., 215–216
Bor, S., 106
Bor, T. C., 156*f*
Boutin, A., 128–129
Boutorabi, M., 120–121

Branch, B., 274
Branch, B. A., 275*f*
Brandes, H. G., 140–141
Brandi, J., 132*t*
Brearley, A. J., 254*f*
Brey, E. M., 175–176, 176*f*
Brinkman, H. C., 43
Brito, V., 161
Brostow, W., 49–50
Brothers, A. H., 106
Brunauer, S., 178–179
Budd, P. M., 67, 69*f*
Burtscher, E., 183*f*, 184, 184*t*
Bushby, A., 17–18

C

Cabot, A., 82
Cadek, M., 56
Cai, W., 149*f*
Calderon, H. A., 171*f*
Cao, J., 7, 145–147
Cao, X., 158*f*
Carlisle, K., 208
Carlisle, K. B., 145*t*, 161
Carlson, N. N., 141*f*
Cejka, J., 124, 126*f*, 127–128
Chae, H. G., 244, 244*f*
Chang, C., 217
Chang, J. W., 79, 80*f*
Chapman, K. W., 128, 128*f*
Chawla, K., 27*f*, 208
Chawla, Krishan K., 2, 5, 23–24, 26, 30, 35*f*, 47*t*, 61–62, 96, 145*t*, 154, 239–241
Chawla, N., 27–28, 27*f*, 176–178, 177*f*, 178*f*, 218–219
Chen, H., 158*f*, 159*f*
Chen, J., 147–148
Chen, L., 61–62
Chen, W., 161*f*
Cheng, S., 215–217
Chew, H. B., 249
Chmutin, I., 207
Cho, I. S., 9, 10*f*
Choi, Y. H., 244, 244*f*
Chou, T., 159*f*
Chou, Y. T., 49–50
Choudalakis, G., 49–50
Choy, W. J., 267–268, 269*f*
Christodoulou, I., 201–202
Chu, H., 243–244
Chung, T. S., 154*f*
Chupas, P. J., 128, 128*f*
Circone, S., 261
Coble, R., 34
Coble, R. L., 26
Cogan, S. F., 37
Cohn, J., 24
Collins, C. L., 168*f*
Colombo, P., 96, 111*f*, 117–119, 118*f*, 221–222
Cooper, A. I., 64–65, 68–70, 68*f*
Cooper, P. W., 241–242, 242*f*
Cordi, S., 220
Cordier, P., 254*f*
Cordova, K. E., 66–67
Coudert, F., 128–129
Cox, M. E., 132*t*

D

Darby, J. R., 42–43, 44*f*
Darolia, R., 37, 254–256, 256*f*, 258*f*
Dattelbaum, D. M., 273–274, 275*f*
Davidovits, J., 220
Dawson, R., 64–65, 67, 68*f*
De Volder, M. F. L., 244
Defouw, J. D., 106
Del Rio, J., 49–50
Delaney, G., 161
Deng, X., 27*f*
deSaja, J. A., 64*f*
Deshpande, V., 199
Di, J., 158*f*
Diebold, U., 263–264
Dimitrov, N., 72
Ding, B., 108, 157, 162*t*

Ding, H., 147–148
Dirac, P. A. M., 172
Divandari, M., 120–121
Dobsak, P., 102
Donev, A., 161
Dong, Y., 7, 7*f*, 89
Du, B., 12*f*
Du, H., 100–101
Du, N., 162*t*
Du, Y., 261*f*
Du, Y. Q., 110*f*
Du Fresne Von Hohenesche, C., 132*t*
Du Prez, F. E., 141, 143
Duan, Y., 11–12, 12*f*, 201–202
Dudziak, N., 37
Dumas, J., 201–202
Dumon, M., 64*f*
Dunand, D. C., 82–83, 86, 87*f*, 88*f*, 89*f*, 106, 132*t*
Dunstan, D., 17–18

E

Eceolaza, S., 49–50
Egashira, M., 145, 146*f*
Egorov, V., 259–260
El Mel, A., 54, 80–83
EL-Ashkar, N. H., 168*f*
Eldrup, M., 173–174, 174*f*, 175*f*
Eliášová, P., 126–127
Elices, M., 47*t*
Elizetxea, C., 244
Elola, M. C., 244
Elsayed, H., 117–119, 118*f*
Elzatahry, A. A., 263–265, 265*f*, 266*f*
Emets, S. V., 141*f*
Emmett, P. H., 178–179
Erdeniz, D., 86, 87*f*, 88*f*, 89*f*
Erlebacher, J., 72
Etxeberria, A., 49–50
Eucken, A., 216–217
Eyre, B. L., 21–22
Eyring, H., 43

F

Fan, H. J., 77, 80, 83*f*
Faucher, S., 54
Fehr, M., 263, 265*f*
Feliczak-Guzik, A., 59
Felske, J., 217
Feng, C., 162*t*
Feng, Y., 210*f*
Ferry, J. D., 45
Finger, F., 265*f*
Flanigen, E. M., 125–126, 125*f*
Fleck, N., 199
Fleisher, A., 115*f*
Flory, P. J., 45
Fox, J., 45
Franchin, G., 221–222
Freitag, B., 171*f*
Fricke, J., 106–107, 253
Friedrich, M., 113*f*
Frisch, K., 231
Frisch, K. C., 204*f*
Fritsch, D., 69*f*
Fu, J., 155, 157*f*
Fu, K., 18
Fu, Z., 99
Fuji, M., 7
Fujinami, M., 49–50, 172
Fujishima, A., 264
Fukuhara, M., 106
Fung, T., 106

G

Ganesh, V., 27*f*
Gao, N., 130*f*
Garzón-Alvarado, D. A., 112–113
Gbureck, U., 268*f*
Geisendorfer, N. R., 12–13, 13*f*
Geller, D. E., 148–150, 150*f*
Ghanem, B. S., 69*f*
Gibson, L. J., 12, 96, 197, 203–206, 222–223
Gidley, D. W., 172
Giridhar, G., 181, 183–184

Gladysz, G., 145*t*, 193, 208, 215
Gladysz, G. M., 2, 5, 61–62, 96, 141*f*, 145*t*, 161, 239–240
Gohardani, O., 244
Gokmen, M. T., 141, 143
Goldfinger, G., 49
Gorbunov, A. A., 181
Gorni, D., 208
Gösele, U., 77, 83*f*
Gotsis, A. D., 49–50
Grousson, S., 168
Guan, C., 147–148
Guan, J., 158*f*
Guiang, C., 125
Guo, X., 7, 7*f*
Guo, Z., 109*f*, 130*f*
Gupta, N., 197, 208, 242–243
Gupta, S. K., 208

H

Hadiko, G., 7
Hagg-Lobland, H. E., 49–50
Hajjar, J. F., 96, 225
Halder, G. J., 128, 128*f*
Han, F., 106
Han, Y. S., 7
Harpale, A., 249
Hart, A. J., 244
Hartman Kok, P. J. A., 49–50
Haruta, M., 71–72
Harvey, E., 130–131
Hashin, Z., 217
He, T., 46–47, 47*t*
He, X., 155–156
He, Y., 157*f*
Heinl, P., 117–119, 120*f*
Heli, H., 80
Henrist, C., 132*t*
Herring, C., 33–34
Heuerding, S., 148–150, 150*f*
Hill, A. J., 49–50, 50*f*
Hill, J. M., 49–50, 50*f*
Hoa, S. V., 243–244
Holzapfel, B. M., 117–119, 130–131, 131*f*, 266–267, 268*f*
Honda, K., 264
Hong, J. H., 89
Hoo, K., 215
Hoo, K. A., 141*f*
Hou, B., 12*f*
Hou, N., 12*f*
Hrubesh, L. W., 170, 171*f*
Hsu, C., 217
Hsu, H., 217
Hu, R., 161*f*
Hu, X., 208
Hu, Y., 89
Hu, Z., 4
Huang, L., 162*t*
Huang, N., 66–67
Huang, S., 7, 7*f*
Huang, Y., 4, 49–50, 50*f*, 208
Huang, Z., 155–156
Hulbert, S., 130–131
Hutmacher, D. W., 268*f*
Hutzler, S., 161
Hyodo, T., 145–147, 146*f*
Hyun, S., 99–101
Hyun, S. K., 100–101, 100*f*, 101*f*

I

Ichihashi, T., 158–160, 243
Iijima, Sumio., 158–160, 243
Ikeda, T., 100–101
Inglis, C., 28
Inoue, A., 106
Ionita, A., 275*f*
Irausquín, I., 106
Iriarte, M., 49–50
Ishizaki, K., 99
Iskandar, F., 140–141
Islam, M. S., 243–244

J

Jacob, D., 254*f*
Jacobs, M., 159*f*
Jakkali, V., 176–177, 177*f*, 178*f*, 218–219
Jakus, A. E., 12–13, 13*f*
Jana, S., 79, 80*f*
Jannotti, P., 8–9
Jean, Y., 173
Jelle, B. P., 5
Jemmott, G., 151–152
Jerro, H. D., 208
Jha, B., 59–60
Jha, N., 106
Jia, Q. M., 49–50
Jiang, D., 66–70
Jiang, G., 161*f*, 162*t*
Jiang, H., 57*f*
Jiang, J., 68–71
Jiang, L., 158*f*, 176–177, 177*f*, 178*f*, 218–219
Jiang, X., 89
Jiang, Y., 100–101
Jiao, Y., 145*t*
Jimenez, W. C., 160
Jin, H., 261*f*
Jin, L., 145*t*
Jin, S., 68–70
Jin, Y., 243–244
Jin Fan, H., 82–83
Jing, D., 210*f*
Jinno, S., 49–50, 172
Jinschek, J. R., 171*f*
Jourlin, M., 168

K

Kalajzic, I., 268–269, 271*f*
Kang, Y. C., 89–90
Karma, A., 72
Károly, Z., 145–147
Kashihara, M., 100–101, 100*f*
Kauppinen, E. I., 57*f*
Kearsley, A. T., 249–251
Kee Paik, J., 245*f*
Kelly, A., 17
Kennedy, A. R., 120–121
Kennedy, L., 160
Kern, W., 101
Khan, S. A., 84*f*
Khlobystov, A. N., 57–58, 59*f*
Kim, D., 9, 10*f*
Kim, J. H., 89
Kim, S., 142*f*
Kim, S. H., 71–72, 73*f*
Kim K. K., 147–148
Kingery, W. D., 26
Kingston, H. J., 69*f*
Kirby, S. H., 261
Kirihara, S., 273
Kirkendall, K. O., 35–37, 77
Kistler, S. S., 249
Kittel, C., 24
Kivistö, S., 57*f*
Klein, J. D., 37
Klempner, D., 204*f*, 231
Klett, J., 216
Ko, Y. N., 89–90
Kobayashi, H., 7, 106
Kobayashi, T., 53, 71–72
Kobi, T., 100–101, 100*f*
Koike, R., 120–121
Kondo, S., 112–113
Kongdee, A., 183*f*, 184, 184*t*
Koopman, M., 27*f*, 145*t*, 161, 162*t*, 208, 263–264
Körner, C., 117–119
Kotaki, M., 155–156
Krishna, P., 14, 14*f*
Kuettner, L., 275*f*
Kujime, T., 100–101
Kumar, S., 244, 244*f*
Kumar, V., 62–63, 190
Kumbar, S. G., 268–269, 271*f*
Kurabayashi, K., 48
Kurtis, K. E., 168*f*

L

Labrincha, J. A., 220–221, 220*f*, 221*f*
Lai, S., 145*t*
Laksmana, F. L., 49–50
Lancheros, Y., 112–113
Landel, R. F., 45
Landi, E., 132*t*
Langenhorst, F., 254*f*
Lapidus, S. H., 128–129, 128*f*, 129*t*
Larson, J. C., 175–176, 176*f*
Leach, A., 216–217
Léaux, F., 106
Lee, C., 86*f*
Lee, J., 100–101, 142*f*
Lee, J. M., 68–70
Lee, J. S., 153, 153*f*
Lee, K. Y., 142*f*
Lee, S., 86*f*
Lefebvre, S., 201–202
Leitlmeier, D., 99–100
Lemaitre, J., 199
Leng, J., 243–244
Lenggoro, I. W., 140–141
Lepage, G., 3–4, 4*f*
Leroux, H., 252–253, 254*f*
Levin, D. A., 249
Levshov, D. I., 56
Lewis, M., 145*t*
Lewis, P. L., 12–13, 13*f*
Li, B., 162*t*
Li, D., 158*f*
Li, J., 155–156
Li, J. B., 49–50
Li, J., 4
Li, L., 7, 7*f*, 54
Li, W., 82–83
Li, X., 100–101, 148*f*, 157*f*
Li, X. N., 90*f*
Li, Y., 12*f*, 148*f*, 162*t*
Liang, W., 243–244
Lightbody, D., 173
Lim, K. S., 97, 98*f*
Lin, Y., 217
Lind, A., 132*t*
Lindahl, J., 117–119
Lindén, M., 132*t*
Lindquist, P., 82–83
Lippmann, H., 199
Lips, K., 265*f*
Liu, G., 161
Liu, Q., 83
Liu, S., 149*f*
Liu, W., 109*f*, 130*f*
Liu, X., 100–101, 145*t*
Liu, Y., 145*t*, 243–244, 273
Liu, Z., 61–62
Llorca, J., 47*t*
Lopez, V. H., 120–121
Lou, X. W., 157, 162*t*
Louzguine-Luzgin, D. V., 106
Lu, B., 89
Lu, L., 72
Lu, Y., 144, 145*t*
Luan, D., 157, 162*t*
Luiten-Olieman, M. W. J., 155, 156*f*
Lula, J., 145*t*, 193, 208
Luo, P., 145*t*
Luo, Q., 145*t*
Lyle, S. J., 66–67

M

Ma, C. Y., 90*f*
Ma, J., 208
Ma, L., 247*f*
Ma, W., 145*t*
Ma, X., 145*t*, 147–148
Ma, Y., 147
Ma, Z., 199–200, 199*f*
Maca, K., 102, 103*f*
Machado, C., 238
Mackenzie, J. K., 25–26
Macmillan, N. H., 17
Maharsia, R., 208
Maharsia, R. R., 208

Mahian, O., 210*f*
Maire, E., 178
Manepalli, R. K. N. R., 181
Manivannan, J., 55*f*
Manjuladevi, M., 55*f*, 56
Mann, U., 141*f*
Manonukul, A., 106
Manoukian, O. S., 268–270, 271*f*
Mao, X., 97–98, 99*f*, 132*t*
Marenduzzo, D., 14–15
Marín, J. M., 264–265
Mark, H., 49
Martínez, J., 201–202
Matlock, D. K., 35–37
Matsoukas, T., 7, 145–147
Matsumaru, K., 99
Matthews, J. R., 21–22
Maurer, F. H. J., 173–174, 174*f*, 175*f*
Mceachen, G., 145*t*, 193
McGeary, R. K., 161, 212
Mcginity, J. W., 147–148
McKeown, N. B., 69*f*
Medalist, R. M., 208
Mensah, P., 197
Merchan-Merchan, W., 160
Merewether, R., 151–152
Merlin, G., 3–4, 4*f*
Meyer, J. C., 170–172
Meyers, M. A., 23–24, 30, 47*t*
Mi, Y., 145*t*
Michaelides, E. E., 210*f*
Micheletti, C., 14–15
Michielsen, B., 157, 159*f*
Middelkoop, V., 159*f*
Miller, D., 62–63, 190
Mills, N., 96
Minami, H., 7, 106
Minsky, M., 167–168
Minus, M. L., 244, 244*f*
Miura, T., 112–113
Mo, J., 89
Mobbs, R. J., 269*f*
Mohammadi-Jam, S., 182*f*
Molitch-Hou, M., 110–112, 119*f*, 121*f*, 122*f*
Morris, R. E., 124, 126*f*, 127–128
Moshoeshoe, M., 59–61
Mou, F., 158*f*
Msayib, K. J., 69*f*
Mueller, B. J., 208
Muenya, N., 106
Mullens, S., 159*f*
Müller, F. A., 117–119
Müller, L., 117–119
Müllner, P., 82–83
Murakami, K., 101*f*
Murakami, M., 145, 146*f*
Murphy, S., 161
Myers, K., 113–116

N

Nabarro, F. R. N., 33–34
Nadiye-Tabbiruka, M., 59
Nagai, A., 68–70
Nagorny, R., 208
Naik, N. N., 168*f*
Nairn, K. M., 49–50, 50*f*
Nakajima, H., 99–101, 100*f*, 101*f*
Nakamura, R., 54, 80–81
Nakanishi, H., 173
Narkis, M., 49–50
Nasibulin, A. G., 57*f*
Nelson, R., 184
Ni, Z., 54–55
Nieh, T. G., 49–50, 145*t*
Nielsen, L. E., 211–212
Nijmeijer, A., 156*f*
Nika, D. L., 48
Niu, K., 81–82, 81*f*
Nolas, G., 24
Notario, B., 61–64, 64*f*, 189
Nöth, U., 268*f*
Novais, R. M., 220–221, 220*f*, 221*f*
Nunes, D., 53
Nutt, S. R., 208

O

Obuseng, V., 59
Ochoa, F., 27*f*
O'Donnell, P. B., 147–148
O'Dwyer, C., 259–260
Oh, Y. J., 89, 141–143, 142*f*
Ohashi, K., 101*f*
Ohji, T., 268–269
Oka, T., 49–50, 172
Oktay, E., 106
Okubo, M., 7, 106
Okuno, K., 211
Okuyama, K., 140–141
Oliver, D. W., 184
Olson, D. L., 35–37
Olsson, M., 151–152
Orlandini, E., 14–15
Orowan, E., 23
Orsini, V. C., 33
Ortona, A., 111*f*
Osswald, T. A., 113*f*
Ostwald, W., 84–85, 161–162
Ota, K., 101*f*

P

Pack, D. W., 147–148
Padilla, E., 176–177, 177*f*, 178*f*, 215*f*, 218–219, 219*f*
Pan, S., 247*f*
Pandey, D., 14, 14*f*
Pandey, G., 269
Park, B., 89–90
Park, G. D., 89
Park, J., 81–82, 81*f*, 89
Park, J. K., 153, 153*f*
Park, J. Y., 46
Park, M., 82–83
Park, S., 89
Park, S. B., 89–90
Parr, W., 269*f*
Parsons, G. N., 84*f*
Pasechnik, V. A., 181
Patterson, B. M., 275*f*
Paul, D. R., 46
Pavón, J. J., 106
Paz y Puente, A. E., 82–83, 86, 87*f*, 88*f*, 89*f*
Peabody, C., 259, 260*f*
Peng, H., 172
Peng, N., 154*f*
Peng, Q., 82–83, 84*f*
Peng, Y., 59–60, 157*f*
Perrier, G., 3–4, 4*f*
Perry, B., 145*t*, 193
Pethrick, R. A., 14–15
Phan, K., 269*f*
Pierotti, R., 53, 178–180
Pinisetty, D., 242–243
Pinto, J., 61, 64*f*, 189
Poco, J. F., 170, 171*f*
Pollard, C., 61–62
Ponce de León, C., 259
Porfiri, M., 208
Poulikakos, D., 215–216
Pu, Y., 148*f*
Pullar, R. C., 220–221, 220*f*, 221*f*

Q

Qin, W., 145*t*
Qiu, L., 210*f*

R

Raaijmakers, M. J. T., 156*f*
Rackwitz, L., 268*f*
Ragab, T., 55–56
Ramakrishna, S., 155–156
Ramakrishnan, N., 25, 27–28
Rao, P., 249
Rasooli, A., 120–121
Rawtani, D., 269
Raza, C., 268–270, 272*f*
Rech, B., 265*f*
Reichert, J. C., 268*f*
Reischl, M., 101
Ren, Z., 159*f*

Restrepo, G., 264–265
Rezek, Y., 208
Rhim, T., 142*f*
Riaz, H. A., 268–269, 272*f*
Ribitsch, V., 101
Ricci, W., 161, 242
Rietmeijer, F. J., 254*f*
Rietzel, D., 113*f*
Rioux, R. M., 79, 80*f*
Roberts, J. T. A., 37
Rodríguez, J. A., 106
Rodríguez-Pérez, M. A., 61, 64*f*, 189
Rong, G., 162*t*
Rosca, I. D., 243–244
Rose, R. M., 37
Rouquerol, J., 53, 178–180
Rudraiah, S., 268–269, 271*f*
Ryshkewitch, E., 221–222

S

Sakoda, K., 273
Salganik, R. L., 28
Salomão, R., 132*t*
Sandreczki, T., 173
Sano, H., 71–72
Santoliquido, O., 111*f*
Saquing, C., 84*f*
SasidharaRao, V., 243–244
Satcher, J. H., Jr, 170, 171*f*
Saveliev, A. V., 160
Sawant, S. S., 249
Scarritt, S., 27*f*
Schafer, B. W., 96, 225
Schantz, J., 268*f*
Scheinecker, M., 183*f*, 184–186, 184*t*
Scheunemann, R., 106
Schlögl, S., 101
Schnegg, A., 265*f*
Sears, J. K., 42–43, 44*f*
Seeder, W. A., 43
Seo, J., 120–121, 123*f*
Seo, J. Y., 142*f*
Serdyuk, N., 99–100
Serp, P., 58
Shabde, V., 215, 217
Shabde, V. S., 140, 141*f*, 144
Shah, R. N., 12–13, 13*f*
Shakeel, N., 268–269, 272*f*
Shanmugam, K., 55*f*
Shao, L., 147–148
Shapovalov, V., 99–100
Shapovalov, V. I., 99–100
Sharma, A. K., 243–244
Sharma, S., 173
Sherwood, J., 173
Shi, A., 199–200, 199*f*
Shi, F. G., 49–50
Shi, W., 89, 158*f*
Shi, X., 12*f*, 199–200, 199*f*
Shim, D., 120–121, 123*f*
Shimai, S., 97–98
Shimizu, Y., 145, 146*f*
Shunmugasamy, V. C., 242–243
Shutov, F. A., 1, 3–4, 191–192
Sieradzki, K., 72
Silva, S. R. P., 56
Silverstein, M. S., 144*f*
Simmons, R. O., 22
Sing, K. S., 179
Singer, R. F., 117–119
Singh, D., 59–60
Slack, G., 24
Smallman, R. E., 33
Smått, J., 132*t*
Smigelskas, A. D., 35–37, 77
Smith, B. H., 96, 225
Snijkers, F., 159*f*
Soleimani Dorcheh, A., 107, 170*f*
Solorzano, E., 64*f*
Solovyova, L. Y., 181
Song, H., 147–148
Song, X., 148*f*
Song, Y., 158*f*
Spagnola, J. C., 84*f*

Spear, K., 61–62
Speight, J. G., 140
Spontak, R. J., 84*f*
Srivastava, R., 217–218
Srivastava, V., 217–218
Stachiw, J., 151–152
Steckel, H., 140–141
Stern, L. A., 261
Stocchi, A., 247*f*
Strååt, M., 207
Striegel, A. M., 183–184
Su, L. H., 22
Subhash, G., 8–9
Subhash, S., 8–9
Suh, N. P., 62
Sun, D. X., 106
Sun, R., 144
Sun, X., 84*f*
Sun, Y., 243–244
Sun, Z., 158*f*, 184
Sung Kim, G., 245*f*
Swaddle, T. W., 25
Szépvölgyi, J., 145–147
Szyniszewski, S., 96, 225

T

Tagliavia, G., 208
Takahashi, M., 7
Tan, P. J., 201–202
Tane, M., 100–101
Tang, B., 108
Tang, Y., 86*f*
Taniguchi, N., 53
Tao, W., 217
Tappan, B. C., 105*f*
Tattershall, C. E., 69*f*
Tawfick, S. H., 244
Tay, T. E., 243–244
Teller, E., 178–179
Thallapally, P. K., 262–263, 264*f*
Tharmavaram, M., 269
Thayamballi, A. K., 245*f*
Thielmann, F., 182–183
Thijs, I., 159*f*
Thomas, G., 45
Thomson, W., 179–180
Thornton, A. W., 49–50, 50*f*
Thostenson, E. T., 159*f*
Tian, Y., 57, 57*f*
Tillotson, T., 106–107, 253
Timmermans, M. Y., 57*f*
Torres, Y., 106
Trousselet, F., 128–129, 129*t*
Tu, K., 77
Turing, A., 112–113

U

Uma Rani, R., 243–244
Unger, K. K., 132*t*
Uriarte, C., 49–50

V

Vachon, R., 215–217
Vaikhanski, L., 208
Vais, R. D., 80
Valencia, S., 264–265
Vallery, R. S., 172
Van Der Voort Maarschalk, K., 49–50
Van Wijk, W. R., 43
Velasco, M. A., 112–113, 114*f*, 115*f*
Velbel, M. A., 254*f*
Verma, P., 243–244
Vita, M. C., 273
Vromans, H., 49–50

W

Wachtman, J. B., 25
Wadley, H. N., 96
Walker, E., 125
Waller, P. J., 66–67
Walsh, F. C., 259
Walsh, W. R., 269*f*
Wang, A., 144
Wang, B., 99
Wang, F., 243–244
Wang, G., 132*t*, 149*f*

Wang, J., 65–66, 119–120, 121*f*, 122*f*, 147–148, 155–156
Wang, M., 108, 157, 162*t*
Wang, N., 158*f*
Wang, P., 66–67
Wang, Q., 90*f*
Wang, R., 161*f*
Wang, S., 59–60, 97–98, 158*f*, 173
Wang, S. H., 35–37
Wang, S. Q., 90*f*
Wang, X., 108, 130*f*, 147–148, 149*f*, 157, 161*f*, 162*t*
Wang, Y., 161, 173
Wang, Z., 157, 162*t*
Wang, Z. L., 162*t*
Warshaw, G., 37
Waters, K. E., 182*f*
Weaire, D., 161
Weers, J., 148–150, 150*f*
Wei, W., 147
Wen, D. H., 90*f*
Wen, X., 109*f*
Wessling, M., 156*f*
Westmacott, K. H., 33
Weston, S., 151–152
Wilcox, B., 267
Wilkinson, D. S., 33
Williams, M. L., 45
Winberg, P., 173–174, 174*f*, 175*f*
Winnubst, L., 156*f*
Wirsig, T. B., 262–263, 264*f*
Withers, P. J., 178
Woldesenbet, E., 197, 208
Wong, S., 208
Woodhams, R., 211
Wouterson, E. M., 208
Wu, L., 247*f*
Wu, Z., 149*f*

X

Xi, X., 161*f*
Xia, Q., 243–244
Xia, Y., 158*f*
Xiao, G., 145*t*
Xie, E., 157*f*
Xie, G., 106
Xie, J., 100–101
Xie, Y., 162*t*
Xing, Z., 184
Xiong, J., 246, 247*f*
Xu, B., 145*t*
Xu, C., 261*f*
Xu, C. Z., 49–50
Xu, H., 68–70
Xu, J., 84, 86*f*
Xu, W., 89–90, 90*f*, 145*t*
Xu, Y., 68–71, 89–90

Y

Yaghi, O. M., 66–67
Yamada, N., 71–72
Yan, Y., 130*f*
Yang, B., 243–244
Yang, C., 145*t*
Yang, D., 162*t*
Yang, J., 99, 247*f*
Yang, M., 132*t*
Yang, P., 82–83
Yang, Q., 199–200, 199*f*
Yang, X., 161
Yang, Y., 145*t*, 273
Yang, Z., 86*f*
Yao, Y., 89–90
Ye, B., 106
Ye, R., 208
Yi, R., 145*t*
Yin, J. S., 162*t*
Yin, Y., 77, 79–81
Yonetani, H., 100–101, 100*f*
Yoon, H. J., 142*f*
Yost, A. R., 86, 87*f*, 88*f*, 89*f*
Yu, J., 108, 157, 162*t*
Yu, M., 147
Yu, X., 7, 7*f*, 79, 89

Yuan, B., 82–83
Yucelen, E., 171*f*

Z

Zacharias, M., 77, 83*f*
Zafran, J., 208
Zeltmann, S. E., 242–243
Zeng, Y., 72
Zhai, C., 162*t*
Zhang, C., 4
Zhang, G., 199–200, 199*f*
Zhang, H., 145*t*, 162*t*
Zhang, J., 157*f*, 261*f*
Zhang, L., 208
Zhang, Q. Y., 90*f*
Zhang, R., 263–265, 265*f*, 266*f*
Zhang, W., 86*f*, 259, 261*f*
Zhang, Y., 119–120, 121*f*, 122*f*, 148*f*, 155–156
Zhang, Z., 123, 243–244, 257–258
Zhao, C., 157*f*
Zhao, D., 263–265, 265*f*, 266*f*
Zhao, H., 149*f*
Zhao, S., 261*f*
Zhao, X., 12*f*
Zhao, Y., 106, 158*f*
Zhao, Y. Y., 106
Zheng, H., 81–82, 81*f*
Zheng, J., 79, 147, 148*f*
Zheng, M., 49–50
Zheng, P., 82–83
Zhou, J., 82–83, 89
Zhou, X., 184
Zhou, Y., 89, 161*f*, 162*t*
Zhu, F., 4
Zhu, N., 210*f*
Zhu, P., 147
Zhu, R., 145*t*
Zhu, T., 17–18
Zhu, Y. C., 49–50
Zhu, Z., 57*f*
Zhuang, S., 65–66
Zhurkov, S., 46
Zikry, M. A., 33
Zolensky, M., 252
Zou, R., 117*f*
Zyla, G., 210*f*

Subject Index

Note: Page numbers followed by "*f*" and "*t*" refer to figures and tables, respectively.

A

Ablation, 246–247
Ablative materials, 246–247
  binder phase materials for, 248–249
  microstructure of, 247–248, 248*f*
  reinforced voids, 247–248
  testing, 249
  unreinforced voids, 247–248
Ablative resistance, 246–247
Acoustic damping, 4
Additive manufacturing (AM), 9–13
  binder jetting, 113–116, 115*f*
  of cellular structures, 109–123
  directed energy deposition, 120–121
  material extrusion, 117–119, 117*f*
  material jetting, 112–113, 114*f*, 115*f*
  sheet lamination, 119–120
  subtractive manufacturing methods, 110*f*
  vat polymerization, 110–112, 112*f*
Aeration, 97
  chocolate bar, 97, 98*f*
Aerogels, 105–107 *See also specific aerogels*
  aging, 107
  alumina, 106–107, 170, 171*f*
  carbon, 106–107
  defined, 106–107
  drying, 107
  materials, 106–107
  pore volume, 107
  preparation, 107
  silica, 106–107, 169, 170*f*
  surface area, 106–107
  titania, 106–107
Aerospace, 243–256
  applications, 243
  carbon nanotubes (CNT), 233–243
  heat shields, 246–249
  honeycombs, 245–246
  material, 243–244
  science, 243–244
  silica aerogel for comet dust collector, 249–254, 252*t*, 253*f*, 255*f*
  thermal barrier coatings (TBCs), 254–256
  thermal protection systems, 246–249
Aging
  of aerogel, 107
  density, 49
  free volume, 50*f*, 67
  physical, 49
  polymer chain mobility, 49
  of polymers, 49
  time dependent reduction of free volume, 49
AgO hierarchical porous film, 89–91, 90*f*
AgO nanorods, 89–90
Ag_2O with hexagonal structure (h-Ag2O), 90–91
Airbus A380-800, 245–246
  components of, 246*f*
Alkenes, 65
Alkynes, 65
Alloys *See also specific alloys*
  atomic radii, 19–20
Alumina, 25
  spheres, 151–152
Alumina aerogels, 106–107
  intentional voids in, 170
Aluminum
  diffusion, 33
  infiltration, 116*f*

Aluminum (*Continued*)
vacancies, 22
Aluminum foam, 202–203, 203*f*
stress-strain relationship, 203*f*
thermal conductivity, 216
Ambient temperature drying, 107
Ammonium bicarbonate, 142*f*
decomposing, 141–143
Amorphous materials, 44–45
Amorphous networks, 65
Amorphous polymers, 42*f*, 48
Analcimes, 59–60
Analogous sedimentary rock formation, 9
Arc discharge, 160
Arc jet, 249, 250*f*
Aromatic monomers, 65
Ashby plots, 208
syntactic foam, 208–210, 209*f*
Assemble-disassemble-organization-reassembly (ADOR) method, 126–128, 126*f*, 127*t*
Assembly methods, 72
ASTM D2662 standard, 192
Asymmetric hollow particles, 81–82, 81*f*
Atomic number contrast, 169
Atomic radii, 19–20
Atomic vacancies, 25
Atomizer, 140
Autonomous underwater vehicle (AUV), 237
AVCOATs, 248–249
Avogadro's number, 21

B

Backscattered electrons (BSEs), 169
Bathyscaphe Trieste, 234*f*
Beam theory, 204–205
BET. *See* Brunauer-Emmett-Teller analysis (BET)
Binder jet process, 113–116, 113*f*, 115*f*
three-phase syntactic foam, 116*f*
Binder phase materials
for ablative materials, 248–249
silicon, 248*f*
Bioactive materials, 117–119, 266–267
Biocompatibility, 71–72, 266–267
Biodegradable, 266–267
Biomaterials, 108, 130–131
evolution of, 231–233, 268*f*
and healthcare
drug delivery, 272
introduction, 265–272, 268*f*
nerve regeneration, 268–271, 269*f*, 271*f*
scaffold, 267–268
Biomimetic, 268*f*
Blowing agent, 6. *See also* Chemical blowing agent; Physical blowing agent
Bohr atomic model, 2
Boltzmann constant, 33–34, 217
Bond-coat layer, 255
Bonding, 107–108
covalent, 41, 66
of fibers, 107–109
hollow macrospheres, 151, 153
hydrogen, 41
of particles, 107–109
of powders, 107–109
of spheres, 107–109
van der Waals, 41
Bones
pore structure, 130–131
scaffold, 269
scaffolding, 130
synthetic, 130
Bore fluid, 154
composition, 155
flow rate, 155
temperature, 155
Bottom-up process, 110–112, 111*f*
Brazing, 31
Brunauer-Emmett-Teller analysis (BET), 179
BSEs. *See* Backscattered electrons (BSEs)
Bulk diffusion process, 79

C

Calixarenes, 261
Carbon
  aerogels, 106–107
  nanotube, 54
  polymers based on, 41
Carbon dioxide, 99
Carbon fibers, 244
  carbon nanotubes in, 244*f*
Carbon microballoons (CMBs), 193–194, 193*f*
Carbon nanotubes (CNTs), 5, 54, 158–160, 243–244
  absorbance spectra, 57*f*
  aerospace, 233–243
  applications, 243–244
  arc discharge creating, 160
  in carbon fibers, 244*f*
  chemical vapor deposition of, 160
  CNT diameter, 56
  CO_2 parameters, 57
  diameter of, 55–56
  double walled, 158–160
  equation, 56
  example of void, 56
  flame synthesis, 160
  methods for making, 55*f*
  methods to make, 160
  as nano-reactor, 57–58, 58*t*, 59*f*
  optical properties, 57
  properties of, 160
  pulsed laser vaporization, 160
  Raman spectra, 56
  RBM peaks, 56
  total surface area of, 56
Casting. *See also* Robocasting
  porosity from, 30–31
  voids during, 30–31
Cation exchange capacity (CEC), 59–60
Cation packing, 20*t*
Celgard material, 260*f*
Cell faces, 3–4, 3*f*
  closed cell, 5
  open cell, 3–4
Cells. *See also* Closed cell; Open cell
  ceramic foam, 96
  cylindrical, 257
  form factors, 257
  lithium ion battery, 257
  microbial fuel, 3–4
  pouch, 257
  prismatic, 257
Cell size, 190–191
  diameter, 190
  distribution and regularity, 199–202, 200*f*
  foam, 190
Cell structure
  ceramic foam, 97–98
  examples, 4*f*
  idealized, 3*f*
Cell struts, 3*f*, 4
  deformation characteristics of, 206
  length of, 203–204
  open cell, 4
Cellular structures, additive manufacturing of, 109–123
  binder jetting, 113–116, 115*f*
  directed energy deposition, 120–121
  material extrusion, 117–119, 117*f*
  material jetting, 112–113, 114*f*, 115*f*
  sheet lamination, 119–120
  subtractive manufacturing methods, 110*f*
  vat polymerization, 110–112, 112*f*
Cellulose acetate, 154
Ceramic foam, 31, 231
  applications, 231, 232*f*
  cells, 96
  cell size, 97–98
  cell structure, 97–98
  chemical blowing agent, 101–102
  creating, 96
  hollow particles, 5–6
  porosity, 232*f*
  sacrificial pore former for, 106
  slurry processing, 97–98
  thermal conductivity, 215
Chabazites, 59–60

Challenger Deep, 233
Chemical blowing agent, 96, 101–105
  cell size expansion during, 106
  ceramics, 101–102
  cross linking reaction, 101
  density, 103
  example, 101
  fumed silica, 101
  mechanisms, 101
  sacrificial pore former, 106
  strength, 103
Chemical dealloying, 72
Chemical nanoscience, 57–58
Chemical vapor deposition (CVD), 145–147
  carbon nanotubes made by, 160
Chitosan, 269
Chocolate bars, 97
  aeration of, 97, 98*f*
  X-ray microcomputed tomography of, 98*f*
Chromatographic porosimetry, 180–186
  final component, 180–181
  inverse gas chromatography (IGC), 181–183, 182*f*
  inverse size exclusion chromatography, 183–186
    accessible pore volume, 184
    advantages of, 184
    elution in, 183–184
    mean pore sizes, 184*t*
    raw data generated by, 184
    representative chromatograph for, 185*f*
    size separation in, 183*f*
    stationary phase in, 184
  inverse technique, 181
  mobile phase, 181
  solid stationary phase, 181
  types of, 181
Clathrates, 261
Closed cell
  faces, 5
  high magnification, 4*f*
  for insulation, 5
  mechanical properties, 12
  open cell compared with, 191–192
  porosity, 3–5
  strength of, 5
Clustering of voids
  creep, 34
  from diffusion, 33
CMBs. *See* Carbon microballoons (CMBs)
CNTs. *See* Carbon nanotubes (CNTs)
Coaxial blowing nozzle
  hollow macrospheres made by, 152–153
  schematic, 152*f*
Cobalt oxide (CoO) particles, 78
Cobalt (Co) particle, 78
Coble creep, 33
  equation, 34
  temperature, 34
Coefficient of thermal expansion (CTE), 45
Color centers, 21
Comet dust collector, silica aerogel, 249–254, 252*t*, 253*f*, 255*f*
Composite foam, 2, 5
Composite syntactic foams, 238–239
Compression
  microballoon, 210
  stress-strain relationship, 202–203
Compressive strength, 199
  of syntactic foam, 209*f*
Co nanoparticle, 80–81
Conduction electrons, 24
Confocal microscopy, 168
Conjugated microporous polymers (CMP), 68–71
  and analogues, 65
Continuous zone melting method, 100*f*
Controlled porosity, 132*t*
Conventional foams, 202–207
Coquina, 8–9
Cork, 212
Covalent bonds, 41, 66
Covalent organic frameworks (COFs), 65–67
Covalent triazine frameworks, 65, 67

- Cracks. *See also* Microcracks
  - creep, 33
  - extension force, 30
  - propagation path, 220*f*
  - stress causing, 29–30
  - stress causing elliptical, 29*f*
- Creep, 31
  - clustering, 34
  - Coble, 33
  - cracks, 33
  - defined, 33
  - diffusional, 33
  - failure and, 34
  - Nabarro-Herring, 33–34
  - rate equation, 33–34
- Creep strain
  - homologous temperature for, 33
  - total longitudinal, 34
- Crude oil pipelines, 239–240
- Crush
  - strain, 199
  - zone for high velocity projectiles, 199
- Crystalline materials, 18–25 *See also specific crystalline and semicrystalline materials*
  - arrangements of atoms, 23
  - hexagonal close packed, 19
  - intrinsic voids in, 78
  - Kirkendall effect in, 35–37
  - mechanical properties of, 23–24
  - octahedral void in, 19
  - packing efficiency in, 14
  - radius ratio for, 18
  - strength, 23–24
  - symmetry, 23
  - tetrahedral voids in, 19
  - theoretical density of, 20–21
  - theoretical strength of, 23–24
  - void types in, 19
- Crystalline networks, 65
- CTE. *See* Coefficient of thermal expansion (CTE)
- Cubic AgO (c-AgO,), 90–91
- Cubic voids, 19*t*
- Cuboids, 139
- $Cu(OH)_2$ nanorods, 84
- CVD. *See* Chemical vapor deposition (CVD)
- Cylindrical cells, 257

D

- Dead time, 182
- Dealloying, 72
- Decomposition reaction, 103
- Deep-sea buoyancy, 233–238
- Deepsea Challenger, schematic of, 236*f*
- Deep-sea thermal insulation, 239–240
- Defects, 21–25
  - benefits, 2
  - in hollow macrospheres, 151
  - point, 21–22, 33
  - in real materials, 24
  - real properties, 20–21
  - Schottky, 21–22
  - voids as, 2
- Deformation
  - of cell struts, 206
  - density changes, 210
  - plastic, 30
  - syntactic foam, 210
- Density, 20–21
  - aging, 49
  - calculation, 20–21
  - chemical blowing agent, 103
  - crystalline materials, 20–21
  - deformation changing, 210
  - detonation velocity, 240, 242*f*
  - equation, 195–196
  - explosives, 240–242
  - foam, 95, 195–196
  - hydrostatic strength, 238–239
  - measurement, 167
  - microballoon, 196*f*
  - relative, 196–197
  - sintering, 103

Density (*Continued*)
space travel requirements, 247–248
syntactic foam, 209*f*
theoretical, 20–21, 25, 129*t*
vacancies, 25
Depletion gilding, 72
Desorption, 179
Detonation pressure, 240
estimation, 241
Detonation velocity, 241
analyzing, 241
density, 240, 242*f*
pentaerythritol tetranitrate, 242*f*
Detrimental process, 78
1,4-dicyanobenzene, 67
Dielectric constant, 206–207
defined, 206–207
electric fields, 206
perfluoroethylene-polypropylene, 207*f*
Dielectric loss, 207
Diffusion, 21. *See also* Vacancy diffusion
activation energy, 33
aluminum, 33
coefficient, 32–37
creep, 33
distance, 32–33
examples of, 32
Fick's first law, 32
Fick's second law, 32–33
Knudsen, 179
phase separation induced by, 154
self, 33
time, 32–33
of vacancies, 32–34
vacancy, 77
Diffusional flux, 32
Diffusion coefficients, 79
Directed energy deposition, 120–121
Dirichlet tessellation, 11
Dissipation factor
perfluoroethylene-polypropylene, 207*f*
as quality control, 207
Dope, 154
composition of, 155
flow rate of, 155
temperature of, 155
Double walled carbon nanotubes (DWCNT), 54–55, 158–160, 243
Drill riser buoyancy module, 239*f*
Drug delivery, 58, 272
hierarchical porosity, 147–148, 161–162
Drying. *See also* Spray drying
aerogels, 107
ambient temperature, 107
freeze, 107
supercritical, 107
Dry–wet spinning, 154
heat treatment following, 155
Dust particles, 129–130

E

ECM. *See* Extracellular matrix (ECM)
Elastic beam theory, 204–205
Elastic constants
foam, 203–206
syntactic foam, 211–212
Electrochemical dealloying, 72
Electrochemical energy storage with porous metals, 259–260
Electron beam melting (EBM)
apparatus, 122*f*
Ti6Al4V alloy foam fabrication, 123*f*
Electron–hole pair, 264
Electron microscopy, 168–172
high resolution transmission, 170–172, 171*f*
scanning, 99*f*, 168–169
transmission, 169–172
Electrospinning, 54
collection substrate, 156
components, 156–157
examples, 155
fibers, 108, 109*f*
power supply, 156

process, 157*f*
spinneret, 156
syringe pump, 156
webs, 109*f*
Ellipsoids, 139
hollow particles, 161
Empirically derived vacancy diffusion (EVD), 49
equation, 49
Energy, 257–263
activation, 33
electrochemical energy storage with porous metals, 259–260
guest–host complexes, 261–263
lithium-ion battery, 257–259
phonons in, 48
point defect, 33
Schottky defect, 21–22
shock wave, 240
solar power, 263
strain release rate, 29–30
Energy absorption
example, 198
foam, 95, 197–199
idealized, 198*f*
Energy storage, highly porous structures for, 89–90
Epoxies, 41
Epoxy, 117–119
Equilibrium free volume, 49, 50*f*
Equilibrium glass transition temperature, 45
Equivalent thermal conductivity, 213
Ethanol, 107
Ethylene glycols, 184–186
Excess free volume, 45–46
Exit wave reconstruction, 170–172
Exploiting chemically selective weakness in solids, 124–129
Explosives
density, 240–242
hollow microspheres in, 240, 242
Extracellular matrix (ECM), 130–131, 130*f*
Extrinsic constraints, 17–18

F
Fabrication
membranes, 155
microballoon, 107–108
wollastonite-diopside-based bioceramic scaffolds, 117–119
Failure
creep, 34
diffusion, 33
intrinsic void, 2–3
F-centers, 21
FEA. *See* Finite element analyses (FEA)
Felts, 108
FEM. *See* Finite element methods (FEM)
Fibers. *See also* Hollow fibers; Nonwoven fibrous structures
carbon, 244, 244*f*
electrospinning, 108, 109*f*
glass, 29–30
metal, 108
morphology, 155
nanofibers, 108
polyacrylonitrile, 244
porous, 154–160
random orientation, 108
Fick's first law, 32, 77, 79
equation, 32
Fick's second law, 32–33
equation, 32
Finite element analyses (FEA), 217–219
conceptual understanding, 218
simulation using, 218
software for, 218
Finite element methods (FEM), 178, 217–219
First conjugated microporous polymer, (CMP-1), 68–70, 69*f*
Flame synthesis, 160
Flexure, 208
Foams *See also specific types of foams*
cell size, 190
commonalities of, 96

Foams (*Continued*)
commonalities of forming, 96
components of, 213
composite, 2, 5
conventional, 202–207
density, 95, 195–196
elastic constants, 203–206
energy absorption, 197–199
gasar metallic, 99–100
general characterization, 190–202
geopolymer, 220–222, 221*f*, 223*f*
high velocity projectiles captured by, 198
lotus-shaped, 99–101, 101*f*
metallic, 222–225
multiphase, 108
noise control, 95
nonstochastic, 9–12
polymer, 96
polyurethane, 5
properties, 190
relative density, 196–197
reticulated, 3–4, 4*f*
signal loss, 207
single phase, 2, 5
software packages to simulate, 217–218
stiffness, 95
stochastic, 9–12
strength, 95
stress, 197
stress–strain behavior in compression, 202
three-phase, 108, 194*f*, 217
two-phase, 108
vibration control, 95
Voronoi, 9–12, 12*f*
Fourier's law, 214
Fracture mechanics, 30
Fracture strength, of polymer, 46
Free volume, 17
aging, 49, 67
aging plot, 50*f*
defined, 42
determination, 14–15
equilibrium free volume, 49
excess, 45–46
graphical definition, 43*f*
hole specific, 43
interstitial specific, 43
measurement, 49–50
molecular motion, 43
molecule hosted, 261
organic solids, 262–263
physical implications, 43
polymers, 42
rigid polymers, 45
thermomechanical behavior, 44–46
time dependent reduction of, 49
Freeze drying, 107
Fullerenes, 54, 59*f*
Fumed silica, 101
Functionality-structure relationships, 129
Fused deposition modeling, 12

G

Gas
decomposition, 101
introduction, 97–105
kinetic theory, 24, 48
latent, 97, 99, 101, 106
mixing, 96–98
porosity, 30–31
pycnometer, 192
universal constant, 33
Gas adsorption, 178–180
analysis types, 179
characterization, 178–179, 180*f*
isotherm, 179
method, 178–179
Gasar metallic foam, 99–100
forming, 99–100
stages of, 100–101
Gas chromatography (GC), 181
inverse, 182*f*
Gasoline buoyancy, 233
Ge-containing zeolites, 128

Geopolymer foams, 220–222, 221*f*, 223*f*
forms of, 220–221
Glass
borosilicate, 99
fibers, 29–30
hollow microspheres, 2–3, 108
microballoon, 5, 6*f*, 7, 107–108
Glass microballoons (GMB), 233, 235*f*
Glass transition temperature, 44–45
defining, 44–45
equilibrium, 45
of semicrystalline thermoplastic polymers, 41
Glassy/rubbery system, 44*f*
Gold nanoparticles, 71–72
Grain boundary cavitation, 33, 35*f*
Grain boundary sliding (GBS), 34, 35*f*
Graphene, 54–55, 160, 170–172, 171*f*
Griffith theory of brittle fracture, 29–30
Guarded heat flowmeter technique, 214, 215*f*, 217
Guest–host complexes, 261–263
applications, 261–263
burning, 263*f*
chemical formula, 261
defined, 261
examples, 261
stability, 261–262

H

Hadal zone, 238
Halloysite nanotubes, 270*f*
Halpin–Tsai theories, 56
Hard spheres, 18
Hard templates, 72
hollow particles, 147
HCP. *See* Hexagonal close packed crystals (HCP)
Heat conduction, 214
Heat shields, 246–249
arc jet testing, 249, 250*f*
honeycombs, 250*f*
solar tower testing, 250*f*
Heterogeneous polymerization, 140
particle size, 141
Hexagonal close packed crystals (HCP), 19
Hexagonal COF skeleton, 66, 66*f*
HGMS. *See* Hollow glass microspheres (HGMS)
Hierarchical design, 95, 129–131
in nature, 129–130
Hierarchical porosity, 231–233, 267
drug delivery, 147–148, 161–162
photocatalysis, 161–162
porous hollow particles, 147–148
High internal phase emulsion (HIPE), 143
schematic, 144*f*
High resolution transmission electron microscopy (HRTEM), 170–172, 171*f*
High velocity projectiles
crush zone, 199
foams for capturing, 198
kinetic energy of, 198
Hindrance, 14–15
Hole specific free volume, 43
Hollow cadmium sulfide (CdS) nanospheres, 82
Hollow ceramic microspheres, 113–116
Hollow cobalt selenide (CoSe), 78
Hollow cobalt sulfide (CoS) nanoparticles, 78
Hollow composite macrospheres, 238–239
applications of, 238–239
Hollow fibers, 154–160
applications, 155
bioreactor technology, 155
defined, 154
diffusion induced phase separation to produce, 154
examples of, 154
polymers for, 154
requirements for, 154
spinneret for, 154*f*
surface area, 155
Hollow glass microspheres (HGMS), 2–3, 107–108, 233, 240*f*

Hollow glass spheres (HGSs), 107–108
Hollow macrospheres, 150–153
  bonded, 151, 153
  buoyancy applications, 151–152
  coaxial blowing nozzle to create, 152–153
  commercially available, 151
  composite, 238–239
  defects, 151
  defined, 150
  example, 150
  seamless, 151
  size, 152
Hollow microspheres
  commercial examples, 6
  design options, 242
  in explosives, 240–242
  glass, 2–3, 107–108
  phenolic, 6
  reinforced voids, 242
  unreinforced voids, 242
Hollow mullite microspheres, 113–116
Hollow nanoaluminum nitride (AlN) particles, 79
Hollow particles, 139–153. *See also* Nonspherical hollow particles; Porous hollow particles
  applications, 144
  ceramic, 6
  characteristics, 144, 145*t*
  commercially available, 144
  common, 143
  defined, 139
  ellipsoid, 161
  formation, Kirkendall effect, 78–82
    asymmetric hollow particles, 81–82, 81*f*
    bulk diffusion process, 79
    Ni–Zn hollow nanoparticles, 79, 80*f*
    surface diffusion of core material, 79
    symmetric hollow particles, 80–81, 81*f*
  hard templating, 147
  Kirkendall effect for, 147
  manufacturers, 145*t*
  Ostwald ripening, 161–162
  plasma processing, 145–147, 146*f*
  processing, 145*t*
  self assembly, 147
  soft templating, 147
  spray drying, 145
  types of, 144, 145*t*
Hollow polymeric nanostructures, 139
Hollow silica nanospheres, 7*f*
Honeycombs, 245–246
  advantages, 245
  in aerospace industry, 245–246
  anisotropic behavior of, 246
  applications, 245–246
  drawbacks, 246
  in heat shield, 250*f*
  as structural components, 245
  three dimensional, 245*f*, 246
HRTEM. *See* High resolution transmission electron microscopy (HRTEM)
Human-occupied vehicle (HOV), 233
Hybrid remotely operated vehicle (HROV), 237, 237*f*
Hydrochlorofluorocarbons, 99
Hydrostatic pressure, 31
Hydrostatic strength, 238–239

I

Ideal materials
  compared with real materials, 17
  packing efficiencies, 18
  properties, 18–21, 19*t*, 23*f*
  reasons for studying, 17
  size dependencies, 17
  theoretical strength, 17–18
IGSC. *See* Inverse gas-solid chromatography (IGSC)
Immiscible liquids, 141
Incompatibility, 143
Insulation
  closed cell foam, 5

of crude oil pipelines, 239–240
deep-sea thermal, 239–240
syntactic foam as, 212
Insulator, 206
Intentional voids
in alumina aerogels, 170
classification, 2
defined, 2
length scale, 15*t*
processing induced, 15
International Union of Pure and Applied Chemists (IUPAC), 53, 178–179
Interstitial specific free volume, 43
Intrinsic constraints, 17–18
Intrinsic void
atomic length scale, 2–3, 3*f*
in crystalline materials, 78
defined, 2
failure from, 2–3
length scale, 15*t*
nanometer length scale, 2, 3*f*
processing induced, 15
thermoplastic polymers, 42
thermoset polymers, 41
Inverse gas chromatography (IGC), 181–183, 182*f*
Inverse gas-liquid chromatography (IGLC), 181
Inverse gas-solid chromatography (IGSC), 181
carrier gas, 182–183
conditions, 181
dead time, 182
mobile phase in, 182
solid material, 182
Inverse-sieving technique, 183–184
Inverse size exclusion chromatography (ISEC), 181, 183–186
accessible pore volume, 184
advantages of, 184
elution in, 183–184
mean pore sizes, 184*t*
raw data generated by, 184
representative chromatograph for, 185*f*
size separation in, 183*f*
stationary phase in, 184
Ionic crystals, 19–20
ISEC. *See* Inverse size exclusion chromatography (ISEC)
Isotropic diffusion model, 79

K

Kelvin Problem, 11
Kelvin radius, 179–180
Kelvin tessellation, 11
Killed steels, 30–31
Kinetic theory of gases, 24
thermal conductivity, 48
Kirkendall effect, 17, 54, 77, 249, 259–260
AgO hierarchical porous film, 89–91, 90*f*
coalescence of voids, 34*f*
in crystalline materials, 35–37, 36*f*
in Cu–Zn system, 78*f*
examples, 35–37
experiment, 35–37
filamentary superconductors, 37
hollow particles, 147
hollow particles formation, 78–82
asymmetric hollow particles, 81–82, 81*f*
bulk diffusion process, 79
Ni–Zn hollow nanoparticles, 79, 80*f*
surface diffusion of core material, 79
symmetric hollow particles, 80–81, 81*f*
Ni–Cu laminate, 36*f*
porous and hollow structures, 85–91
schematic, 148*f*
tubes, 82–85, 83*f*, 84*f*
coaxial $ZrAl_2O_4$ microtubes, 85*f*
micrometer-scale coaxial $ZrAl_2O_4$ microtubes, 84*f*
shell of, 84
Kirkendall-type diffusion, 77
Knudsen diffusion, 179

L

Laser scanning confocal microscopy (LSCM), 167–168
  applications of, 168
Latent gas
  activation of, 97
  agents, 99
Lattice holes, 18
  common types, 19*t*
  filling, 24
  size, 18, 24
  temperature, 24
  thermal conductivity limitation, 24
Light-activated polymerization, 110–112
Linear polymers, 42*f*, 143
Linear voids, 19*t*
Linkages, 68–70
  length of, 70–71
Lithification, 9
Lithium ion battery
  anode, 257–258
  applications, 257–259
  cathode, 257–258
  cells, 257
  chemistries, 257
  components, 257
  electrolyte, 257
  separator, 257–258, 258*f*
  stress on, 259
Lithium-ion diffusion, 89–90
Lotus, 129–130
  foam shape, 99–101, 100*f*, 101*f*
  macrometer scale, 128*f*
  micrometer scale, 128*f*
  nanometer scale, 128*f*
  pore structure, 128*f*
Low dimensional sample geometries, 18
LSCM. *See* Laser scanning confocal microscopy (LSCM)

M

Machining process, 109–110
Macropores, 53, 179
Macroscale voids, 120–121
Magnesium (Mg), 100–101, 106
Magnetic resonance imaging (MRI), 176*f*
Marshmallows, 97
Mass customization, 110–112
Materials *See also specific materials*
  aerogels, 106–107
  amorphous, 44–45
  bioactive, 117–119, 231–233, 266–267, 268*f*
  design, 117–119
  versus digital design of voids, 12–13
  extrusion, 117–119, 117*f*
  hierarchical porosity, 132*t*
  jetting, 112–113, 114*f*, 115*f*
  modulus, 25–28
  nonspherical hollow particles, 162*t*
  powder processed, 25, 31
  sacrificial core, 7, 105–106
  space travel requirements, 247–248
Materials science and engineering, 62*f*
Matrix nanoporosity, 6*f*
Mechanical properties
  closed cell, 5
  crystalline materials, 23–24
  syntactic foam, 208
  voids, 25–30
Mechanical stretching, 96, 123
  pore structure by, 123
Mechanism, 77
Melting
  continuous zone method, 100*f*
  point, 44–45
  semicrystalline thermoplastic polymers, 41
Membranes, 257–258
  fabrication of, 155
  inorganic, 155
Menger sponges, 272–274
mer making ethylene, 42*f*
Meshing
  example, 218–219
  of solder, 219*f*
Mesopores, 53, 66–67, 179

Metal-based foams, 96, 231
Metallic foams, 222–225
Metals, 96 *See also specific metals*
applications, 231, 232*f*
degassing, 31
fibers, 108
intrinsic properties of, 259
irradiation causing vacancies, 37
porosity in, 30–31, 232*f*, 259
sacrificial pore former, 106
thermal conductivity, 24
vacancies, 21
Methane hydrates, 262*f*
applications, 261–263
burning, 263*f*
chemical formula, 261
stability, 261–262
MFM. *See* Multiphoton fluorescence microscopy (MFM)
Micro air vehicles (MAVs), 244
Microballoon
carbon, 193–194, 193*f*
compression, 210
density, 196*f*
fabrication of, 107–108
glass, 5, 6*f*, 7
reinforcing effect, 211
Microbial fuel cell, 3–4
Microcellular materials, 62–63
Microcracks, 28
effect on Young's modulus, 28
Micro-electro-mechanical-systems (MEMS), 243
Micrometer-scale coaxial $ZrAl_2O_4$ microtubes, 84*f*
Micrometer-scale metal fibers, 108
Micrometer scale voids, reinforced, 6–7
Micropores, 53, 179
Microscopy, 167–172. *See also* Electron microscopy; Scanning electron microscopy (SEM); Transmission electron microscopy (TEM)
confocal, 168
electron, 168–172
scanning electron microscopy, 168–169
transmission electron microscopy, 169–172
laser scanning confocal, 167–168
multiphoton fluorescence, 176*f*
optical, 167–168
photoacoustic, 176*f*
positron reemission, 172
transmission electron, 169–172
transmission positron, 172
Microstructure
ablative materials, 247–248, 248*f*
porous hollow particle, 149*f*
PulmoSphere, 272*f*
syntactic foam, 6
Mixing. *See also* Rule of mixtures
in food industry, 97
gas, 96–98
temperature control during, 97
Modulus, 25–28, 202*f*
effect of porosity, 25
equations, 26
materials, 25–28
prediction, 25
reinforced void, 194–195
spherical voids, 25–26
syntactic foam, 212
unreinforced void, 194
void volume fraction, 26*f*
Young's, 23, 25–30, 27*f*, 204–205, 223–225, 226*f*
Molecular motion, 43
Mousse, 98*f*
MRI. *See* Magnetic resonance imaging (MRI)
Multiphase foam, 108
Multiphoton fluorescence microscopy (MFM), 176*f*
Multiwalled carbon nanotubes (MWCNT), 54–55, 243
Multiwalled nanotube (MWNT), 158–160

N

Nabarro-Herring creep, 33
  creep rate, 33–34
  temperature, 33–34
NaCl. *See* Sodium chloride (NaCl)
Nanocellular materials, 62–63
Nanocontainment, 57–58
Nanofibers, 108
NanoFOAM process, 103, 104*f*
  general reaction synthesis of, 105*f*
  series of events in, 103, 104*f*
  used for, 105
Nanomaterials, 53, 139
  macrostructures, 53
Nanometer-scale (solid) aluminum particles, 79
Nanometer-scale porous structures
  nanoporous noble metals, 71–72
  nanoporous organic networks, 64–71
    conjugated microporous polymers, 68–71
    covalent organic frameworks, 65–67
    covalent triazine frameworks, 67
    different classes of, 65
    polymers of intrinsic microporosity, 67, 69*f*
  nanoporous polymers, 61–64
    fabrication techniques, 61
  nanotubes, 54–58
  solid material, 54
  zeolites, 59–61
    basic formation of, 60*f*
    chemical formula, 59, 60*t*
    naturally occurring, 58*t*, 59
Nanometer scale voids, reinforced, 7
Nanoparticles, 53, 71–72
Nanoporosity, 53
Nanoporous gold film, 72, 73*f*
Nanoporous materials, 54
  applications, 54
Nanoporous noble metals, 71–72
Nanoporous organic networks, 64–71
  conjugated microporous polymers, 68–71
  covalent organic frameworks, 65–67
  covalent triazine frameworks, 67
  different classes of, 65
  polymers of intrinsic microporosity, 67, 69*f*
Nanoporous polymers, 61–64
  fabrication techniques, 61
Nanoporous triazine network, 68*f*
Nanotechnology, 53
Nanotubes, 53–58
  formation with hierarchical porosity, 85–91
Natural axon nerve repair, 268–269
Natural materials, porosity in, 7–9
Natural polymers, 54
Nereus, 237–238
Nereus hybrid vehicle, 151–152
Nerve regeneration, 268–271, 269*f*, 271*f*
Nickel based superalloys, 37
Nickel nanoparticle, 79
Nickel–zinc hollow nanoparticles, 79, 80*f*
NiTi microtubes, 87*f*
NiTi tube structure, 89*f*
Nitrogen, 53, 99, 103, 104*f*, 105, 105*f*
Noble metals, nanoporous, 71–72
Node point, 68–70
Nodes, 218–219
Noise control, 95
Nonsolvent-induced phase separation, 154
Nonspherical hollow particles, 160–162
  characteristics, 162*t*
  close packing, 161
  functionality, 161–162, 162*t*
  materials, 162*t*
  Ostwald ripening, 161–162
  processing methods, 162*t*
  shapes, 162*t*
  sizes, 162*t*
  surface area, 161–162
  types, 162*t*
Nonstochastic foams, 9–12
Nonwetting surfaces, 109

Polyacrylonitrile fibers, 244
Poly(aryleneethynylene)-based CMPs
  chemical structures of, 70*f*
  pore size for, 71*f*
Poly(aryleneethynylene) polymers, 70–71, 70*f*
Polydimethylsiloxane (PDMS), 101
  chemical reaction with cross-lining agent, 102*f*
  silica reinforced, 105–106
  temperature, 173–174, 174*f*
Polyetherimide (PEI) foams, 190
Polyethylene, 41, 47, 67, 154
  forms, 47*t*
  processing, 47
  tensile properties, 47*t*
(poly)ethylene glycol (PEG) molecules, 184–186
Polyethylenimine (PEI), 62–63
  mechanical properties of, 63*f*
Polyimide, 154
Polylactic acid (PLA), 221–222
Polylactide-co-glycolide (PLGA), 12–13, 13*f*
Polymer-based two-phase syntactic foam, 240*f*
Polymer chains, 41
  aging and mobility, 49
  alignment, 47
  hindrance, 14–15
  preferential orientation, 48
  theoretical strength, 46
  thermal vibrations, 48
Polymeric foams, 189
Polymerization
  heterogeneous, 140
  light-activated, 110–112
  plasma, 7, 145–147
  vat, 110–112, 112*f*
Polymers *See also specific types of polymers*
  aging, 49
  amorphous, 42*f*, 48
  branched, 42*f*
  carbon based, 41
  cross linked, 42*f*
  expansion with temperature, 45
  foams, 96
  fracture strength, 46
  free volume, 42
  hollow fibers, 154
  homogenous solution, 154
  kinetic strength theory, 46–47
  linear, 42*f*, 143
  maximum strength, 46–47
  occupied volumes, 45–46
  rheological properties, 101
  silicon based, 41
  structure, 41–43
  tensile properties, 47
  thermal conductivity of, 48
  thermal transport, 48
Polymers of intrinsic microporosity (PIMs), 65, 67, 69*f*
Polymethylmethacrylate (PMMA), 63–64
  thermal conductivity, 64*f*
Polyolefin polymer, 124*f*
Polypropylene, 41, 67
  separators, 258*f*
Polystyrene (PS), 5, 41, 238
Polysulfone, 154
Polytetrafluorethylene, 154
Polyurethane foam, 5
Polyvinyl alcohol (PVA) fiber, 83
Polyvinyl chloride, 41
Pore formation, 96, 101–102. *See also* Sacrificial pore former
  porogens, 141–143
Pore structure, 98*f*
  bone scaffolding, 130–131
  lotus leaf, 128*f*
  mechanical stretching, 123
  titania, 264
Porogen, 106–107
  liquid phase, 143
  placeholder, 141–143
  pore formation, 141–143

solid, 141–143
types, 143
Porosity. *See also* Hierarchical porosity
calculations, 129*t*
in casting, 30–31
ceramic foam, 232*f*
closed and open cell, 3–5
controlled, 132*t*
gas evolution, 30–31
hydrostatic pressure, 31
International Union of Pure and Applied Chemists standards, 178–179
Kirkendall effect, 89
macroscale, 130*f*
in metals, 30–31, 232*f*, 259
micrometer scale, 104*f*, 130*f*
modulus, 25, 29*f*
multiple length scale, 147–148
nanometer scale, 104*f*, 130*f*
in natural and synthetic materials, 7–9, 8*f*
open cell, 192
optimization, 15
powder processed materials, 31
rule of mixtures, 26
schematic, 130*f*
shrinkage, 30–31
solders, 27–28
Young's modulus, 27*f*
Porous and hollow structures, Kirkendall effect, 85–91
Porous bioscaffold, 112–113
Porous fibers, 154–160
Porous foil current collector, 259, 261*f*
Porous hollow particles, 147–150
defined, 139
example, 148–150
hierarchical porosity and, 147–148
microstructure, 149*f*
porous macrospheres, 150–153
Porous halloysite/chitosan composite, 271*f*
Porous macrospheres, 150–153
pseudo-double-emulsion technique to fabricate, 153
schematic, 153*f*
water-in-oil-in-water technique, 153*f*
Porous polylactic acid (PLA) structures, 119–120
Portland cement, 220
Positron annihilation lifetime spectroscopy (PALS), 14–15, 49–50
schematic, 173*f*
Positronium, 14–15, 22
types, 172
Positron, 14–15, 22
charge, 172
generation, 172
lifetime, 49–50
prediction, 172
Positron reemission microscopy, 172
Pouch cells, 257
Powder bed fusion, 117–120, 119*f*
Powder processed materials, 25
porosity, 31
Powders
bonding, 107–109
hollow particles, 140
porous particles, 140
Primary building units (PBU), 60–61
Prismatic cell, 257
PS. *See* Polystyrene (PS)
Pseudo-double-emulsion technique, 153
PulmoSphere, 148–150
diameter of, 272
microstructure, 272*f*
spray drying, 150*f*
Pulsed laser vaporization, 160

Q

Quality control, 207

R

Radiative thermal conductivity, 217
Radius ratio, 197
cation packing, 20*t*
constant, 18
crystalline materials, 18

Radius ratio (*Continued*)
formula, 18
Raman spectra, 56
Raman spectroscopy, 55–56
Ramie, 109, 109*f*
Random close packing, 161
Reaction-diffusion model, 112–113
Real materials, defects, 24
Reinforced void, 5–7, 193–195
ablative materials, 247–248
examples, 5
hollow microspheres, 242
macrometer scale, 7
micrometer scale, 7
modulus, 194–195
nanometer scale, 7
strength, 194
volume fraction of, 193
Relative density, 62–63, 196–197
Resonant Ultrasound Spectroscopy, 27–28
Reticular chemistry, 66–67
Reticulated foam, 3–4, 4*f*
Reverse osmosis, 155
Rigid polymers, 45
Robocasting, 157
Rocket nozzles, 246–247
r-type cavity, 34*f*
r-type voids, 33
Rule of mixtures, 26

S
Sacrificial core material, 7, 106
Sacrificial pore former, 105–107
ceramic foams, 106
chemical blowing agent differentiated from, 106
metal foams, 106
process of, 105–106
Sandia National Laboratory's Solar Tower Facility, 249
Scanning electron microscopy (SEM), 99*f*, 168–169
components, 169
imaging modes, 169
Scattering mechanisms, 48
Schottky defect, 21
energy formation for, 21
Secondary block unit (SBU), 60–61
Secondary electrons (SEs), 169
Segmentation, 218–219
Self consistent field theory, 216–217
Self diffusion, 33
Self-sustaining combustion synthesis reaction, 103
Self-templating process, 78
Semiconduction, 21, 264–265
Semicrystalline thermoplastic polymers, 44–45
glass transition temperature, 41
melting, 41
Separators
dry process, 257–258
in lithium ion battery, 257–258, 258*f*
wet process, 257–258
Serial sectioning, 175–176
Service induced voids, 30–31
SEs. *See* Secondary electrons (SEs)
Shear modulus, 224
Sheet lamination process, 119–120, 121*f*
diagram of, 121*f*
porous PLA structure, 122*f*
Shock wave energy, 240
Sievert's law, 30–31, 100
Signal loss, 207
Silica
chemical blowing agent and fumed, 101
hollow nanospheres, 7*f*
polydimethylsiloxane reinforced, 105–106
Silica aerogels, 169, 170*f*
for comet dust collector, 249–254, 252*t*, 253*f*, 255*f*
density, 106–107
surface area, 106–107
Silicon
based polymers, 41

binder phase, 248*f*
Silicone-based preceramic polymer ink, 117–119
Silicones, 41, 117–119
Single nanometer porosity, 54
Single-phase foam, 2
examples, 5
Single walled carbon nanotubes (SWCNT), 54–56, 58, 243
Single walled nanotube (SWNT), 158–160
armchair, 159*f*
orientation, 158–160
types, 159*f*
zigzag, 159*f*
Sintered fibrous media, 108
Sintering, 102–103, 103*f*, 106
density, 103
strength, 103
Site of contortion, 67
Size exclusion chromatography (SEC), 181
Skutterudites, 24
Slitlike microstructure, 124*f*
Slit, voids, 28
Slurry-processing, 97–98
Sodium chloride (NaCl), 106
Soft templates, 72
hollow particles, 147
Solar energy, 259
Solar power, 263
degradation, 263, 265*f*
efficiency, 263
Solar tower, heat shield testing, 249, 250*f*
Sol–gel process, 106–107
Solders
meshing, 219*f*
porosity, 27–28
voids, 31
Solid freeform fabrication, 109–110
Solidification, 97
unidirectional, 100–101
Solid prototype, 109–110
Space travel material requirements, 247–248
Specific volume
defined, 43
function of temperature, 44–45
graphical definition, 43*f*
total, 43
Spheres. *See also* Hollow macrospheres; Hollow microspheres; Hollow silica nanospheres; Porous macrospheres
alumina, 151–152
bonding, 96, 107–109
hard, 18
Spherical voids, modulus and, 25–26
Spinneret, 154
air gap, 154
electrospinning, 156
geometry, 155
hollow fibers, 154*f*
requirements, 154
temperature, 155
Spray drying, 140
essential parts, 141*f*
hollow particles, 145
process, 140
PulmoSphere, 150*f*
schematic, 141*f*
water-in-oil-in-water emulsion for, 142*f*
Spray nozzle, 140
Staebler-Wronski effect, 263
Stardust aerogel, 251–253, 251*f*, 254*f*
Steel-based foams, 96
Steel wool, 108
Stochastic foams, 9–12
Strain energy release rate, 30
Strength
chemical blowing agent, 103
closed cell, 5
compressive, 194, 199, 209*f*
effect of voids, 28–29
of foam, 95
fracture, 46
hydrostatic, 238–239
ideal materials, 17

Strength (*Continued*)
  kinetic theory, 46–47
  polymer chains, 46
  size effect, 17–18
  syntactic foam, 211
  tensile, 23
Stress
  for crack propagation, 29–30
  foam, 197
  lithium ion batteries, 259
  plateau region, 197
  theoretical, 23, 23*f*
Stress-strain relationship, 18
  aluminum foam, 203*f*
  in compression, 202–203, 210–212
  foam, 202
  syntactic foam, 210–212, 210*f*
Subtractive manufacturing methods, 110*f*
Superalloy substrate, 255
Supercritical drying, 107
Superhydrophobic surfaces, 108–109, 129–130
Support material, 113*f*
Surface area
  aerogels, 106–107
  hollow fibers, 155
  nonspherical hollow particles, 161–162
  silica aerogels, 106–107
Surface diffusion of core material, 79
SWNT. *See* Single walled nanotube (SWNT)
Symmetric hollow particles, 80–81, 81*f*
Sympathetic implosions, 238
Synthetic bone, 130
Syntactic foams, 2, 5–6, 208–212
  applications, 233–243
  Ashby plots, 208–210, 209*f*
  composite, 238–239
  composite syntactic foams, 238–239
  compressive strength, 210–212, 210*f*
  deep-sea buoyancy, 233–238
  deep-sea thermal insulation, 239–240
  defined, 107–108, 212
  deformation, 210
  density, 209*f*
  elastic constants, 211–212
  explosive formulations, 240–242
  growth and performance, 208–210
  hollow composite macrospheres, 238–239
  insulation, 212
  matrix, 107–108
  mechanical properties, 208
  microstructure, 6
  schematic of, 108*f*
  schematic
  shear modulus, 212
  strength, 5, 211
  stress-strain relationship, 210–212, 210*f*
  thermal properties, 212–217
  three-phase, 6–7, 6*f*, 192–193, 193*f*, 194*f*, 195*f*, 196*f*, 208, 210, 217
  two-phase, 6
  Young's modulus, 212
Synthetic materials, porosity in, 7–9
Synthetic polymers, 54
Synthetic zeolites, 60–61, 125
  three-phase foam, 217

T

TEM. *See* Transmission electron microscopy (TEM)
Template, 64–65, 96, 105–107
  hard, 147
  hollow particles, 147
  soft, 147
Template-free process, 78
Tensile properties
  polyethylene, 47*t*
  polymers, 47
Tensile strength, 23
Test capacitor, 206–207
Tetrahedral void
  crystalline materials, 19
  defined, 19*t*
  geometry, 14*f*, 23*f*

hexagonal close packed crystals, 19
Theoretical close packing, 161
Theoretical density, 129*t*
calculation, 20–21
crystalline materials, 20–21
vacancies, 25
Theoretical strength
crystalline materials, 23–24, 23*f*
ideal materials, 17
polymer chains, 46
size effect, 17–18
Thermal barrier coatings (TBCs), 254–256
Thermal conductivity
aluminum foam, 216
amorphous polymers, 48
ceramic foam, 215
kinetic theory of gases, 48
lattice holes limiting, 24
measurement technique, 213–214
metals, 24
models, 214–216
in nanoporous polymers, 63–64
polymers, 48
polymethylmethacrylate (PMMA), 64*f*
radiative, 217
resistance in series model, 216
self consistent field theory predicting, 216
three-phase foam, 217
tuning, 24
Thermal decomposition, 47
Thermal expansion
coefficient of, 45
microcracks formation, 28
Thermally grown oxide (TGO) layer, 254
Thermal properties, 24–25
syntactic foam, 212–217
Thermal protection systems, 246–249
Thermal transport
polymers, 48
primary means, 24
Thermomechanical behavior, 44–46
Thermoplastic polymers, 41
examples, 41
intrinsic void and, 42
Thermoset polymers, 41, 152
examples, 41
intrinsic void, 42
network structure, 41
Three dimensional imaging, 174–178
destructive, 175
nondestructive, 175–176
spatial resolution, 176*f*
3D-Printed Habitat Challenge, 249
3D-printed porous Nickel-stainless steel electrode, 262*f*
Three-dimensional voids, 17, 33
Three-phase foam, 193
designs, 194*f*
thermal conductivity, 217
Three-phase syntactic foam, 116*f*, 192–193, 193*f*, 194*f*, 195*f*, 196*f*, 208, 210, 217
by binder jet AM process, 116*f*
tin-based solder containing voids, 219*f*
tin-based solder joint tested in shear, 220*f*
Titania, 155, 263–265
applications, 263–265
electron-hole pair creation, 264
mechanism of, 266*f*
nanometer-scale porosity in, 265*f*
pore structure, 264
water splitting, 264, 266*f*
Titanium, 238, 266–267, 268*f*
Titanium hydride (TiH_2), 120–121
Titanization, 86
Tobramycin, 148–150, 272
Topcoat layer, 254
Top-down process, 110–112, 111*f*
Total surface area, 56
Transmission electron microscopy (TEM), 169–172
components, 169
Transmission positron microscopy, 172
Trigonal planar voids, 19*t*
Tubes, Kirkendall effect, 82–85, 83*f*, 84*f*

Tubes, Kirkendall effect (*Continued*)
coaxial $ZrAl_2O_4$ microtubes, 85*f*
micrometer-scale coaxial $ZrAl_2O_4$ microtubes, 84*f*
shell of, 84
Turbine blade, schematic of, 256*f*
Two-phase syntactic foam, 116*f*, 211

U

Ultrasonic vibrator, 145–147
Ultrasound (US), 176*f*
Universal gas constant, 33
Unmanned aerial vehicles (UAVs), 244
Unreinforced voids, 5–7, 193–195
ablative materials, 247–248
applications, 5
hollow microspheres, 242
modulus, 194
strength, 194
volume fraction of, 193
Unsaturated polyesters, 41

V

Vacancies, 17
aluminum, 22
atomic, 25
direct methods, 22
equations for, 22
indirect methods, 22
irradiation, 37
metals, 21
movement, 33
parameters influencing, 21–22
temperature, 21–22
theoretical density, 25
Vacancy diffusion, 33–34, 77
Van der Waals bonds, 41
Van der Waals forces, 66
Vaporization, 97
pulsed laser, 160
Vat polymerization, 110–112, 112*f*
Vibration
control of foam, 95
polymer chain, 48
ultrasonic, 145–147
υ-induced syneresis, 143
Vinyl esters, 117–119
Viscosity, 43
Voids, 96 *See also specific types of voids*
casting, 30–31
defect, 1–2
defined, 2–7
effect on mechanical properties, 25–30
effect on strength, 28–29
formation, 30
hierarchical design with, 129–131
Kirkendall, 78–79, 81, 83, 86–88, 88*f*
length scale, 13–15
linear, 19*t*
macroscale, 120–121
materials *versus* digital design of, 12–13
r-type, 33
service induced, 30–31
slit, 28
solders, 31
three dimensional, 17, 33
Volume *See also specific volumes*
fraction, 56
occupied, 43
total specific, 43, 43*f*
Voronoi cells, 201*f*
Voronoi decomposition, 11
Voronoi foams, 9–12, 12*f*
Voronoi partitioning, 11
Voronoi tessellation, 11
Vulcanized rubber, 41

W

Water
open cell absorption capacity, 4
titania splitting, 264, 266*f*
Water-in-oil-in-water, 141–143
porous macrospheres, 153*f*
spray drying, 142*f*

Wavelength, 168–169
Webs, 108
Wollastonite-diopside-based bioceramic scaffolds, 117–119, 117*f*
Woods Hole Oceanographic Institution (WHOI), 237
w-type cavity
  schematic, 35*f*

X

Xinc chloride ($ZnCl_2$), 67
χ-induced syneresis, 143
X-ray
  benefits, 178
  imaging depth, 175–176
  solders, 176–177
  synchrotron source, 176
  types, 176
X-ray microcomputed tomography, 175–176

Y

Young's modulus, 23, 25–28, 30, 204–205, 223–225, 226*f*
  microcracking, 28
  porosity, 27*f*
  syntactic foam, 212
Yttria stabilized zirconia (YSZ), 255–256

Z

Zeolites, 59–61, 124–125
  basic formation of, 60*f*
  chemical formula, 59, 60*t*
  general chemical formula, 124–125
  naturally occurring, 58*t*, 59
  pore sizes, 125–126, 125*f*
  synthetic, 125
Zinc aluminum borosilicate glass ceramic, 9, 10*f*
Zn-containing organometallic liquid, 79

Printed in the United States
By Bookmasters